Exterior Home Improvement Costs

New Ninth Edition

The practical pricing guide for homeowners & contractors

RSMeans

Exterior Home Improvement Costs

New Ninth Edition

The practical pricing guide for homeowners & contractors

Copyright ©2004
Reed Construction Data, Inc.
Construction Publishers & Consultants
63 Smiths Lane
Kingston, MA 02364-0800
781-422-5000
www.rsmeans.com
RSMeans is a product line of Reed Construction Data

The managing editor for this book was Mary Greene. The editor was Andrea St. Ours. The editorial assistant was Jessica Deady. The RSMeans engineering editors were Barbara Balboni, John Chiang, John Ferguson, Robert Kuchta, Robert Mewis, Steve Plotner, and Gene Spencer. The production manager was Michael Kokernak. The electronic publishing specialist was Sheryl Rose. Electronic publishing support provided by Kathryn Rodriguez. The proofreader was Robin Richardson. The book and cover were designed by Norman R. Forgit. The cover photographs are copyright Peter Vanderwarker and Nancy H. Fenton, courtesy of Fenton Inc. The illustrations were by Carl Linde, Richard W. Lowre, Robert Megerdichian, and RSMeans. Original contributing authors were George Ghiorse and Timothy Jumper.

Printed in the United States of America

10 9 8 7 6 5 4 3 2

Library of Congress Catalog Number Pending

ISBN 0-87629-742-4

Table of Contents

Acknowledgments

The editors would like to thank the following individuals for their contributions to this publication.

Nathan and Peter Fenton of Fenton Inc. Residential Remodeling in Wellesley, MA, for granting us permission to include photographs of several of their home remodeling projects.

J.P. Gallagher of J.P. Gallagher Construction & Development Co., Inc. in Hanover, MA, for allowing us to photograph homes the company has recently built.

Roscoe Butler of Holway House Design for permission to use photographs of his studio and some of his design projects. Also, Abode Design/Build of Falmouth, MA, who completed the construction work for the Holway House Design projects.

Chrissy and David Hoadley of Hanover, MA, for allowing us to photograph their home.

Craig Kooda and Kooda Exteriors, www.koodaexteriors.com, for allowing the use of photographs of their outdoor landscaping projects.

Robert and Susan Alexander for permission to photograph their home.

Howard Chandler and the Builders Association of Greater Boston for helping us connect with leading home builders who so generously assisted with the photographs.

We would also like to acknowledge:

Scot Simpson, author of *Builder's Essentials: Framing & Rough Carpentry*, the source of select glossary terms in this book.

Richard Connolly, President of Cornerstone Consulting, Inc., in Weymouth, MA, for providing the "Questions for the Building Department" and "Contractor Evaluation" forms.

Introduction

Purpose of the Book

This book and its counterpart, *Interior Home Improvement Costs,* are designed to take some of the mystery out of remodeling projects—by giving you an idea of what they'll cost and what's involved in doing the work.

What You'll Find

The Model Projects

The bulk of the book is made up of 68 model projects with estimates. Each contains a list of materials, along with an estimate of professional labor-hours and total costs. Each project also includes:

- "What's Involved"—the tasks and skills needed to complete the job.

- "Materials & Tools"—guidance on selecting materials, and the tools you'll need.

- "What to Watch Out For"—helpful hints on tricky parts of the work and how to avoid potential problems.

- "Level of Difficulty"—how much skill and experience are required.

These overviews will help you decide which parts of the job you want to take on, if any, and which are better left to a contractor.

All of the project cost estimates include a list of the materials required in a standard renovation. Factored into the "Contractor's Fee, Including Materials" is an estimate of the overhead and profit a contractor is likely to add to the total cost—for time spent visiting your home, planning and estimating the job, and a percentage of the costs of doing business.

Working with Design & Building Professionals

Following these introductory pages is guidance for working with builders, architects, and landscape designers/ contractors, along with questions you may need to ask your local building department before you get started.

Detailed Costs

Following the 68 model projects are costs for individual construction items—from foundations, to roofing and siding materials, to windows and doors, and much more. These costs can be used to modify any of the model projects, or to develop your own individual estimate from scratch.

Adjusting Project Costs to Your Location

Following the Detailed Costs, you'll find factors for over 900 cities and towns, organized alphabetically and by zip code. Multiply the total cost of any model project by the factor given for your location to get the most accurate cost estimate for your project.

Safety & Shopping List

At the end of the book, you'll find important safety precautions that should be followed throughout any construction project. The last section, just before the index, is a shopping list that you can use to help make sure you've thought of all the materials and tools you'll need, and to compare prices from different retailers.

New in This Edition

All of the project materials and costs have been updated, and we've also added eight completely new projects, including an outdoor living area with appliances, a three-car garage, and landscape projects such as lighting, stone walls, fountains, and in-ground sprinkler systems. We've also added icons to quickly identify projects that strongly feature three special attributes:

- Best Return on Investment— improvements that attract buyers when you sell your home.

- "Green Building"—healthier, more comfortable surroundings, and/or lower utility and water bills.

- Curb Appeal—improvements to your home's appearance from the street that enhance its value.

Do it Yourself, or Hire a Contractor?

There's no question that doing your own work can save money and give you the satisfaction that comes from creating something of value—but not everyone can or should undertake every task. Honestly assess your skills and fitness for the tasks before tackling any home improvement project. Also ask yourself if you really want to do the work. If you're doing it just to save money and don't enjoy it, you may not give it your best effort.

Do You Have the Skills?

Throughout the model projects, we refer to three levels of do-it-yourselfer:

- Beginner: a first-time remodeler who has tools and equipment for small repairs and projects.

- Intermediate: a do-it-yourselfer with moderate building skills, experience, and a reasonable collection of tools for various kinds of work.

- Expert: a skilled remodeler with extensive building experience and tools.

Most remodeling projects require accurate layout and precise measuring, cutting, and assembly of materials. Complex preparations, and particularly work involving structural elements, should be handled by qualified tradespeople and experts. When in doubt, call a contractor, and consult an architect or engineer, as needed. The long-term benefits of professional workmanship outweigh a reasonable initial expense.

Electrical, plumbing, structural, and foundation work must meet local building code requirements and are best performed by licensed professionals. Some proposed changes may require a permit and plans approved or certified by an engineer or architect.

How Long Will It Take?

This is a major consideration, especially for disruptive projects, like a room addition or large dormer that involves opening up your living space. Think about whether you really have the time to focus on the project to get it done quickly. If it will take a total of eight days, and you have only weekends free, it will be a month before it's complete. Hiring a contractor may be worth it when compared with weeks of major

inconvenience. Try to plan your work for the daytime hours.

Doing it yourself means saving money, but don't try to also save time by rushing to get the project done. Many do-it-yourselfers have broken windows, scratched finishes, and damaged other materials with improper handling, storage, or installation.

Follow the instructions for tools, materials, appliances, and fixtures. If the paint can says to let the first coat dry for eight hours before applying the second coat, wait eight hours. Sometimes we're in such a hurry to enjoy the fruits of our labor, that we eliminate steps, such as the recommended sanding, final coat, or buffing. Overcome the urge. The work may take longer, but it will last and look better.

Timing

If your work will require shutting off water, gas, or electric service, make sure it's done at a time when you can get emergency repairs or professional assistance. Don't choose a holiday weekend or a Sunday afternoon. Also remember that most specialty stores, like plumbing and electrical, may be closed on Saturday afternoons and Sundays.

Check your local trash pickup or dump rules and schedule. Find out in advance what materials are accepted, and when.

Other Things to Consider

Storing Materials

Plan how you'll transport the materials, which ones will need delivery (and when), and where you'll store them. Is your vehicle large enough to haul the materials and equipment you need? Are gates, doorways, and windows large enough to accommodate large items or equipment?

Tools & Equipment

Assess your collection and make a list of what you need. If you're looking for bargains, make sure the product is complete and of acceptable quality, and that it can be serviced. Keep in mind that the better the quality or grade of tools and materials, the easier they tend to be to work with.

Neighborhood Relations

A big remodeling project can be a strain on homeowners—with endless decisions to be made, disruption, dust, and strangers in your house—but it can also affect your neighbors. To preserve a good relationship, let them know of your plans and what they can expect. Tell them what day the work will begin, how long you expect the project to take, and who they can get in touch with if a problem comes up when you're not available.

Talk to your contractor in advance to find out if the neighbors' property will be affected in any way, so you can explain it to them beforehand and ask their permission. Also let neighbors know if and when big trucks will be on the street. (Talk to your contractor about keeping trucks on your side of the street, and about avoiding noise outside of regular business hours.) Have materials placed in your yard or driveway, to prevent dirt and debris from going into the street.

Try to keep your yard respectable looking, and keep an eye on your next-door neighbors' yards to make sure any stray items, such as stripped-off roofing materials, don't land there. Have dumpsters removed as soon as they're no longer needed. You might even ask your neighbors if they'd like to dispose of anything if you have room left before it's hauled away. When it's all over, be sure to thank your neighbors for putting up with the inconveniences. Invite them in for a look at the remodeled space.

Documenting Your Project

Keep a journal, and track the costs and time involved. Photograph the steps along the way, especially for major, time-consuming projects. This documentation may be useful for future home improvements or for resale.

And Finally...

There's no greater satisfaction than accomplishing a major project ourselves, but if we all could do it ourselves, the professionals would be out of business. Recognize your limitations; plan your project well, and evaluate its impact on your family's comfort; read the safety section at the back of the book; seek expert advice; and most importantly, enjoy the experience.

Additional Resources

All of the projects in this book are for exterior applications. For interior projects, consult *Interior Home Improvement Costs,* which contains another 68 model projects—various kitchens, bathrooms, fireplaces, flooring, stairways, closets, and more.

Whether you're hiring a contractor or doing the work yourself, it's important to have a quality standard for workmanship in mind. We recommend *Residential & Light Commercial Construction Standards,* a compendium of industry standards for quality in construction. This information can help avoid disputes with your contractor and clarify many installation questions. The book is based on building code requirements and guidance from leading professional associations, product manufacturer institutes, and other recognized experts. It's available at the contractor desk of many major home improvement centers, in bookstores, or it can be ordered from RSMeans at 1-800-334-3509 or *www.rsmeans.com.*

Return on Your Home Improvement Investment

Defining "Value"

While most remodeling projects are driven by the desire for a more functional and pleasant space to live in, it's also important to know how your project stacks up in terms of economic return on your investment. There are several ways to measure "cost versus value."

One is to compare the dollars you spend on the project with how much it's likely to add to the sale price. You can also think of return in terms of how the improvement helps you attract buyers and sell quickly—if and when you put your home on the market.

Another dollar value measure is how much money the improvement will save you over time—in the form of lower utility bills. Examples are replacing plumbing fixtures with low-flow models, upgrading light fixtures and using fluorescent bulbs, or adding a whole-house fan to your attic. Some upgrades, such as adding insulation, a more efficient heating and air conditioning system, or thermal replacement windows, may qualify for rebates from your utility company or for energy-saving loans.

If the improvement is converting a garage or attic to living space—or a major room addition, it could even *earn* you money in the sense that it provides another housing option for a family member, or if it can be rented (and local zoning laws allow rentals).

Factors that Determine Value

Keep in mind that there are many other factors that influence "value" as determined by a home's appraised or selling price. These include:

- Quality of workmanship and materials.
- Appropriateness of the improvement in terms of style/design. (Think neutral and conservative if you're planning to sell fairly soon, and always make sure your improvement is in harmony with the rest of your house.)
- Condition of your house overall.
- Selling prices of surrounding homes and desirability of the neighborhood (schools, commuting distance, etc.).
- Number of new and older homes currently on the market in your area.

Which Improvements Tend to Bring the Best Return on Investment?

Studies by realtors and other organizations generally agree that the following improvements typically have the best payback when homes are sold. Projects are listed in descending order, with kitchens having the highest return on investment.

- Room Addition (for family room or master suite)
- New Deck (desirability varies regionally)
- Replacement Windows
- Exterior Painting and Landscape Upgrade (These move to the top of the list if you're selling your home, and it's worn or in dire need of curb appeal.)

The return on investment for many projects is affected by how much they're needed. If your home's façade looks out-of-date or worn, an upgrade will have a lot more impact than if you're just making a change because you want a new color scheme.

Look for the icon throughout the projects in this book, which indicates a high return on investment.

A Final Thought

Return on investment goes beyond just the selling price if and when you decide to sell your home. If a particular space or feature is something you really want, and you'll use and enjoy it every day, it may be worthwhile, especially if you don't plan to sell your house for a long time.

Green Building

Green building is a major movement in new home construction. Builders, architects, and material suppliers are responding to home buyers' interest in healthier, more comfortable houses with lower energy and water bills—not to mention building practices that support a healthy environment. New standards, such as LEED™ (Leadership in Energy and Design), ENERGY STAR® (for appliances and electronics), Greenspec® (for building materials and products), and the Forest Stewardship Council (lumber), have made it easier to identify and specify environmentally friendly building products. (Refer to these organizations' Web sites, listed at the end of this section, for more information.)

Some green building features, such as integrated solar collection systems, super-efficient heating and air conditioning systems, and re-circulating hot water piping, are usually easier to build in as part of a new house. Many others, however, such as low-flow faucets and showerheads, thermal windows, fluorescent light fixtures, or whole-house fans, can easily be incorporated as part of another remodeling project. As mentioned previously in "Return on Your Home Improvement," there may also be rebates available to help cover the cost.

Look for the green building icon throughout the projects in this book, highlighting improvements that have "green" benefits.

Green Materials

Green building materials are those that:

- Are healthy for people—they don't produce indoor air quality problems in the home from harmful fumes or fibers, and they don't harm the workers who manufacture them.

- Are healthy for home use and the natural environment. They're better for people with chemical sensitivities. They don't degrade the environment or deplete scarce resources. They don't generate hazardous by-products or waste.

- Help minimize use of energy or water.

- Require little energy to manufacture.

- Are durable, reusable, recyclable, and/or biodegradable.

- Can be obtained locally to reduce fuel use for long-distance transportation.

What to Look For

Following are some examples of green building materials, products, and design approaches that you might consider incorporating into one of your remodeling projects. Some of these items are easier to find than others, but manufacturers are constantly expanding their offerings of environmentally friendly products. In some cases, green products may be a little more expensive initially, but the savings in energy or water use can pay back the investment very quickly.

- Appliances—Look for ENERGY STAR®-rated models.

- Brick and stone available locally.

- Cabinetry made with formaldehyde-free glues.

- Carpet and padding made of natural or recycled materials.

- Ceilings—for suspended ceilings: acoustic panels that are toxin-free, made of recycled materials.

- "Daylighting"—placing windows to maximize natural light and views of the outdoors—to reduce eye strain and stress. Daylighting also means designing to diffuse or direct the light for best effect, and to avoid glare.

- Drip irrigation systems—which lose less water through evaporation than sprinkler systems.

- Engineered wood products, made with smaller and faster-growing tree species. Products include finger-jointed lumber, glu-lam beams (which offer the advantage of extra strength over a long span), and pre-fabricated wood trusses and joists.

- Fireplaces—consider radiant wood-burning stoves that burn cleaner and more efficiently.

- Flooring made of natural materials, such as bamboo, linoleum, and ceramic tile (made from recycled glass), cork, recycled rubber, or recycled or wool carpeting. (See also "Wood flooring.")

- Floor finishes—water-based urethanes.

- Furniture and furnishings—features to look for include: ergonomic design, nontoxic materials, and sustainably harvested or recycled wood.

- High-efficiency heating, ventilating, cooling, and hot water systems. ENERGY STAR® ventilation/thermostats.

- Insulation with no, or low levels of irritants and pollutants. Cellulose and the new formaldehyde-free fiberglass products are reasonable choices. Purely natural materials include perlite and cotton.

- Lighting—quality light fixtures and bulbs that minimize headaches and eye strain. Fluorescent fixtures and bulbs (including screw-in compact fluorescent bulbs) that maximize energy savings and last much longer. ENERGY STAR®-rated lighting.

- Paints—low or no-VOC paints (volatile organic compounds) to minimize unhealthy air quality.

- Pavement—porous materials, such as gravel or brick-in-sand, that can absorb rainwater and reduce run-off.

- Plants that prevent erosion. (See also "Xeriscaping.")

- Plumbing fixtures and faucets—low-flow models, re-circulating systems, and faucet aerators that use less water and/or energy.

- Ponds—to help manage rainwater run-off.

- Recycled lumber—made from recycled plastic and wood products. It has the additional advantage of insect and moisture resistance and low- to no painting, staining, or maintenance.

- Recycling centers—storage (in a kitchen, pantry, mudroom, garage, or shed) to contain recyclables between pickups or drop-offs.

- Roofing—light-colored to reflect heat, insulated.

- Shingles—wood, plaster, and fiber cement.

- Thermal windows and doors—with the correct solar heat gain, heat loss, and visual transmittance values—both for your climate and for the orientation of the room to the sun.

- Trees—planted with your house's heat gain and loss in mind—for example, a deciduous tree that shades the house in summer, but allows light to filter through in winter.

- Water filtering systems—to improve the quality of your drinking water.

- Wood flooring that is recycled from old buildings, or new-growth (or new "certified") wood.

- Wood sheathing made from recycled wood fibers.

- Xeriscaping—using native plants in your yard that will flourish with a minimum of watering, fertilizing, chemicals, and insecticides.

Also, try to recycle as much as possible of the waste if your project involves demolition. For example, your old cabinets, windows, doors, or other items may be of interest to a salvage yard or community building recycling center.

Visit these Web sites for more information.

LEED™: www.usgbc.org

ENERGY STAR®: www.energystar.gov

Greenspec®: www.buildinggreen.com

Forest Stewardship Council: www.fscus.org

Forest Certification Resource Center: www.certifiedwood.org

Curb Appeal

For decades, realtors have talked about "curb appeal" when rating a house's attractiveness to buyers when they first drive up to it. It's a quality that can be seen from the street, and can either captivate potential buyers or discourage them from even setting foot in the house. Curb appeal is important not just for selling a house, but for welcoming guests to your home, and for making you feel good whenever you return to it.

There are many ways to enhance curb appeal, because many different elements contribute to it. Siding, windows and doors, new trim, driveways and walkways, decorative fences and gates, landscaping and planters, lighting, accessories and furniture, and even your mailbox and house numbers work together to define your home's curb appeal.

You can invest a little or a lot, depending on what your home needs and your budget. Even small improvements can make a big difference.

Look for the curb appeal icon throughout the projects in this book.

Following are some ideas to consider.

Siding

- Repaint if paint is peeling, or if it's a dated or inappropriate color for your home's architectural style. Perhaps you only need to touch up the trim work. While you're at it, make sure window casings are in good shape and free of rot.

Shutters and Gutters

- Usually a purely decorative element, shutters can dress up a home, but in disrepair, they definitely take away from its appearance. The solution, sanding and repainting, is simple, if a bit time-consuming. The work goes better if you remove the shutters and work on them on a flat surface.

Porch

- If your porch floor, railings, or trim are peeling, loose, or damaged, they're detracting from the looks of your home. Like overgrown shrubs, a scruffy porch gives the impression that the house is not well cared for.

  Loose railings and rotted floorboards are also a safety hazard. Loose railings can sometimes be easily fixed by tightening their lag bolts. Railings that are screwed into brick may need to have the screws and their sheaths replaced.

Doors and Windows

- Replace an old, worn, and uninspired door with an attractive new one in wood, steel, or fiberglass. You can also add sidelights or a transom window to bring more light into the house and make the entryway more interesting. If you're happy with your existing door, you might freshen it up with a new coat of paint, which could also provide a color accent. Look around at other homes for color combinations that work well.

- If your windows are in poor condition, consider replacing them rather than performing major repairs. A new style or combination of windows can bring more light into the house and update the look from the outside. New thermal windows can also help pay for themselves in energy savings.

House Numbers

- These must be clearly visible from the street. If your house numbers are recessed into your home's façade (e.g., stone or brick), consider painting them to make them easier to read. Make sure your mailbox numbers are complete and clear as well. While house numbers are a functional item, they can also be attractive, for example, mounted on a tile or stenciled on a glass transom window in a style that matches your house's architecture.

Planters

- Add a large pot, or a pair of planters, overflowing with seasonal flowers and greenery at or near the front door.

Driveway

- This is one of your home's largest features and has a definite impact on curb appeal. While it must be functional, there are ways to soften it and give it some character. Landscaping on either side of the driveway can distract from a straight rectangular shape. An arbor built over part of the driveway can add charm to the front of your house and can even provide shelter for an extra car. A curved driveway is a more natural shape if your yard is suited to it, and a circular drive makes it easy to come and go.

- You can also use a more natural paving material, like brick or gravel (in a number of colors), or various shapes of concrete pavers, rather than asphalt or concrete. To keep the cost down, you can stain an existing concrete driveway.

Lawn

- If it has bare spots or is weed-infested, remove the weeds and re-seed or sod to bring it back to a respectable state. Trim around walkways, trees, and other features.

Flower Beds

- Flowers can enhance curb appeal along the driveway, across the front of the house, clustered near pathways or the doorway, or at the street. Flower beds between properties help

delineate your yard and, with some taller specimens, can give you and your neighbors some privacy. You can incorporate some interesting rocks, or even make a rock garden.

Shrubs

- Many homes are overwhelmed with shrubbery that has been allowed to grow way too large for the front of the house, obscuring windows, robbing the interior of natural light, and even making it difficult to walk down a pathway or get in the doors. Pruning is one of the quickest and least expensive ways to add curb appeal.

  If shrubs have completely outgrown their shape and cannot be attractively pruned, take them out and replace them with well-proportioned foundation plantings that offer a variety of texture and color, and enhance your home's architecture.

Walls and Fences

- Stone walls add tremendous character to a property. They're not inexpensive because of the material cost and the considerable labor involved, but you can reduce the cost if you're able to do the work yourself and have some help. Wood or brick walls can also enhance a yard by defining its borders and providing texture and a backdrop for landscaping.

Lighting

- Landscape lighting can add warmth and highlight special architectural features, trees, or plantings. To enhance security/safety lighting, choose decorative fixtures (such as a post light or wall-mounted lanterns) that complement your house's style in the daytime and after dark.

Consider the curb appeal possibilities for your home. Think about what will add value to your home, but also about features you'll enjoy right now—not just what future buyers may want. Then, look through the projects section of the book and find the costs to help you plan many of these improvements.

Remodeling an Older Home

There's something special about an old house. You get the sense that any work you do should meet with the approval of the craftsmen who originally built it. Whether you're making repairs to maintain an old structure or introducing modern conveniences, you may want to do it in a way that maintains the home's original character. You'll need to decide how "pure" you want to be in materials and design changes.

Houses constructed prior to 1940 have unique characteristics not seen in homes built today—from foundation to finishes. Consider the following:

• Old houses were often built with "true" dimensional lumber. This means that a 2 x 4 is actually 2" x 4", whereas the actual dimensions of a 2 x 4 purchased today are 1-1/2" x 3-1/2". This difference is consistent for all framing lumber produced today. In old houses, the boards used for both rough and finished carpentry may be a full 1" thick, compared to stock used today that is actually 3/4" thick.

• If you're pursuing an exact restoration, you'll need to find a mill that produces the exact dimensional lumber you need. If exact reproduction isn't important, you'll have to introduce shims to build up narrower new members to match what's already in place. Where new members (both cosmetic and structural) will be exposed, true dimensional lumber will need to be used, and can be costly.

• Some older homes are "balloon-framed," which means that the vertical structural members (the posts and studs) are continuous pieces from sill to roof plate. The intermediate floor joists are supported by ledger boards "spiked to," or "let into," the studs. Temporary bracing and floor support can be critical to working on both interior and exterior walls that have been balloon-framed. Adding new wires and utilities may actually be easier in a balloon-framed wall, because the studs run uninterrupted from sill to roof plate, with the exception of fire stops.

• Old houses usually have higher ceilings that require longer lengths of lumber for wall framing. If the walls are over 8' high, you may incur the cost of greater waste when purchasing wall finish materials. You may also find that the partition walls are load-bearing, requiring more material when re-framing. In some cases, the original builder (or a subsequent remodeler) may have placed additional blocking or bracing within the wall structure. This may obstruct the placement of new pipes or wiring, which could lead to additional demolition and re-framing.

• Old house doors and windows are often unique shapes and sizes. Replacement may involve the services of a custom millwork company. Re-working old doors and sashes can be very time-consuming. There may also be a long lead time for ordering custom units. The different dimensions of framing members may result in varying wall thicknesses. Window and door frames may require extension jambs, adding to the cost and time of construction. Door and window frames that are part of the structural components may require greater care during removal.

• Cabinets and bookcases in old homes are often built in place. Removing them may involve a lot of patching and matching of finishes. Lath and plaster is generally thicker than drywall used today. It can be difficult to patch this wall covering so that it matches existing surfaces, if you want to achieve a perfect finish. It may be necessary to reconstruct and finish an entire wall.

• To match interior and exterior moldings, you may need to have material custom-milled. When removing existing moldings and trim, try to determine the nailing pattern and fastening method that was used originally. Try to save as much material as possible. Even small pieces may come in handy for patching and replacing sections of trim, siding, moldings, or other details. When you're trying to match existing materials, remove all layers of paint to establish true dimensions.

• Matching new exterior trim to the old can also be difficult, in the sense that many details become obscured or disappear under numerous coats of paint. Hand-planed moldings, and those of the same shape but different size, may be costly to reproduce, but many stock moldings can be used in combination to produce a detail very similar to an original design.

• When rebuilding cornices and overhangs, dismantle them carefully to understand how they were put together, and to make patterns from the original pieces. Carefully document the lengths of overhangs and flashing details so you can reproduce the look. Taking things apart carefully also lets you fabricate parts and do some of the assembly on the ground, rather than on a scaffold. We recommend priming and applying first coats of paint prior to attaching any exterior trim or finish materials.

- If your renovation involves replacing siding, mark the spacing of the old siding on the cornerboards. Then, you can apply the new material to match the exact layout of the existing siding.

- Many companies specialize in reproducing building components, light fixtures, bath and kitchen fixtures, cabinetry, and hardware. Using reproductions can help you satisfy current building codes, while ensuring your safety.

- Owners of old homes should consider working with a professional to address structural concerns, fireplace and chimney rebuilding, electrical, plumbing, and abatement of hazardous materials (such as lead paint or asbestos pipe covering). You can also seek guidance from your local historical society.

- Working on an old house is special. You may do it for sentimental reasons, to preserve its aesthetic character, or out of respect for the craftsmanship. Whatever your motivation, be patient and enjoy the satisfaction that comes from knowing you've added your signature to something that will endure because you cared.

Part One
Working with Design and Building Professionals

Working with an Architect

Many of the features we admire in beautiful homes were designed by professional architects. To achieve a unified space that meets your needs—functional and aesthetic—you might consider consulting an architect who can define your remodeling project on paper—providing you with a "total picture"—before construction even begins.

For major home projects, such as master bath and kitchen/family room remodels, additions, updates, and garages that change the basic "footprint" of your home by adding or removing walls and other structural elements, hiring an architect is a sound investment. You may even save money in the long run, as architects are professionally trained to translate your ideas into a fully balanced and functional new space—built efficiently and within your budget.

Architects not only design your project in detail, but can give you creative ideas to enhance space or storage, choose the most suitable products and finishes, save energy, maximize your budget, and tie your project into the rest of your home. As your representative, an architect can also make sure your project complies with local building codes; help secure planning, zoning, and building department approvals, if needed; oversee contractors; and monitor quality throughout the construction process.

Before You Meet

Prior to your first meeting, ask yourself the following questions:

- Why do you want to add to or renovate your house?

- What is your lifestyle? Are you at home a great deal? Do you work at home? Do you entertain often?

- What activities will take place in the new space?

- What do you think the addition or renovation should look like?

- Do you have strong ideas about design styles?

- How much can you realistically afford to spend?

- How soon would you like to be settled in your new space? Are there rigid time constraints?

- Do you plan to do any of the work yourself?

- How much disruption can you tolerate as you renovate or add on to your home?

Answers to these questions will help focus your discussion with prospective architects.

Tips for Selection

As you search for a suitable architect, get advice and referrals from other homeowners who have undertaken similar projects. Ask who designed the homes and remodeling projects you admire. Many local chapters of the American Institute of Architects maintain referral lists and can help you identify architects in your area who specialize in certain types of projects.

It's recommended that you interview a few architects before making your final decision. This way, you can get a sense of the range of design possibilities for your project. You also get the chance to compare approaches and prices, as well as personal chemistry, so that your final choice is someone you're comfortable with and confident in. When interviewing architects, ask them to show you "before and after" projects that are similar to yours. Also, ask for references you may contact. Once you select an architect, be sure to address the schedule, scope of work, and costs, and retain a contract in writing.

For more information, consult the American Institute of Architects (AIA), which offers additional guidance on selecting an architect and local referrals. Visit AIA's Web site at http://www.aia.org

Working with Landscape Professionals

Quality landscaping can add considerable value to your home, and can make a big difference in both the price and time on the market if and when you sell your home. This section will guide you in deciding whether to tackle these projects yourself or hire a professional designer or contractor.

Designers

A landscape designer can help make your yard more functional, attractive, and harmonious, particularly large yards with disjointed trees and plantings. A landscape designer's goal is to create a sense of continuity, balance, and structure—and to make it look natural. He or she should be professionally trained to lay out the features of your property, integrating plantings, walkways, walls or fences, lighting, patios, and water features to create a usable, attractive space.

Landscape designers are familiar with the shape, growth, and requirements of various types of plants, and can help you choose what is best for your yard. They can suggest species to enhance privacy, provide year-round bloom, or even help you save on heating and cooling costs (with summer shade and winter sun). A designer can help you define watering requirements, design sprinkler or irrigation systems, and incorporate plants that withstand insects and diseases.

Designers also offer plans to give your house's entryway new curb appeal, with ideas for container gardening, walkways, accessories, and furniture.

Here are some tips if you're considering hiring a designer:

- Take inventory of your yard and needs. Sketch your whole property. Decide which plants, trees, hills, or rock outcroppings you want to keep.
- Have in mind what you'd like to accomplish. Do you want more privacy? A barrier to reduce street noise? Will your yard be used for large gatherings and outdoor dining? Do you want a level, open area for children to play in, visible from inside the house? Do you want a vegetable, perennial, or cutting garden? Formal or informal?
- Define your budget. Landscape work can get involved and expensive quickly. Prioritize your needs and "wish list," so that the designer can focus on what's most important to you.
- Interview several landscape designers before making a decision. Take a look at their "before and after" plans and photographs. Be sure to have a written contract that details the work involved, the schedule, and the payments. Ask for references.

Contractors

Whether or not you design your own plan or hire a designer, you'll need to bring it to fruition. Depending on the scope of your project, it may be a good idea to hire a landscape contractor. Much like designers, contractors are knowledgeable about various plant species, as well as the best ways to install and care for them.

If you're hiring a landscape contractor on your own, here are some tips to get started:

- Determine the services you'll expect from the contractor (e.g., planting seed, shrubs, trees, and other plants; building walkways, decks, or patios; installing retaining or stone walls; or installing irrigation systems or lighting).
- Research the company, their certification, and their expertise. Ask for references from other customers whose projects were recently completed. Inquire about insurance—both liability and Worker's Compensation.
- Ask about warranties and guarantees. Does the landscaper or nursery have a replacement policy for their plants and materials?

Maintenance

The more complex and sizable your landscape features are (such as gardens, gazebos, and ponds), the more time-consuming regular maintenance will be. Professional landscape maintenance can be a sound investment to ensure the health and beauty of your plantings, and a neat appearance to walkways and lawns. When selecting a service, be sure to establish a maintenance schedule, including regular mowing, seasonal fertilization, and any additional tasks such as weed control, aeration, or de-thatching. Find out which chemicals, if any, will be used in your yard, particularly if you have children or pets.

Working with a Contractor

Some home improvement projects clearly require professional help—for part or all of the work. An addition, for instance, could call for at least seven mini "projects," such as removing old cabinets and installing new ones; patching and painting walls and ceilings; installing new flooring, countertops and appliances; and often replacing windows and doors. If you're not comfortable tackling all aspects of the work on your own, hire a contractor to get the job done. In most cases, an electrician and plumber are required for any work involving wiring or piping.

Many people put more time and energy into choosing a contractor than they do choosing a physician! Because of the close relationship you have with your home, you need to be completely satisfied with a contractor's commitment to the project.

This section includes suggestions for selecting and working with a contractor. It doesn't cover every detail or situation that might arise, but can help you think about the relationship you'll have if you hire someone to work on your home.

Selecting a Contractor

Hiring a licensed contractor guarantees that he or she has satisfied the state's requirements to perform a certain type of work. It does not, however, guarantee that the work will be high quality or performed on time. You want a contractor who has a proven track record. Here are some suggestions for finding one:

- Ask friends and colleagues for recommendations when looking for a carpenter, plumber, electrician, painter, or other contractor. Ask if they were satisfied with the quality of the work, the schedule, and if there were any problems.

- Pay attention to the construction going on in your neighborhood. Look for signs and names on trucks, or ask one of the workers for a business card.

- Ask at your local lumberyard, building supplier, or town building department. They don't usually make specific recommendations, but may give you some good general leads.

- Ask a contractor you trust about another tradesperson. For example, ask your carpenter to recommend a plumber or electrician.

- Check with your local professional builder's association or other trade associations. The National Association of the Remodeling Industry (NARI), for example, can provide a list of members in your area.

- Look for a contractor who works with a qualified team of subcontractors and suppliers. Contractors who do repeat business with subcontractors and suppliers receive discounts on materials and labor cost savings that can be passed on to you.

Once you've put together a list of potential contractors, interview each of them. Carefully review their credentials, ask to see their previous projects, and follow up with the references they provide. Be sure to address things like the project schedule, warranties of products and workmanship, and cost estimates.

(See the "Contractor Evaluation" forms at the end of this section for more on what to cover with contractors during interviews.)

Establishing a Relationship

There is no reason to believe that hiring a contractor will be anything but a positive experience. As long as you both agree to the terms of the work and follow through on your obligations, the relationship should be productive and mutually profitable. Here are some guidelines:

- Don't be shy about asking a lot of questions. Make sure they're answered to your satisfaction and that you and the contractor both understand the terms of the work to be done. Communicate your needs clearly, especially if you want specific materials or products used.

- Your contractor should be able to gauge the approximate length of time the project will take, and explain both the nature of disruptions to your daily life and the impact of the work on existing plumbing, heating, and electrical systems—and on your landscaping.

- Request a written proposal that clearly defines the scope of the work and the time it will take.

A professional contractor will provide you with a schedule of the job so you can track its progress. The schedule should include deadlines for you too, such as when paint colors or materials must be selected. Keep in mind that the forces of nature (rain, snow, and other weather conditions) and extenuating circumstances (illness, availability of materials) can delay the project. Be firm but fair in your expectations.

- Agree on a payment schedule, and adhere to it.

- Be reasonable in your requests. Establish the terms of the work to be done before it begins; try not to change the terms of the agreement after the job is under way. If you do request a change, realize that it will probably involve additional time and expense.

Contractual Arrangements

Even a small job requires a legally binding contract that spells out the terms and conditions of your agreement. A professional contractor will probably insist on a detailed contract with all aspects of the job in writing. A fair contract will address the needs of both parties. There are several types of contractual arrangements that can be used for construction work. The following is a partial list providing basic information; consult a legal professional for further advice.

- A *lump sum contract* stipulates a specific amount as the total payment due to the contractor for performance of the work.

- A *cost-plus-fee contract* provides for payment of all costs associated with completion of the work, including direct and indirect costs as well as a fee for services, which may be a fixed amount or a percentage of costs.

- A *cost-plus-fee contract with not-to-exceed clause* is similar to the cost-plus-fee contract, but the profit is a set amount. The contractor guarantees that the project cost will not exceed a set amount.

- In a *labor-only contract*, the contractor is paid only for the hours that he and his crew work. The materials are purchased by the owner or others.

- In a *construction-management-fee-only contract*, the owner pays all subcontractors on the project, and pays a supervision fee to a general contractor in a lump sum or as a percentage of the work involved.

In addition to the price of doing the work, make sure the contract clearly states the materials to be used, warranties, and the schedule, even for small projects like painting or flooring. Only once you've read the contract and are fully satisfied with it should you sign it. Be sure to retain a copy for yourself. *(See the "Points to Cover Before Signing a Contract" section of the "Contractor Evaluation" form.)*

What to Watch Out For

- Don't pay any amount that does not seem justified or that was not agreed to beforehand without asking for an explanation. You should not be required to pay for a job in full before the work is complete. On the other hand, most contractors will require at least one partial payment before and/ or during the course of the work.

- Ensure that the contractor you intend to work with is fully insured and licensed for the work he or she will perform. (You can contact your state's department of business regulations or licensing for information in your area.) Contractors should carry Worker's Compensation and general liability insurance.

- Be sure to ask for all warranties, manuals, and other literature related to new equipment or materials being installed. Request documentation of brands being specified.

- If you intend to do some of the work yourself, you and your contractor should both understand where (and when) your work ends and the contractor's work begins.

- Maintain a businesslike relationship with the contractor. If you become too casual, the terms of the agreement

may become blurred, time may be wasted, and disagreements may ensue.

- Make sure you understand who will be doing the actual work. The person you make arrangements with may not be the same person who will do the physical work. Again, ask plenty of questions.

Ethical Considerations

- When getting more than one estimate for the job, allow each contractor to visit your home. This ensures a more complete estimate and an even playing field.

- Once the decision has been made to hire a certain contractor, let the other bidders know. (Keep in mind that you may want to hire them in the future for another job.)

- As much as possible, stay out of the workers' way while they're doing their jobs. Be available to answer questions or make last-minute decisions, but try not to create distractions. You don't want to be responsible for accidents or delays.

- If you've agreed to do certain prep work yourself, make sure it's complete and cleaned up before the workers arrive to do their jobs.

- Respect safety issues and job site conditions. If a contractor asks you to move all breakable items and keep small children and pets out of the area, there's good reason to do so.

- If you know of any condition or situation that could pose a problem during the course of the work, let the contractor know.

Closing Out the Project

When you decide to complete a home improvement project on your own, you have less of an obligation to finish the project by a certain date, and can often take your time. If you hire a contractor, however, you assume that all of the work will be completed as scheduled. The following items should not be overlooked.

- Inspections and sign-off by your local building department, if needed.

- Occupancy permits obtained, if necessary.

- All warranties, manuals, and other literature associated with new equipment or materials that have been installed—these should be in your possession.

- Materials left over from the job. Ask your contractor for extra materials, such as tiles or flooring, that you can store for future repairs. If you have no need for additional materials and you've paid for them in advance, ask about credit for materials not used.

- Ensure that any remaining tasks at the end of the project (referred to in the industry as a "punch list") have been completed.

Project Forms

The following pages contain forms that can be helpful as you begin making decisions about your home improvement project. The first, "Questions for the Building Department," could be completed over the phone for a small project. For larger jobs, you'll probably have to make an appointment with your local building department. Many homeowners wrongly assume that permits and inspections are not required for small projects, or that the permitting process will cause more aggravation than it's worth. These are false assumptions. The purpose of establishing and enforcing building codes is to promote public safety and adherence to minimum building standards. When in doubt, don't hesitate to ask your local building officials. Obtaining answers to the questions on the form will enhance your understanding of the building process.

The second set of forms, the "Contractor Evaluation Sheet," covers important information you should have for every contractor you consider hiring, such as licensing, experience, contact information, and points to consider before signing a contract. You may need to make several copies if you're interviewing multiple contractors.

Contractor Evaluation Sheet

Part 1: Telephone Interview

Company: _____

Address: _____

Town: _____

State: _____ Zip: _____

Telephone: _____

Cell phone: _____

Main contact: _____

If a post office box is given for the main address, ask for a residential address.

Address: _____

Town: _____

State: _____ Zip: _____

	Yes	No
POINTS TO COVER WITH A CONTRACTOR:		
1. Are you required to have a license?	☐	☐
If yes, what type of license? _____		
If yes, what is your license number? _____		
2. Are you required by the state to be registered?	☐	☐
3. Do you carry Worker's Compensation?	☐	☐
4. Do you carry Liability Insurance?	☐	☐
5. Can your insurance company send me a Certificate of Insurance?	☐	☐
6. Is this business your sole means of support?	☐	☐
If no, what percentage of your time is dedicated to your other interests? _____		
If no, of the remaining time, how much will be dedicated to my project? _____		
7. How long have you operated your business? _____		
8. How many years of experience in the industry do you have? _____		
9. Do you belong to any professional organizations?	☐	☐
If yes, what organizations? _____		
10. Do you subscribe to and read any professional publications?	☐	☐
If yes, what publications? _____		
11. Are you involved personally with performing the work?	☐	☐
If yes, what portions? _____		
12. Will you be involved in other projects while this one is underway?	☐	☐
If yes, what percentage of the time? _____		

	Yes	No

13. Will you be able to start and complete the work before you undertake another project? ☐ ☐

14. Does your work require you to respond to emergencies? ☐ ☐

    If yes, what percentage of the time? _____

15. What are your daily starting and ending times? _____

16. What are your operating and office hours? _____

17. Will you or your employees take vacations during the project? ☐ ☐

18. Do you or your employees work on weekends? ☐ ☐

19. Do you or your employees work on holidays? ☐ ☐

20. In the event you cannot be at the work site as scheduled, how will I know? _____

21. What is your policy on changes I request that may add or subtract from the work? _____

22. What is your policy on changes initiated by you that may add or subtract from the work? _____

23. What is your policy on my giving instructions to your employees? _____

24. What is your policy on cleaning up the work that you or your workers do? _____

25. What is your policy on being a resource to me for work outside your expertise? _____

26. Do you provide warranties for your workmanship? ☐ ☐

    If so, please explain your warranty coverage. _____

27. Do you provide warranties from manufacturers? ☐ ☐

    If yes, please explain how these warranties are implemented. _____

28. Are you or your employees required by any manufacturer to have special training in order to install its product? ☐ ☐

    If yes, what products? _____

    If yes, is the warranty affected without this training? _____

29. Do you provide me with a Release from Payment from your suppliers? ☐ ☐

30. Do you allow me to provide my own materials? ☐ ☐

31. Do you subcontract any of the work? ☐ ☐

    If yes, what parts? _____

    If yes, do you provide me with a Release from Payment from your subcontractor(s)? ☐ ☐

32. What is your policy regarding actions I take that have a negative effect on you, such as a delay in making a decision?

33. Are your workers made aware of your policies and procedures? ☐ ☐

34. When are payments made? _____

35. What form of payment do you accept? _____

36. Are all checks for payment made out to your company? ☐ ☐

    If no, what is your Social Security number? _____

37. Are you willing to be paid according to a bank's schedule? ☐ ☐

		Yes	No

38. Have you previously done work similar to this project? ☐ ☐

    If no, how are you qualified to work on this project? _____

39. Do we agree on who is supplying which materials under what circumstances? ☐ ☐

40. Do you have any other policies or procedures I need to be aware of? ☐ ☐

    If yes, what policies? _____

    If yes, what procedures? _____

41. As references, please provide me with the names and telephone numbers of three of your former customers.

    Name _____ Telephone _____ Time to Call _____

    Name _____ Telephone _____ Time to Call _____

    Name _____ Telephone _____ Time to Call _____

42. As references, please provide me with the names and telephone numbers of three of your professional contacts.

    Name _____ Telephone _____ Time to Call _____

    Name _____ Telephone _____ Time to Call _____

    Name _____ Telephone _____ Time to Call _____

43. As references, please provide me with the names and telephone numbers of three of your suppliers.

    Name _____ Telephone _____ Time to Call _____

    Name _____ Telephone _____ Time to Call _____

    Name _____ Telephone _____ Time to Call _____

Part 2: Points to Cover Before Signing a Contract

	Yes	No

44. Do we agree on the basic costs of the work? ☐ ☐

    If no, what costs need to be resolved? _____

45. Do we agree on the optional costs of the work? ☐ ☐

    If no, what optional costs need to be resolved? _____

46. Do we agree on those items of work that are not included? ☐ ☐

    If yes, what are those items of work that are not included? _____

47. Do you have any miscellaneous charges of which I need to be aware? ☐ ☐

    If yes, what charges? _____

48. Please indicate your payment schedule:

    Payment 1:   Payment 2:   Payment 3:

    Payment 4:   Payment 5:   Payment 6:

49. Please indicate what sections of the work each payment covers:

    Payment 1:   Payment 2:   Payment 3:

    Payment 4:   Payment 5:   Payment 6:

		Yes	No
50.	Are there any unresolved issues?	☐	☐
	If yes, what issues are unresolved? _____		
	If yes, whose responsibility is it to resolve each issue? _____		
51.	Do you have two copies of the specifications?	☐	☐
52.	Have you attached to your contract a copy of the specifications?	☐	☐
53.	Does your contract read: "Refer to Attached Plans and Specifications?"	☐	☐
54.	Have you signed both copies of the contract?	☐	☐
55.	Do you have my signature on both copies of the contract?	☐	☐
56.	Do I have a copy of the signed contract?	☐	☐

FOR THE HOMEOWNER:

The above points were discussed with the contractor as a basis of understanding and do not represent contractual commitments by either party.

Signed _____

Signed _____

Date _____

FOR THE CONTRACTOR:

The above points were discussed with the client as a basis of understanding and do not represent contractual commitments by either party.

Signed _____

Title _____

Date _____

Questions for the Building Department

	Yes	No
1. What general information do you require on permit forms? _____		
2. Do I need to submit plans?	☐	☐
If yes, how many copies do I need to file for permit? _____		
Do the plans need to be prepared and/or stamped by an architect or engineer?	☐	☐
3. Do you require the following on the blueprints?		
1/4" scale floor plan?	☐	☐
Exterior elevations?	☐	☐
Floor framing?	☐	☐
Roof framing?	☐	☐
Interior door schedule?	☐	☐
Exterior door schedule?	☐	☐
Window schedule?	☐	☐
Interior elevation?	☐	☐
Wiring schematic?	☐	☐
Lighting plan?	☐	☐
Heating or air conditioning plan?	☐	☐
Labor specifications?	☐	☐
Material specifications?	☐	☐
4. Do I need a certified plot plan by a registered engineer?	☐	☐
If yes, how many copies? _____		
Is the plot plan required to include septic design?	☐	☐
Site utility locations?	☐	☐
What other elements need to be shown? _____		
5. Is a septic plan and evaluation inspection required?	☐	☐
Is one or the other required for the building permit?	☐	☐
6. Are there fire alarm requirements?	☐	☐
If yes, do I obtain that information from the fire department?	☐	☐
7. How much advance notice is required to schedule an inspection? _____		
8. Does the town require soil compression tests?	☐	☐
If yes, by whom is the testing done? _____		

	Yes	No
9. Does the town require an as-built drawing upon completion of the foundation?	☐	☐
10. Does the town require that the foundation be sited by a registered engineer before pouring of the concrete?	☐	☐
11. Does the town require the foundation to be a specific height above the crown of the street?	☐	☐

If yes, what is the height? ————————————————

	Yes	No
12. Does the town require a footing for the foundation?	☐	☐
If yes, does the town require a footing inspection?	☐	☐
13. Does the town require a foundation inspection before backfilling?	☐	☐
14. Does the town require a post-footing inspection?	☐	☐
15. Does the town require that the rough electrical and plumbing work be completed before framing inspection?	☐	☐
16. Does the town require an insulation inspection?	☐	☐
17. Does the town require fireproofing?	☐	☐
If yes, solid blocking?	☐	☐
If yes, mineral wool?	☐	☐
If yes, fireproof caulking for electrical and plumbing holes?	☐	☐

If an attached garage is being installed are there fire-related material requirements?

	Yes	No
For drywall on wall between garage and house?	☐	☐
For drywall on ceiling of garage?	☐	☐
Fire-rated garage door?	☐	☐

If a new heating system is being installed are there fire-rated material requirements?

	Yes	No
For drywall on walls in system area?	☐	☐
For drywall on ceiling in system area?	☐	☐

18. What are the requirements for a Certificate of Occupancy? ——————————

19. What is the fee for the building permit? ————————————————

20. How long does it take to get the building permit? ——————————————

21. How long does the building permit remain in effect? ————————————

Part Two
Exterior Projects with Estimates

How to Use the Model Projects

Each model project in this book includes a number of components to help you plan your project, purchase appropriate materials, assess the work involved, and accurately budget your costs. The text, illustration, and cost estimate all work together to provide a realistic picture of the project and its level of difficulty.

The text gives you an overview of the work, materials and tools, hints on how to avoid problems, and the level of difficulty. For do-it-yourselfers, the "Level of Difficulty" is your key to determining whether or not it's a good idea to build all or a portion of the project yourself. The "What to Watch Out For" section lets you know how to cope with tricky procedures. In some cases, an "Options" section suggests alternate or additional materials, or ways to enhance the project.

The illustrations can help you visualize a project of this type. Inspiration photographs are sometimes included to help you get a sense of what the finished product might look like, but are not intended to exactly duplicate the estimated project.

The cost estimate for each project lists all the needed materials and their quantities, along with professional installation times, and corresponding costs.

The "Detailed Costs" section following the model projects provides costs for individual construction items—helpful if you want to add or substitute components in one of the model plans to customize your project. Tips for using the Detailed Costs can be found at the beginning of that section.

CAUTION: The projects outlined in this book are intended as a general overview only. Consult a skilled professional to learn proper, complete, and detailed procedures for all home improvement work.

Key to Abbreviations

C.L.F. – 100 linear feet
C.Y. – cubic yard
Ea. – each
Gal. – gallon
Lb. - pound
L.F. – linear foot
Pr. – pair
Sq. – square
S.F. – square foot
S.Y. – square yard
V.L.F. – vertical linear foot

Project Size: The basic dimensions of each model project, where applicable. The cost estimates are based on these measurements.

Quantity/Unit: How much of each item to purchase, and its standard unit of measure. *(See the Key to Abbreviations, also on this page.)*

Description: List of the materials and components in the model project. *(Some projects include a separate "Alternate Materials" box, with costs per square foot, linear foot, etc., so you can, for example, substitute marble for ceramic tile, or hardwood instead of vinyl flooring.)*

Labor-Hours: The estimated time it takes a professional tradesperson to perform the installations in the project. The total number of contractor's hours required for a complete project is given at the bottom of this column.

Material: An estimated cost for each item in the "Description" column. Each cost is derived by multiplying the price of a single unit by the required number of units shown in the Quantity/Unit column. The prices reflect the national average of what you can expect to pay at a building supply retailer. The total material cost for the whole project is at the bottom of this column. (Note that sales tax has not been included in these figures and will have to be added for a more accurate total.) To tailor these prices to your specific location, refer to "Adjusting Project Costs to Your Location" at the back of this book.

Contractor's Fee, Including Materials: The total price a contractor would charge to purchase and professionally install all of the materials in the project. Factored into this cost is an estimate of the overhead and profit a contractor is likely to add to the total cost. (Overhead includes items such as the contractor's time spent visiting your home, planning and estimating the job, and a percentage of the miscellaneous costs of doing business.) Refer to "Adjusting Project Costs to Your Location" at the back of this book to adjust this national average cost to your location.

Attached Single Garage, 12' x 22'

Description	Quantity/ Unit	Labor-Hours	Material
Site clearing, layout, excavate for footing	1 Lot	7.0	
Footing, 8" thick x 16" wide x 44' long	2 C.Y.	5.6	242.40
Foundation, 6" concrete block, 3'-4" high	264 S.F.	23.2	529.06
Edge form at doorway	10 L.F.	0.5	3.24
Floor slab, 4" thick, 3000 psi, with sub-base, reinforcing, finished	265 S.F.	5.6	340.26
Wall framing, 2 x 4 x 10', 16" O.C.	38 L.F.	6.1	162.34
Headers over openings, 2 x 8	14 L.F.	0.7	13.94
Sheathing for walls, 1/2" thick, 4' x 8' sheets	384 S.F.	4.4	285.70
Ridge board, 1 x 8 x 12'	12 L.F.	0.3	17.57
Roof framing, rafters, 2 x 6 x 14', 16" O.C.	280 L.F.	4.5	181.44
Joists, 2 x 6 x 22', 16" O.C.	220 L.F.	2.8	142.56
Gable framing, 2 x 4	45 L.F.	1.3	18.36
Sheathing for roof, 1/2" thick, 4' x 8' sheets	352 S.F.	4.0	261.89
Sub-fascia, 2 x 8 x 12'	24 L.F.	1.7	23.90
Felt paper, 15 lb.	340 S.F.	0.7	12.24
Shingles, asphalt, standard strip	4 Sq.	6.4	194.40
Drip edge, aluminum, 5"	52 L.F.	1.0	15.60
Flashing, aluminum	21 S.F.	1.2	16.38
Fascia, aluminum	21 S.F.	1.2	53.93
Soffit, aluminum	38 S.F.	1.4	46.06
Gutters and downspouts, aluminum	44 L.F.	2.9	65.47
Load center, circuit breaker, 20A	1 Ea.	0.7	10.26
Non-metallic sheathed cable, #14, 2-wire	40 L.F.	1.3	8.16
Switch, w/ 20' #14/2 type NM cable	1 Ea.	0.5	9.72
Duplex receptacle, 20A, w/ 20' #12/2 type NM cable	1 Ea.	0.6	22.80
Lighting outlet box and wire	1 Ea.	0.3	11.28
Light fixture, canopy type, economy grade	1 Ea.	0.2	31.20
Wood, window, standard, 34" x 22"	1 Ea.	0.8	250.80
Garage door, wood, incl. hardware, 9' x 7'	1 Ea.	2.0	528.00
Siding, cedar, rough-sawn, stained	400 S.F.	13.3	1,444.80
Sheetrock, 5/8" fire resistant	484 S.F.	8.0	162.62
Painting, door and window trim, primer	85 S.F.	1.7	5.10
Painting, door and window trim, 2 coats	85 S.F.	1.7	5.10
Totals		113.6	$5,116.58

Contractor's Fee, Including Materials: **$13,758**

Decks

A deck's size, materials, and design should be determined by its intended use, the style and size of your house, and your budget. Following are some tips that may help get you started.

- Be sure to integrate the new deck with the house and your landscaping. For example, posts and railings for a new deck should coordinate with the house trim.

- When deciding on a size for the deck, add a few more feet than you think you'll need. Most people underestimate the space they'll need for furniture and activities.

- When shopping for deck materials, try to select the decking boards; avoid those with excessive knots, voids, twisting, warping, or cupping.

- The factory coloring of pressure-treated deck material will gradually change to a natural wood color over several years. If you want to apply a finish treatment to your deck, transparent stain looks good and requires little upkeep. If you use a heavy-bodied stain, regular maintenance will be needed to preserve its appearance. Finishing your deck will generally ensure a consistent aging process; in addition, finishing allows you to use different materials. If you're adding to an existing deck and wish to match it, painting or staining may be necessary.

- Premium length stock is more expensive, but the fewer the joints, the better.

- If the deck will be 2'-6" or more above the ground, the railing must be strong enough to prevent a 200-pound person from falling, and its rails must be spaced close enough to prevent a 4" ball from passing through it. Consult your local building department for spacing of balusters or decorative rail components.

- Framing anchors and hangers to secure deck members may be specified by local building codes.

- In damp areas, decks can develop a mossy surface. This can be removed with a mild bleach solution.

- Exposure to extreme weather conditions dries wood out, making it susceptible to water retention and rot unless it's treated annually with sealer-preservative.

- The most common materials are cedar, pressure-treated pine, or composites made from recycled plastic or other materials. Comparison-shop, ask questions about structural requirements and fastening systems, and don't be afraid to mix composites with traditional framing lumber, such as using pressure-treated framing material with composite material decking.

Elevated Deck

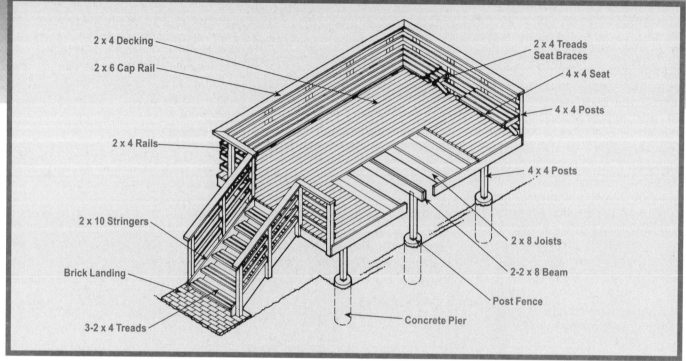

- 2 x 4 Decking
- 2 x 6 Cap Rail
- 2 x 4 Rails
- 2 x 10 Stringers
- Brick Landing
- 3-2 x 4 Treads
- 2 x 4 Treads Seat Braces
- 4 x 4 Seat
- 4 x 4 Posts
- 4 x 4 Posts
- 2 x 8 Joists
- 2-2 x 8 Beam
- Post Fence
- Concrete Pier

This model project is a 10' x 16' deck with a stairway with safety railing and a two-sided bench railing.

Materials

The 10' x 16' deck is constructed from pressure-treated wood. Consider the attributes of the various types of deck materials, including mahogany, fir, vinyl, and composites, before deciding which is best for your home. Wood-polymer decking is a popular choice, as it doesn't warp or require a preservative, and resembles wood in appearance. Be sure to arrange delivery (often free of charge), as the lumber requires trucking. The support system and frame for the pressure-treated deck plan consist of six 4 x 4 posts set in 3' concrete footings. The 2 x 8 joists are spaced 2' on center, with doubled support joists at 6' intervals, and doubled 16' headers. The foundations

require 8" diameter tube forms. The decking in this plan provides an attractive and durable surface, but boards of other widths can also be used. Changing the pattern of the decking will affect the cost.

There are many design options for the railing, bench, and stairway. Standard 4 x 4 posts and 2 x 4 rails are the most economical, but 2 x 2 balusters can be installed at added expense to enhance the railing design or make it child-safe. The built-in bench can be altered to suit personal taste or practical needs. Costs can be cut by eliminating the bench altogether, or by reducing its length and limiting it to one side of the deck. The railing, an essential safety feature in any deck this high off the ground, consists of 2 x 4 rails, and 4 x 4 fitted and lagged posts.

What's Involved

It's crucial that you lay out the footing and post locations accurately, since the rest of the structure depends on their alignment. Double check all measurements as you go. Before you begin constructing the frame, review the layout template to be sure the doubled support joists will be correctly aligned when they are placed. Square the four corners before locating the post positions, and make sure you follow the plan. Economizing by eliminating structural components will cost you more money and aggravation in the long run because of sagging, loosening of the structure, and other conditions that require repair.

The deck frame must be plumb, level, and square. Before fastening the joists, review the accuracy of the layout to be sure the doubled joists are precisely aligned with the 4 x 4 support posts. This alignment must be exact to ensure safe bearing of the structure's load. Before laying the decking, double-check the frame for level and square. Then tie the outside corners temporarily at right angles while you place the decking.

For conventional decking, be sure to measure accurately and make precise saw cuts; the cut-off pieces will be used later to fill out the courses. Place the cupped side of the decking boards down, and use enough nails based on professional guidance. To keep the rows of nail heads straight, lay the entire deck with minimum fastenings, and then complete the nailing by following chalk lines. Use a spacer to keep the opening between courses at about 1/8" or slightly wider. This space is important, as it allows for drainage of rainwater, helps to ventilate the structure, and prevents the build-up of moisture-laden matter between the deck boards.

Seek assistance from a qualified person if you have not installed a stairway before. Safety is a primary consideration for any set of stairs, and consistent vertical spacing from tread to tread is one way of ensuring it. Proper support and correct placement of stringers and stairway railings are also critical safety issues.

Level of Difficulty

The pressure-treated deck can be built by most people with a reasonable level of skill in home improvement projects. Its 10' x 16' size makes it a major project, but the work can be spread out over a few weekends. Basic carpentry knowledge is needed for all the deck's components. Beginners can tackle this project if they get expert instruction in advance and during construction. If you're a novice, plan to triple the labor-hours in the estimate. Intermediates

should add 40% to the professional time for the framing and decking tasks and 75% for the footing, post, railing, bench, and stairway installations. Experts should add 10% to the estimated labor-hours for all parts of this project. If you're installing boards diagonally, this will add to your time.

What to Watch Out For

If a deck is 4' or less above grade level, installing the supports is fairly easy; but for decks higher than 4' or on sloping terrain, the job is more difficult. Plan to use a stepladder for some of the railing

work, as much of the layout and fastening can be approached most efficiently from outside the structure.

If you're concerned about the safety of preservatives in pressure-treated lumber, you should know that the substance in question, CCA (chromated copper arsenate), is no longer allowed as a wood preservative for residential applications as of 2004, and has been replaced.

See also:

Ground-Level Deck, Elevated L-Shaped Deck, Multi-Level Deck, Roof Deck, Patio Doors, Outdoor Living Area.

Elevated Deck, 10' x 16'

Description	Quantity/Unit		Labor-Hours	Material
Layout, excavate post holes	0.50	C.Y.	0.5	
Concrete, field mix, 1 C.F. per bag, for posts	6	Bags		41.76
Forms, round fiber tube, 1 use, 8" diameter	18	L.F.	3.7	32.62
Deck material, pressure-treated posts, 4 x 4 x 4'	24	L.F.	0.9	38.88
Post base, 4 x 4	6	Ea.	0.4	39.24
Post cap, 4 x 4	6	Ea.	0.4	19.22
Headers, 2 x 8 x 16'	64	L.F.	2.3	76.80
Joists, 2 x 8 x 10'	120	L.F.	4.3	144.00
Decking, 2 x 4 x 12'	576	L.F.	16.8	338.69
Stair material, pressure-treated stringers, 2 x 10 x 10'	20	L.F.	0.8	35.76
Treads, 2 x 4 x 3'-6", 3 per tread	70	L.F.	2.0	41.16
Landing, brick on sand, laid flat, no mortar, 4.5 brick per S.F.	16	S.F.	2.6	47.81
Railing material, pressure-treated posts, 4 x 4	48	L.F.	1.7	77.76
Railings, 2 x 4 stock	240	L.F.	7.0	141.12
Cap rail, 2 x 6 stock	48	L.F.	1.5	44.35
Bench material, seat braces, 2 x 4 stock	108	L.F.	3.1	63.50
Joist and beam hangers, 18 ga. galvanized	12	Ea.	0.5	8.35
Nails, #10d galvanized	15	Lb.		25.02
Bolts, 1/2" lag bolts, 4" long, square head, with nut and washer	28	Ea.	1.6	12.10
Totals			50.1	$1,228.14

Contractor's Fee, Including Materials:	**$4,596**

Elevated L-Shaped Deck

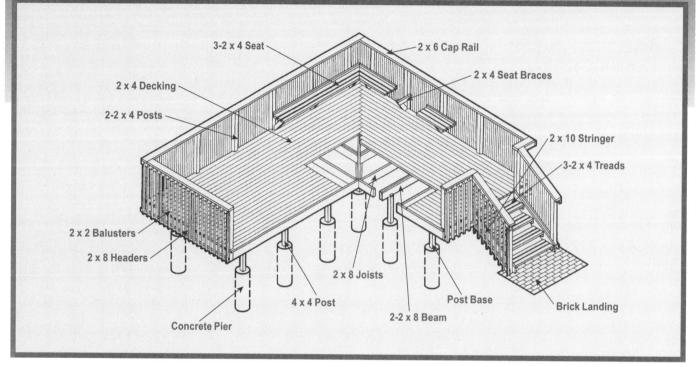

3-2 x 4 Seat

2 x 4 Decking

2-2 x 4 Posts

2 x 2 Balusters

2 x 8 Headers

Concrete Pier

4 x 4 Post

2 x 8 Joists

2-2 x 8 Beam

Post Base

2 x 6 Cap Rail

2 x 4 Seat Braces

2 x 10 Stringer

3-2 x 4 Treads

Brick Landing

One of the fun things about deck design is the variety of configurations that can be used to maximize the space, based on the style of the house and the surrounding terrain. This L-shaped plan consists of two separate decks tied into one structure, bordering two sides of the house. The same modular construction method can also be used for U-shaped, angled, and multi-level deck designs. With proper instruction before they start, and guidance along the way, many do-it-yourselfers can tackle this project with professional results.

Materials

The lumber for this deck is standard, pressure-treated pine. The legs of the L are formed from two separate, but attached, deck frames, each with its own 4 x 4 post-and-footing support system. More time is required to lay out the footings for this deck because there are more of them, but the process is the

same as with simpler decks. The most important aspect of the footing and post placement is the establishment of a square and level support system for the rest of the structure. If you are inexperienced in this procedure, get some help from a knowledgeable person and take the job in steps, building the frame for the large 8' x 16' section first before laying out the footing and post support for the 8' x 12' section. This way, you can adjust for small alignment errors and reduce materials waste resulting from layout errors.

What's Involved

The frame consists of 2 x 8 joists and headers that are assembled as two separate units. As in the layout for the footings and support posts, be sure to double check all measurements and make sure the corners of each section are properly squared. Since this deck is built only 3' off the ground, most of the frame for the 8' x 16' section should be

built in place. If you have room on the ground next to the deck, you could lay out and start the frame on the ground, then position it on the 4 x 4s to be finished in place. The 8' x 12' frame can be built in the same way.

The decking consists of 2 x 4s, but 2 x 6s may also be installed on frames with 24" joist spacing. If you want to use thinner deck boards, the joists will have to be placed at closer intervals and their number will be increased, so plan accordingly. Measure and cut the boards accurately, so that the shorter, cut-off pieces can be used to fill out the courses as the decking is placed. When laying the boards, place the cup side down and leave a space between courses. Also, keep a check on the gain at both ends and the middle as you proceed, to be sure that the decking is square to the frame. After all of the boards have been placed and fastened, snap a chalk line and saw the ragged ends to an even finished length.

The railing, bench, and stairway design offer the opportunity to add an imaginative touch to the deck. This plan, for example, suggests a corner bench design. The railing can be modified to suit your taste and your home's style. In both cases, the cost of the project will be affected, depending on the alterations. A balustered railing design is attractive and safe, but the material expense and installation time for these are considerable. The same is true for more elaborate or additional built-in benches.

The ship's ladder stairway design enhances the open motif of the deck, but you can also use conventional stringers and treads for the same material cost. If you're not experienced in stairway building, seek instruction before you start. Accurate layout of tread height is a must.

Level of Difficulty

The carpentry skills required for this project are the same as for standard straight, square, or rectangular decks. Aligning the two sections is a little more complicated, but if you take the project in steps, the layout is simply a matter of more time, not more skill.

Beginners who have ample instruction can do much of the work on this project, but will need time to proceed slowly and should seek advice as they go. Intermediates and experts should also allow additional time for the layout and placement of the support system for this deck. Beginners should triple the professional time estimates for most tasks. Intermediates should add about 60% to the estimate; and experts about 20%. If the deck will be built more than 6' off the ground, more time will be required for everyone.

What to Watch Out For

Take some time to draw up the plans and do some reading on the subject before you decide on a deck design, as this project is a big investment of both time and money. If you're concerned about the safety of preservatives used in pressure-treated lumber, you should know that the substance that has been most questioned is CCA (chromated copper arsenate), which producers have replaced with other treatments. CCA is no longer allowed for use as a wood preservative for any residential uses.

Options

Multi-level decks are not that difficult to build, particularly if they are close to ground level. U-shaped, angled configurations, and even hexagonal or octagonal designs, where appropriate, can be built to professional standards by do-it-yourselfers who are willing to learn first and then work slowly.

See also:
Elevated Deck, Ground-Level Deck, Multi-Level Deck, Roof Deck, Patio Doors, Garden Arbor, Outdoor Living Area.

Alternate Materials
Elevated L-Shaped Deck
Cost per Square Foot, Installed

2 x 6 composite	$4.27
2 x 4 treated pine	$3.26
2 x 6 treated pine	$3.18
1 x 4 redwood	$7.10
2 x 6 cedar	$12.00

Elevated L-Shaped Deck, 8' x 12' + 8' x 16'

Description	Quantity/Unit		Labor-Hours	Material
Layout, excavate post holes	1	C.Y.	1.0	
Concrete, field mix, 1 C.F. per bag, for posts	12	Bags		83.52
Forms, round fiber tube, 1 use, 8" diameter	42	L.F.	8.7	76.10
Post base, 4 x 4	14	Ea.	0.9	91.56
Post cap, 4 x 4	14	Ea.	0.9	44.86
Deck material, pressure-treated posts, 4 x 4 x 3'	42	L.F.	1.5	68.04
Headers, 2 x 8 x 16'	80	L.F.	2.8	96.00
Joists, 2 x 8 x 8'	96	L.F.	3.4	115.20
Joists, 2 x 8 x 12'	72	L.F.	2.6	86.40
Decking, 2 x 4 x 12'	720	L.F.	20.9	423.36
Stair material, pressure-treated stringers, 2 x 10 x 6'	12	L.F.	0.5	21.46
Treads, 2 x 4 x 3'-6", 3 per tread	60	L.F.	1.7	35.28
Landing, brick on sand, laid flat, no mortar, 4.5 brick/S.F.	16	S.F.	2.6	47.81
Railing material, pressure-treated posts, 4 x 4	56	L.F.	2.0	90.72
Balusters, 2 x 2 stock	540	L.F.	14.4	194.40
Cap rail, 2 x 6 stock	72	L.F.	2.3	66.53
Bench material, pressure-treated seats & braces, 2 x 4 stock	96	L.F.	2.8	56.45
Joist and beam hangers, 18 ga. galvanized	18	Ea.	0.8	12.53
Nails, #10d, galvanized	20	lb.		33.36
Bolts, 1/2" lag bolts, 4" long, square head, with nut and washer	40	Ea.	2.3	17.28
Totals			72.1	$1,660.86

Contractor's Fee, Including Materials: **$6,441**

Multi-Level Deck

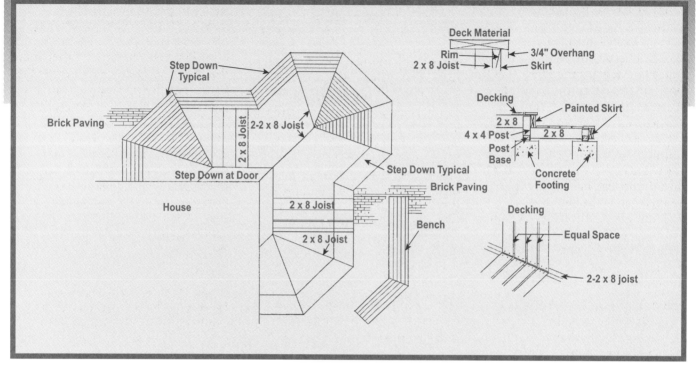

The introduction of wood-polymer lumber has opened up new possibilities for decking. What was once thought of as a structure that will need annual cleaning, staining, and repairs has, with this material, become a long-lasting, low-maintenance, valuable outdoor living space. This project is an advanced, multiple-level design that wraps around the corner of the house. It requires good carpentry skills, and up-front time spent selecting materials and planning the layout. Multi-level decks are easily adapted to sloped yards and allow you to eliminate stairs by using step-down areas. This layout can also nicely conceal an exposed foundation. The step-down in this project consists of a second platform that extends around the perimeter of the upper level.

Materials

This model project uses synthetic/ recycled decking made from plastics and/or wood products. Wood-polymer decking and lumber have a wood-like appearance, do not warp, are resistant to moisture and insects, are usually slip-resistant, and never require preservative treatments, staining, or sealing. The materials can be sawed, sanded, nailed, screwed, and drilled, and an experienced do-it-yourselfer should have no problem working with them. Because all synthetic/recycled decking and lumber is not alike, it's important to read the specific manufacturer's recommendations on where and how to use the different-sized members, recommended joist spans and space between boards, and material performance in extreme weather conditions.

The framing is accomplished with pressure-treated lumber per building code requirements. The skirt that covers the riser at the step-downs could be synthetic lumber, or as in this case, painted wood trim. Because synthetic decking does not chip, split, or crack, you should be able to minimize waste by reversing the cut-off board pieces as you go around the deck. Consider purchasing decking in lengths that work well for using up the entire piece.

You may want to wait until you've installed most of the decking before purchasing just enough additional material to complete the job.

Most deck fasteners work well with wood-polymer lumber, though stainless steel and hot-dipped galvanized fasteners are recommended. Because of the longevity of synthetic decking and lumber, high-quality fasteners and anchors should be used for the deck framing and anchoring systems. This plan requires more framing anchors and joist hangers than typical wood rectangular or square decks because of the greater weight of the material and the more demanding design. Check with your material supplier for custom hangers that may not be on the shelf, but can be special-ordered. Planning ahead to use the most appropriate materials can save a lot of aggravation for the inexperienced builder, and ensures a high quality job for the more advanced do-it-yourselfer.

What's Involved

The design features an octagonal extension at the corner of the deck, and 45° angles at the other two outer corners. The center of the octagonal section is located off of the corner of the house by bisecting the right angle created by extending the exterior wall lines from the corner (this should be a 90° angle). The pattern of the decking is designed so that no ends are showing. Each piece of decking requires careful measuring and cutting. The key to this design is to perfect the layout so that when the decking material is applied, the angle cuts at the joints are equal, or 22.5°. Any error in the layout will cause the lengths of the angle cut at the end of the decking to be unequal, which would detract from the appearance of the pattern.

Although wood-polymer lumber does not noticeably swell or shrink in wet or dry conditions, changes in temperature can cause some expansion and contraction. The boards should be spaced appropriately—both between the decking boards and where end joints meet, using removable spacers.

Because of the deck pattern, it's best to begin at the outside edges of the deck and work towards the center to ensure a proper and consistent overhang. (The outside piece of decking should overhang the skirt.) Synthetic/recycled plastic decking requires joist spacing similar to that used for traditional lumber. When using 5/4 x 6 or 2 x 6 decking boards, the joist spacing should be 16" on center.

Level of Difficulty

This project requires an ability to visualize the end result while planning the job, laying it out, and building it. Working on a multi-level design is different than working at one elevation. Care must be taken to maintain proper riser heights (7" is recommended) between levels. You also need a basic understanding of the geometry of bisecting angles, and the ability to maintain straight and true lines.

While not strictly necessary, a builder's level might be easier to use than a standard level because it allows you to set multiple elevations from one point. The elevation of the top of the footings in relation to the top of the finished deck and step-downs is critical for safety. Cutting the decking and minimizing waste is somewhat difficult for a beginner. Constructing this type of deck requires two people—for pulling and snapping chalk lines, and for holding the decking pieces at the correct angle for nailing. Beginners should add 100%, and intermediates and experts 50% and 20%, respectively, to the estimated labor-hours.

What to Watch Out For

Although this deck is large, don't assume that it can carry greater loads.

Consult construction professionals or your town building inspector if you expect heavier than normal loads (such as large groups of people, a hot tub, or landscape planters) on the deck. Make sure the structure complies with local ordinances for setbacks from property lines. Don't underestimate the fastening required to safely attach the deck to the house. Take time to locate the fasteners so that they won't interfere with the joist layout. Find out the specific installation, performance, and maintenance expectations for different synthetic decking materials. Research the proper cleaning procedures and what to expect in terms of fading and aging. Plan for access to outlets or wiring for lighting, speakers, or appliances.

See also:
Ground-Level Deck, Elevated Deck, Elevated L-Shaped Deck, Roof Deck, Outdoor Living Area, Garden Arbor.

Multi-Level Deck, 800 S.F.

Description	Quantity/Unit		Labor-Hours	Material
Layout, excavate post holes	3	C.Y.	3.0	
Concrete, field mix, 1 C.F. per bag, for posts	26	Bags		180.96
Forms, round fiber tube, 1 use, 8" diameter	78	L.F.	16.1	141.34
Post base, 4 x 4	26	Ea.	1.6	170.04
Deck material, pressure-treated posts, 4 x 4 x 3'	61	L.F.	2.2	98.82
Headers, 2 x 8 x 12'	396	L.F.	14.1	475.20
Joists, 2 x 8 x 10',12',14',16'	640	L.F.	22.8	768.00
Decking, wood-polymer lumber, 5/4 x 6 x 16'	1,540	L.F.	38.5	5,008.08
Wood skirt board, 1 x 8 x 10'	240	L.F.	8.5	449.28
Drilling anchors, 3/4" diameter, 4" deep	20	Ea.	3.6	4.32
Expansion anchors, 3/4" diameter, long	20	Ea.	2.5	90.72
Lag screws, 3" long, 1/2" diameter	20	Ea.	1.2	9.12
Bolts & hex nuts, 1/2 x 4	40	Ea.	2.8	17.28
Joist hangers, heavy duty 12 ga., galvanized, 2 x 8	96	Ea.	4.7	77.18
Joist hangers, heavy duty 12 ga., galvanized, two 2 x 8	21	Ea.	1.1	63.76
Connector plates	6	Ea.	1.9	198.00
Paint skirt	240	L.F.	3.0	17.28
Stainless steel nails, plain	25	Lb.		172.50
Totals			127.6	$7,941.88

Contractor's Fee, Including Materials: **$18,517**

Ground-Level Deck

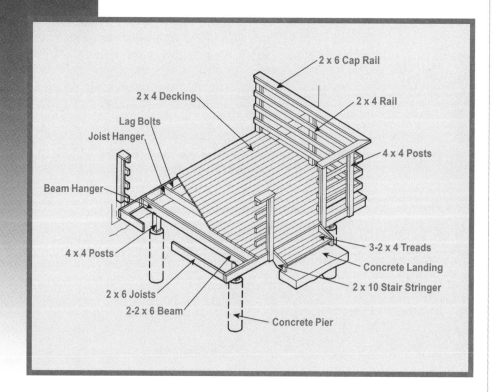

2 x 6 Cap Rail

2 x 4 Decking

2 x 4 Rail

Lag Bolts

Joist Hanger

4 x 4 Posts

Beam Hanger

3-2 x 4 Treads

4 x 4 Posts

Concrete Landing

2 x 10 Stair Stringer

2 x 6 Joists

2-2 x 6 Beam

Concrete Pier

Adding a deck is one of the easiest and most economical ways to increase your home's value and living space. Single-level decks with standard decking patterns are usually manageable for do-it-yourselfers, even those with limited building experience. This plan shows how an attractive and functional structure can be built for a modest investment of both time and money. The project also provides an excellent opportunity for inexperienced do-it-yourselfers to improve their construction skills.

Materials

Most decks built at foundation or ground level have the same basic components. These vary in size and number, according to the type of wood, the method of installation, the dimensions of the structure and its height above grade, and the complexity of the design. The materials used in this model project are basic decking products available at most building supply outlets. Since they vary somewhat from one retailer to another, it's best to do some comparison-shopping. When you purchase the

lumber, select each piece yourself, if possible, looking for straight deck boards and sound support members. Clarify the conditions of delivery at the time of the purchase. Most suppliers will make curb deliveries free of charge for a complete deck order.

All of the wood products for this deck are designed for use in exterior structures. Pressure-treated deck materials are more costly than conventional, untreated lumber, but the additional expense is well worth the investment to guarantee the deck's longevity. Consider the attributes and costs of the various other types of deck materials as well, including mahogany, fir, vinyl, and composites, before deciding which is best for your home.

The support system for this 8' x 10' ground-level deck consists of footings, 4 x 4 posts, and the frame, which is comprised of 2 x 6 joists and headers. Cylindrical tube forms for the supports are available at most building supply centers and lumberyards.

The decking in this plan consists of 2 x 4s laid across the frame and fastened

with galvanized nails. Decking materials of another thickness, width, or edge type can also be installed. Remember that the decking is the most visible component of the structure, so additional expense for aesthetic reasons may be a worthwhile investment.

This model project includes a basic post-and-rail design for the railing system. It's one of the simplest railings to install, and it creates an attractive border with a 2 x 6 cap, which can double as a convenience shelf or a support for planters. More elaborate railings are also possible with additional materials and installation time. A more extensive railing, with spaces no wider than 4" between rails or balusters, would be required if the deck is elevated at 2'-6" or more above the ground.

The posts for the railings should be constructed from 4 x 4s mortised into the frame and then fastened with lag bolts. Although the fitting of the posts takes a little longer, and the lag bolts are more expensive than nails, the finished installation looks better, is more secure, and will last longer.

What's Involved

Before you begin the excavation for the footings, lay out their locations accurately to ensure a square and level support system for the rest of the structure. Several methods can be employed to establish the location of the support posts and their footings, but one of the most accurate is to approximate the layout with batter boards at the four corners, then establish, by the use of string lines, equal diagonals.

Failure to place the footings and posts precisely at square with one another and with the side of your house will adversely affect the rest of the project. Double-check the post locations before you set them in concrete, and be sure to use cylindrical tube forms for the foundations. Review the layout of the 2 x 6 frame as well, to be sure that the doubled joists are correctly aligned with the 4 x 4 supports.

When building the frame, be sure that it is completely plumb and level. Mortise the 4 x 4 railing posts to the frame and fasten with the lag bolts. Lay the decking on the frame, leaving at least 1/8" between each board, and once they are all where you want them, snap chalk lines as guides for nailing them in place.

Level of Difficulty

This basic deck plan is an approachable project for beginners and intermediates. A few basic carpentry skills, along with some experience in measuring, plumbing, and leveling, are required. Beginners should seek advice before they start and as the project progresses, and should double the professional time for all tasks involved in this project. Intermediates and experts should add 30% and 10%, respectively, to the estimated labor-hours.

What to Watch Out For

This plan is very economical and basic in its design, but it is not child-proof. Safer, more expensive railings should be installed to protect small children. Stairway railings are not included in this plan because of its one-step access, but they should be installed (at extra cost) on deck stairways with more than one step.

Because this deck is built at ground level with limited access to the terrain underneath, precautions should be taken to restrict unwanted vegetation from growing in this area. Some weeds and hardy grasses can establish themselves under the deck, eventually restricting ventilation and creating an eyesore. Decks that are built in sunny locations are particularly susceptible to this problem. To prevent it, remove about 3" to 4" of topsoil from the area to be covered by the deck and then cover it with a layer of polyethylene. Place enough gravel over the plastic to bring the entire area to grade level. This procedure will increase the cost of the job, but it will prevent a problem that is difficult to correct after the deck is in place.

If you're concerned about the safety of preservatives in pressure-treated lumber, you should know that the substance that has been considered hazardous, CCA (chromated copper arsenate), has, as of 2004, been replaced with other treatments.

See also:

Elevated Deck, Elevated L-Shaped Deck, Multi-Level Deck, Roof Deck, Outdoor Living Area, Garden Arbor.

Ground-Level Deck, 8' x 10'

Description	Quantity/ Unit		Labor- Hours	Material
Layout, excavate post holes	0.50	C.Y.	0.5	
Concrete, field mix, 1 C.F. per bag, for posts	4	Bags		27.84
Forms, round fiber tube, 1 use, 8" diameter	12	L.F.	2.5	21.74
Post base, 4 x 4	4	Ea.	0.2	26.16
Post cap, 4 x 4	4	Ea.	0.2	12.82
Deck material, pressure-treated posts, 4 x 4 x 4'	16	L.F.	0.6	25.92
Headers, 2 x 6 x 10'	40	L.F.	1.3	36.96
Joists, 2 x 6 x 8'	72	L.F.	2.3	66.53
Decking, 2 x 4 x 10'	280	L.F.	8.1	164.64
Stair material, pressure-treated stringers, 2 x 10 x 3'	6	L.F.	0.2	10.73
Treads, 2 x 4 x 3'-6", 3 per tread	12	L.F.	0.3	7.06
Landing, precast concrete, 14" wide	4	L.F.	0.1	19.30
Railing material, pressure-treated posts, 2 x 4 x 3'	30	L.F.	0.9	17.64
Railings, 2 x 4 stock	52	L.F.	1.5	30.58
Cap rail, 2 x 6 stock	26	L.F.	0.8	24.02
Joist and beam hangers, 18 ga. galvanized	7	Ea.	0.3	5.63
Nails, #10d galvanized	8	Lb.		13.34
Bolts, 1/2" lag bolts, 4" long, square head, with nut and washer	18	Ea.	1.0	7.78
Totals			20.8	$518.69

Contractor's Fee, Including Materials: **$1,934**

Roof Deck

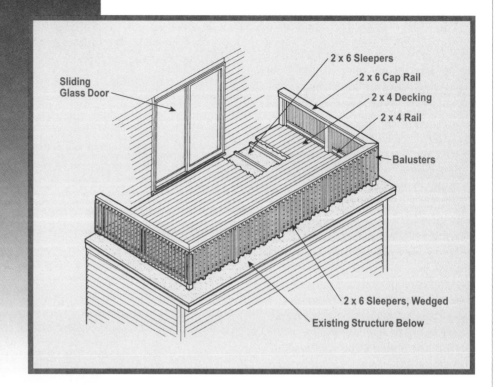

Labels in illustration:
- Sliding Glass Door
- 2 x 6 Sleepers
- 2 x 6 Cap Rail
- 2 x 4 Decking
- 2 x 4 Rail
- Balusters
- 2 x 6 Sleepers, Wedged
- Existing Structure Below

Not all houses can accommodate a roof deck, but those that can have the potential to add a unique outdoor living area. This plan demonstrates how normally unused roof area can be turned into deck space. It's a challenging project for the do-it-yourselfer because of the general inconvenience involved in working at a height and without damaging the roof materials, and because of the many carpentry skills that are required.

Generally, a flat or gently sloping roof is best suited for a deck. A steeper roof can also accommodate a deck, but a different support system must be employed. The critical factor is the existing dormer, window, or exterior wall, which must be able to accommodate a door for access to the proposed deck. If you don't have these basic structural conditions, the project cannot be accomplished without costly preliminary work. If you are planning to build a one-story addition on your home, you might want to consider incorporating a roof deck, provided a flat roof or low-pitched roof fits in with your house design.

Materials & Tools

Conventional deck materials and design are used in this roof deck plan, with the addition of sleepers that function as the deck frame and support system. Pressure-treated lumber is employed for the basic components, including the sleepers, decking, and railing. You'll also need shims for leveling, caulking, and paint or stain if you plan to finish the railing. Many options are available in the choice of doors. Make sure yours is compatible with the design of your house, and the structure and size of the wall. This plan includes the cost of materials for framing a 6' sliding glass door. Tools for this project are standard hand and power tools used for carpentry projects.

What's Involved

If the roof is flat, the 2 x 6 sleepers can be laid on edge as in this model project; but if the roof is slanted, they have to be sawn at an angle to compensate for the slope. Be sure to make the angled cuts precise and consistent for all of the sleepers to ensure stability and equal load distribution. Some shimming will also be needed. Before you place the sleeper system, carefully check the condition of the roof and its rafters. If there are questionable areas, take steps to correct the situation, seeking professional advice and assistance if needed. Take care not to damage or puncture the roofing material during the deck installation process. Caulk all points where the deck is attached to the roof.

After the sleeper frame has been placed, fastened, and squared, the decking and railing are installed as in any other deck. The 2 x 4 boards are laid with staggered end joints and with a space between courses to allow for drainage. The railing consists of doubled 2 x 4 posts, fitted and lagged into the sleeper frame, 2 x 4 rails, 1 x 1 balusters, and a 2 x 6 cap. Great care should be taken in the design of the railing system for the roof deck; safety should be a primary criterion. Check your local building codes and regulations before construction begins to determine the minimum requirements for safety railings on above-ground decks.

The doorway to the roof deck should be installed before the deck is constructed to provide a convenient means of access to the roof site during construction. The preparations will vary depending on the type and size of door you select. Work carefully when cutting the opening to prevent excessive damage to the interior wall and exterior siding. Whenever possible, it pays to use an existing window space as part of the new door opening. You'll have to enlarge and modify the rough window opening, but you'll still be a step ahead in your door installation by using this approach.

Level of Difficulty

Beginners should not attempt to build this deck on their own unless it is to be located on a flat roof. Even then, they should seek detailed professional advice before they begin the job or purchase the materials. Beginners should leave the entire door installation to a professional, as the cutting of the opening and placement of the new unit involves considerable carpentry knowledge and skill.

Intermediate do-it-yourselfers should be able to complete the deck construction in about twice the time given in the estimate for professional work, but they should add more time if the roof area is not easily accessible. Plan to work slowly and seek guidance. Hole-cutting and door installation are not tasks for intermediates who are inexperienced in these areas. Experts who are well-versed in these operations should add about 30% to the professional time for all tasks. They should plan on a bit more time for the hole-cutting and door installation, as hidden problems tend to arise with these tasks, especially in older homes.

If electrical wiring is located in the wall, hire a professional to shut off the supply and rewire before you cut the hole. All do-it-yourselfers should consider the risks and inconvenience of working on a roof before they start this project, and should make plans to rent or otherwise acquire the appropriate equipment for safe and efficient roof work.

What to Watch Out For

Several important factors must be considered before installing a roof deck to avoid overloading the existing roof. Remember that the weight of the 8' x 16' deck will increase when it is in use. (This is known as the "live load.")

Be sure that the existing rafters and roof boards are in good condition and are sturdy enough to support more than your greatest anticipated load. If additional support members are needed, they will have to be installed before you begin construction of the new deck. If you're in any doubt as to how to reinforce the roof support system, definitely consult and hire a contractor to do the work. The additional expense will be well worth the investment to ensure your roof deck not only looks good, but is safe.

See also:
Patio Doors, New Window, Deck Projects.

Roof Deck, 8' x 16'

Description	Quantity/Unit		Labor-Hours	Material
Deck material, pressure-treated sleepers, 2 x 6 x 8'	112	L.F.	3.6	103.49
Header, 2 x 6 x 16'	16	L.F.	0.5	14.78
Decking, 2 x 4 x 12'	408	L.F.	11.9	239.90
Railing material, pressure-treated posts, 2 x 4 x 3', doubled	72	L.F.	2.6	116.64
Railings, 2 x 4 stock	72	L.F.	2.1	42.34
Balusters, 2 x 2 treated pine	288	L.F.	7.7	103.68
Cap rail, 2 x 6 stock	36	L.F.	1.2	33.26
Cut opng. for sliding glass door, flr. sheathing & flooring, to 5 S.F.	1	Ea.	1.6	
Cut wall, sheathing to 1" thick, not including siding	1	Ea.	1.1	
Cut wall, gypsum wallboard to 5/8" thick	1	Ea.	0.3	
Frame for door, 2 x 4 to support header	48	L.F.	0.8	19.58
Header, 2 x 8, doubled	14	L.F.	0.7	13.94
Door, sliding glass, stock, wood frame, 6' wide, economy	1	Ea.	4.0	834.00
Nails, #10d galvanized	10	Lb.		16.68
Bolts, 1/2" lag bolts, 4" long, square head, with nut and washer	24	Ea.	1.4	10.37
Totals			39.5	$1,548.66

Contractor's Fee, Including Materials: $4,443

Dormers

Building a dormer can expand second- or third-floor living space and, in the right style and size, can also enhance the exterior appearance of your house. Consider the following when planning a dormer.

- A dormer is a major visual element and has a huge impact on your home's look from the outside. Its design, size, and spacing relative to other features are crucial and should be given careful consideration.

- The two most commonly installed dormer types are gable and shed dormers. Both require adequate area and ridge height.

- The pitch of the dormer's roof must meet building code requirements, just as a full-size roof must do.

- If the new dormer is to be bedroom space, be sure that at least one window in the space meets building code requirements for egress (a clear opening space of 20" x 24" minimum). If the purpose of adding the dormer is to create new living space (from attic space), be sure to check building code egress requirements for window size.

- Don't assume that your existing ceiling joists will serve as adequate floor joists if your dormer project is part of a conversion of currently unoccupied space (e.g., attic). If the project requires that you increase the depth of your floor joists, don't overlook the effects on stairs, headroom, and window height.

- Assemble all the tools and helpers you'll need before you begin work, to minimize the length of time the house is left open to the elements. For a large or complex dormer, have appropriate protective materials available to cover the opening overnight while the project is in progress.

- Wait for good weather to cut into the house, and try to anticipate how much you can complete in one day.

- The cutting of the roof opening is one of the most critical operations in a dormer project. Accurate measurement, both inside and out, is required to lay out the cuts before they're made.

- The thickness of existing materials must be matched, which can be tricky in older homes.

- Most roof work requires proper staging with adequate working space and platforms.

4' Gable Dormer

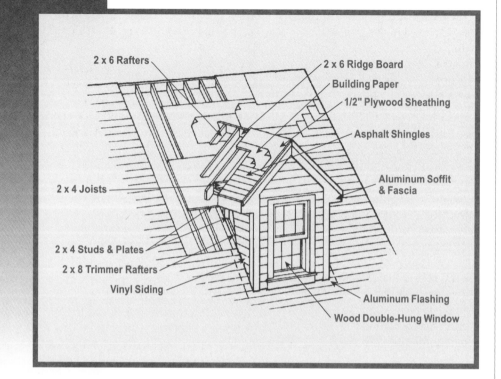

2 x 6 Rafters
2 x 6 Ridge Board
Building Paper
1/2" Plywood Sheathing
Asphalt Shingles
Aluminum Soffit & Fascia
2 x 4 Joists
2 x 4 Studs & Plates
2 x 8 Trimmer Rafters
Vinyl Siding
Aluminum Flashing
Wood Double-Hung Window

A dormer can be a great way to make a house look more interesting and substantial from the outside, while brightening and expanding the useable space in an upstairs room. Dormers are also key features in the conversion of attic space to living space. This project should be high on the list in terms of return on investment, since it provides both curb appeal and living space—valuable assets in any home.

The size and design of a new dormer depends on the style of the house and the amount of area available for the structure. While gable dormers are smaller than shed dormers, their installation still requires significant carpentry know-how and skill.

Materials & Tools

Dormers use the same building materials employed in the exterior wood construction of a complete house, just in smaller amounts. This dormer project consists of a window, a wood-frame support system, wall and roof materials, and other exterior construction products for the finish work, plus gypsum wallboard and paint for the interior. The roofing material suggested in this model plan is standard asphalt shingles; cedar and some other materials will be more expensive. The materials should match as closely as possible the existing roofing. The same applies to the siding, soffit, fascia, and window. The vinyl siding and the aluminum soffit and fascia, for example, should be used only if they are compatible with the materials that are already on the house. When selecting the window, purchase a quality unit that will be energy-efficient and require minimum maintenance.

All of these materials are commonly used in home building and remodeling and are readily available from most lumberyards or home centers. It pays to shop around and compare prices to get the best deal. The materials order for this project is large enough that it should be trucked. Make the delivery arrangements before the purchase is finalized. Standard hand and power tools are used in this project, including those for carpentry, wallboard installation, and painting.

What's Involved

Before the new dormer is constructed, extensive site preparation must be completed. First, you need to precisely determine the location of the dormer on the roof. Then the opening is cut from the outside. Be sure to make the cut adjacent to existing roof rafters and strip the roofing materials carefully in the area of the opening before you saw the roof sheathing. After the rafters in the dormer opening have been cut and the sections removed, the cut ends have to be headed off, and trimmer rafters of the same cross-section added as support to the full length of the existing rafters. These doubled support rafters are important in that they maintain the structural integrity of the roof and help to support the load of the new dormer.

After the opening has been prepared, the dormer can be framed in basically the same order as the framing of a house. The 2 x 4 wall studs are placed, and the 2 x 4 ceiling joists and 2 x 6 rafters are installed. The placing of the framing members for this gable dormer takes time because of the number of angle cuts and the extensive amount of fitting involved in the process. The cutting and placement will probably be done under less than favorable conditions, with some traveling up and down a ladder or in and out of the opening as part of the job.

Once the frame has been completed, 1/2" plywood sheathing is applied to the walls and roof and then covered with roofing felt. Before placing the siding, make sure that the seams between the old roof and the dormer walls have been correctly flashed. This process and the installation of the valley require time and roofing know-how, so seek the help of a professional if you're inexperienced in these tasks. After the dormer has been completed on the exterior, the insulation can be placed in preparation for the interior finish work.

Level of Difficulty

All of the dormer projects in this section require moderate to advanced carpentry skills and a knowledge of roof structure and framing. There are, in addition, the challenges of awkwardness, height, and safety measures in any work done on the roof. Specialized equipment, such as roof jacks and staging, may need to be rented or borrowed.

Steep roofs can be particularly difficult and dangerous to work on without the right equipment. Hand-held power tools will have to be operated in awkward and inconvenient positions. For all of these reasons, beginners and do-it-yourselfers with limited skills are not encouraged to tackle dormer installations and should leave the entire project to a contractor. Experts and very capable intermediates can complete most of the work, but they should proceed cautiously on the hole-cutting and all of the roofing and siding tasks. Even experts should consider hiring an

experienced roofer to install the valley and to restore the disturbed sections of the old roofing.

A poor roofing job can cause interior water damage that will exceed the roofer's fee many times over. Experts and intermediates should add 50% and 100%, respectively, to the professional time for most tasks, and more for the roofing and flashing installations if they are inexperienced in these areas.

What to Watch Out For

Because small gable dormers like this one are often installed in pairs, their symmetrical placement on the roof can further complicate the layout process.

If the proposed opening is easily accessible from the inside, the job becomes much easier, especially if you have a helper. With two workers, one inside and one outside the roof, the rafters can be more easily located, the corners of the opening established, and the cutting efficiently accomplished. Remember that you'll need to complete the dormer as quickly as possible in order to restore the roof to its weather-tight condition.

See also:

7' Gable Dormer, Shed Dormers, New Window, Finishing an Attic, Finishing a Half-Attic.

Gable Dormer, 4' x 6'-6"

Description	Quantity/ Unit	Labor-Hours	Material
Removal of roofing shingles, asphalt strip	200 S.F.	2.3	
Roof cutout and demolition, sheathing to 1" thick, per 5 S.F.	5.20 Ea.	6.9	
Framing, rafters, 2 x 6 x 4', 16" O.C.	56 L.F.	1.5	36.29
Ridge board, 2 x 6 stock	8 L.F.	0.3	5.18
Trimmer rafters, 2 x 8 x 18' long	36 L.F.	1.1	35.86
Studs & plates, 2 x 4 stock	72 L.F.	1.3	29.38
Sub-fascia, 2 x 6 stock	12 L.F.	0.3	7.78
Headers, double, 2 x 6 stock	16 L.F.	0.7	10.37
Ceiling joists, 2 x 4, 16" O.C.	20 L.F.	0.3	8.16
Gable end studs, 2 x 4 stock	4 L.F.	0.1	1.63
Sheathing, roof, 1/2" thick plywood, 4' x 8' sheets	64 S.F.	0.7	47.62
Walls, 1/2" thick plywood, 4' x 8' sheets	64 S.F.	0.9	47.62
Building paper, 15 lb. felt	95 S.F.	0.2	3.42
Drip edge, aluminum, 0.016" thick, 5" girth	20 L.F.	0.4	6.00
Flashing, aluminum, 0.019" thick	35 S.F.	1.9	27.30
Shingles, asphalt strip, organic, class C, 235 lb. per square	1 Sq.	1.6	48.60
Soffit & fascia, 1 foot overhang, aluminum, vented	20 L.F.	2.7	35.52
Window, double-hung, 3' x 3'-6", vinyl clad	1 Ea.	0.9	319.20
Siding, vinyl, double 4" pattern	50 S.F.	1.6	33.00
Gypsum wallboard, 1/2" thick, taped and finished	128 S.F.	2.1	38.40
Insulation, fiberglass, 3-1/2" thick, R-11, paper-backed	50 S.F.	0.3	16.20
Fiberglass, 6" thick, R-19	10 S.F.	0.1	5.52
Totals		28.2	$763.05

Contractor's Fee, Including Materials: **$2,553**

7' Gable Dormer

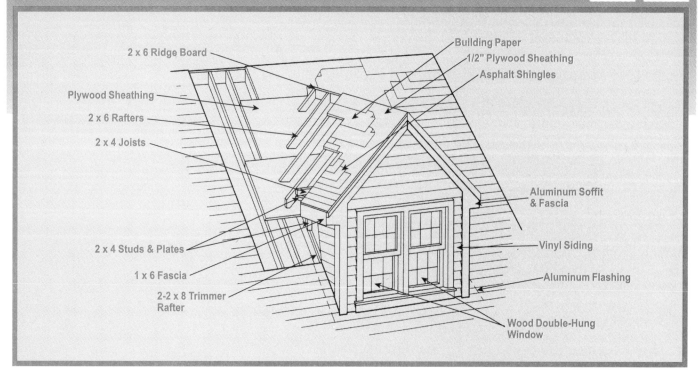

Both this and the smaller gable dormer are fairly major projects—in terms of the work and skill involved, but also the return on investment. You'll gain new living space, enhanced curb appeal, and a good chance of added resale value. The key dormer elements are style, size, and location. A dormer is a prominent feature, so proper planning and workmanship are essential. If you're planning to install two or three gable dormers, take time to carefully plan their location on the roof in relation to one another, as this is crucial to the appearance of your house.

Materials & Tools

The dormer's support system and framed walls require 2 x 4s and 2 x 6s. You'll also need siding, roofing materials, insulation, a window, and exterior and interior paint. Other interior items include gypsum wallboard and finishing supplies, and possibly trim for around the window. These are all standard building materials that you can purchase from one retailer.

Consolidating this order might give you a price advantage, in addition to free local delivery. Do compare between different home centers or lumberyards for the best price.

The roofing, siding, and trim materials should match the products already in place on your house. Also, trim and structural features, such as the width of the fascia and the proportions of the soffit, should be duplicated as closely as possible. Your cost for both labor and materials could change from what is shown in this project estimate if you require different materials to match your particular installation. The most important decision in the selection of finish materials is the choice of a window, the design and size of which should approximate or complement, if not match, the other windows in the house. *(See the previous 4' dormer project for more on materials.)* Standard hand and power tools are used in this project, including those for carpentry, wallboard installation, and painting.

What's Involved

The first major procedure in this project is the layout and cutting of the opening in the roof to receive the new structure. Whenever possible, the longitudinal cuts for the roof opening should be made next to existing rafters. The width of the dormer, therefore, may have to be adjusted to fit the rafter spacing on a given roof. The roofs of most houses have rafters spaced at 16" on center, but in some homes they may be spaced at 20", 24", or even larger intervals. In the latter case, plan on increasing the cost of this operation, as additional rafters and supports may have to be placed to convert the roof structure to safe standards.

While the roof is open, check the condition of the other rafters and sheathing. If either of these components is questionable, have an inspection done by a professional. Any problems should be corrected before you proceed with the new installation. Placing a new dormer on a weak or deteriorated roof surface is a waste of time and money and can cause serious structural problems later on.

Once the opening has been made, new support materials have to be placed, even on newer roofs. The most important of these supports are the trimmer rafters, which double the existing rafters next to the opening and span the entire length from the ridge to the wall plate. The ends of the cut rafters must be capped with doubled headers. In both cases, the lumber should match the dimensions of the existing rafter material.

The walls, ceiling, and roof of the dormer are framed and covered after the opening has been fully prepared. Because of the number of angle cuts and the individual fitting required for many of the framing members, this procedure is time consuming. If you're qualified to do the work yourself, allow extra time to double-check the measurements and angles as you work. Remember, too, that the cutting will usually have to be done in an area remote from the dormer site, on the ground or a suitable surface inside the opening, so you'll use up more time going up and down a ladder or climbing in and out of the opening as the cuts are made. Plywood sheathing is placed on the dormer walls and roof after the framing has been completed. Be sure to scribe the angled side pieces accurately, as a tight fit is desirable.

Level of Difficulty

This project requires considerable carpentry skill and know-how. Beginners and intermediates whose remodeling experience is limited should not tackle it. The structure itself is a challenging undertaking, and its inconvenient roof location makes it more complex. Do-it-yourselfers who undertake this project should anticipate the challenges inherent in roof work—height, limited access to the site, and the awkward position for using power and hand tools.

Two steps in the project are critical to the overall success of the installation. One is the adequate restoration of the

roof support with trimmer rafters and headers after the opening has been cut. The other is correct roofing and flashing installation to restore weathertight conditions to the roof. The roof framing and covering, particularly, may require the services of an experienced contractor. Even expert do-it-yourselfers should consult a qualified roofer before attempting this task.

Generally, experts should add 50% to the professional time for all procedures, and more for the finish roofing and flashing jobs. Intermediates should add 100% to the labor-hours for any work they're capable of doing, such as the interior wallboard installation, and hire a professional for the roofing work.

What to Watch Out For

Because of their roof location, dormers are notoriously difficult to paint and maintain. As a result, you should select finish materials that will minimize future painting, window-washing, and other maintenance chores. For example, windows with removable sashes may cost more initially, but the convenience of washing or painting them from inside the house might be worth the extra investment. In most cases, stained surfaces are easier to apply and maintain than painted ones.

See also:
4' Gable Dormer, Shed Dormers, New Window.

Gable Dormer, 7' x 10'-6"

Description	Quantity/ Unit	Labor-Hours	Material
Removal of roofing shingles, asphalt strip	200 S.F.	2.3	
Roof cutout and demolition, sheathing to 1" thick, per 5 S.F.	15 Ea.	20.0	
Framing, rafters, 2 x 6, 16" O.C.	108 L.F.	2.9	69.98
Ridge board, 2 x 8 stock	12 L.F.	0.4	7.78
Trimmer rafters, 2 x 8 x 18' long	36 L.F.	1.1	35.86
Studs & plates, 2 x 4 stock	120 L.F.	2.1	48.96
Sub-fascia, 2 x 6 stock	16 L.F.	0.4	10.37
Headers, double, 2 x 6 stock	24 L.F.	1.1	15.55
Ceiling joists, 2 x 6, 16" O.C.	36 L.F.	0.5	23.33
Gable end studs, 2 x 4 stock	10 L.F.	0.2	4.08
Sheathing, roof, 1/2" thick plywood, 4' x 8' sheets	105 S.F.	1.2	78.12
Walls, 1/2" thick plywood, 4' x 8' sheets	120 S.F.	1.7	89.28
Building paper, 15 lb. felt	210 S.F.	0.5	7.56
Drip edge, aluminum, 0.016" thick, 5" girth	28 L.F.	0.6	8.40
Flashing, aluminum, 0.019" thick	62 L.F.	3.4	48.36
Shingles, asphalt, multi-layered, 285 lb. per square	2 Sq.	4.0	117.60
Fascia, 1 x 6, rough-sawn cedar, pre-stained	28 L.F.	0.9	74.93
Soffit, 3/8" rough-sawn cedar, plywood, pre-stained	14 S.F.	0.3	20.16
Window, double-hung, 5' x 3', vinyl-clad, thermopane	1 Ea.	1.0	333.60
Siding, cedar, beveled, rough-sawn, pre-stained	120 S.F.	4.0	433.44
Gypsum wallboard, 1/2" thick, taped and finished	240 S.F.	4.0	72.00
Insulation, fiberglass, 3-1/2" thick, R-11, paper-backed	120 S.F.	0.6	38.88
Fiberglass, 6" thick, R-19	120 S.F.	0.8	66.24
Totals		54.0	$1,604.48

Contractor's Fee, Including Materials:	**$5,021**

12' Shed Dormer

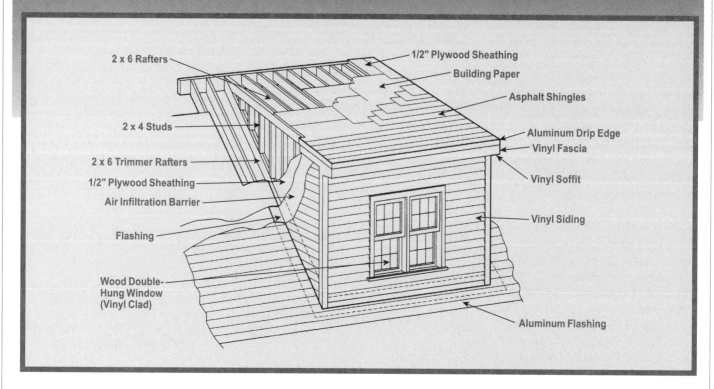

2 x 6 Rafters

2 x 4 Studs

2 x 6 Trimmer Rafters

1/2" Plywood Sheathing

Air Infiltration Barrier

Flashing

Wood Double-Hung Window (Vinyl Clad)

1/2" Plywood Sheathing

Building Paper

Asphalt Shingles

Aluminum Drip Edge

Vinyl Fascia

Vinyl Soffit

Vinyl Siding

Aluminum Flashing

A shed dormer is an efficient way to increase the living space and window area of second- or third-floor rooms without the expense of a full addition. A shed dormer roof often extends from the ridge of the existing roof to a point near or directly over the exterior wall. Because the dormer roof is flat, not gabled or peaked, the amount of usable open space enclosed by the structure is increased, and maximum ceiling height is attained.

Materials & Tools

While the amount of construction materials used for shed dormers is usually greater than for gable dormers, the same basic materials are used, in essentially the same sequence of operations.

As in most exterior house improvements, framing lumber, plywood sheathing, roofing, siding, building paper, flashing, and various finish products are used in this plan. You can save by comparison shopping before buying. For an order this size, the difference can mean substantial savings.

As in other dormer plans, the finish materials should match the products already in place on the rest of the house. This plan includes low-maintenance vinyl siding and aluminum soffit and fascia coverings, but other materials, such as cedar shingles and clapboards and conventional wood trim, may be more appropriate for your house. A gutter and downspout for the shed roof have not been included in the plan, but it's a good idea to add them if the dormer is located above a doorway, or if the face of the dormer is a continuation of the house wall.

The roofing should match the existing roof material, except where prohibited by the pitch of the roof. If the ridge is low, and the pitch of the new shed roof nearly flat, rolled roofing of matching color might have to be used, as shingles are not recommended for roofs with shallow slopes.

Fiberglass batt insulation will be needed between the rafters. When selecting the window unit, consider thermopane sash in the same design of your present windows. The extra expense for a high-

quality window will be returned in energy savings and comfort over the years. These and other changes from the materials list in this plan will affect the cost of the finish work as well as the labor-hours estimate.

The tools you'll need include standard hand and power tools including a measuring tape, level, plumb bob, carpenter's square, drill, hammer, circular saw, reciprocating saw, miter saw, and, for safety, eye protection, a dust mask, and gloves.

What's Involved

Careful removal and reconditioning of the affected roof area is critical to prepare the house for the new structure. The roofing, sheathing, and rafters must be torn out, the structural integrity of the old roof restored, and the rafter support for the dormer installed.

In most cases, it's best to make the longitudinal cuts beside existing rafters, then reinforce them with trimmers. The new supports must span from the ridge to the eaves. The existing rafters are usually spaced at 16" on center, but they may be located at odd intervals in some older houses. Be sure to match or exceed the dimensions of the existing rafters with the trimmer material. Any deficiencies in the existing roof or its support system should be corrected before framing the new dormer.

The frame for the shed dormer consists of 2 x 4 walls and 2 x 6 rafters and ceiling joists. After the front wall of the dormer has been framed, placed, and temporarily braced, the joists and rafters are installed. As in other roof framing operations, the general rule is consistency of length, angling and notching, and spacing of the roof framing members.

The framing of the side walls must be accomplished one piece at a time, with two different angles cut at each end of the studs to fit the roof lines. This process takes some time, so plan accordingly. Once the framing has been completed, 1/2" plywood sheathing is applied and then covered with an air-infiltration barrier, either housewrap or 15-pound felt building paper.

Level of Difficulty

This dormer project is challenging, even for experts and accomplished intermediate-level do-it-yourselfers, and it is out of reach for beginners. Whenever you open a roof and tamper with major structural components like rafters, you have to have the building skill to accomplish the work efficiently and correctly.

Skilled intermediates and experts should be able to complete all tasks involved in the project. They should add 50% and 20%, respectively, to the professional time for all procedures. The new roofing may pose some challenges and require some extra time, especially for flashing and restoring disturbed sections of the old roof. Nevertheless, once the drip edge has been placed and the job correctly laid out, the roofing material should go on smoothly. Remember to allow extra

time and get some help hauling bundles of roofing to the work area.

What to Watch Out For

All dormer installations should be completed or at least closed in as rapidly as possible to protect the inside of the house from the elements. If you're doing the job on your own, arrange several successive days of work in nice weather. Although sudden weather changes can't be predicted, you can reduce the chance of getting caught off-guard by scheduling this project during a warm and drier time of year. Always have an emergency tarp ready to use in the event of a sudden rainstorm, and

then hold off resuming construction until the roof area thoroughly dries. Wet roofs can be slippery and dangerous.

See also:
Gable Dormers, Shed Dormers, New Window.

Shed Dormer, 12' x 8'

Description	Quantity/Unit	Labor-Hours	Material
Removal of roofing shingles, asphalt strip	100 S.F.	1.1	
Roof cutout and demolition, sheathing to 1" thick, per 5 S.F.	20 Ea.	26.7	
Framing, rafters, 2 x 6 x 12', 16" O.C.	144 L.F.	3.9	93.31
Trimmer rafters, 2 x 8 x 18' long	36 L.F.	1.1	35.86
Studs and plates, 2 x 4 stock	192 L.F.	3.4	78.34
Sub-fascia, 2 x 6 stock	14 L.F.	0.4	9.07
Header, double, 2 x 8, 5' long	10 L.F.	0.5	9.96
Header, double, 2 x 12, 20' long	24 L.F.	1.3	46.08
Ceiling joists, 2 x 6, 12' long	144 L.F.	1.8	93.31
Sheathing, roof, 1/2" thick plywood, 4' x 8' sheets	198 S.F.	2.3	147.31
Wall, 1/2" thick plywood, 4' x 8' sheets	90 S.F.	1.3	66.96
Housewrap, spun bonded polypropylene	90 S.F.	0.2	17.28
Building paper, 15 lb. felt	60 S.F.	0.1	2.16
Drip edge, aluminum, 0.016" thick, 5" girth	36 L.F.	0.7	10.80
Flashing, aluminum, 0.019" thick	24 S.F.	1.3	18.72
Shingles, asphalt, strip, 235 lb. per square	6 Sq.	9.6	291.60
Siding, vinyl, double, 4" pattern	90 S.F.	2.9	59.40
Soffit and fascia, aluminum, vented	36 L.F.	4.8	62.21
Window, 3' x 4', double-hung, vinyl clad	2 Ea.	1.8	780.00
Gypsum wallboard, walls, 1/2" thick, taped and finished	192 S.F.	3.2	57.60
Insulation, fiberglass, 3-1/2" thick, R-11, paper-backed	132 S.F.	0.7	42.77
Insulation, fiberglass, 6" thick, R-19	60 S.F.	0.4	33.12
Totals		69.5	$1,955.86

Contractor's Fee, Including Materials: **$6,315**

20' Shed Dormer

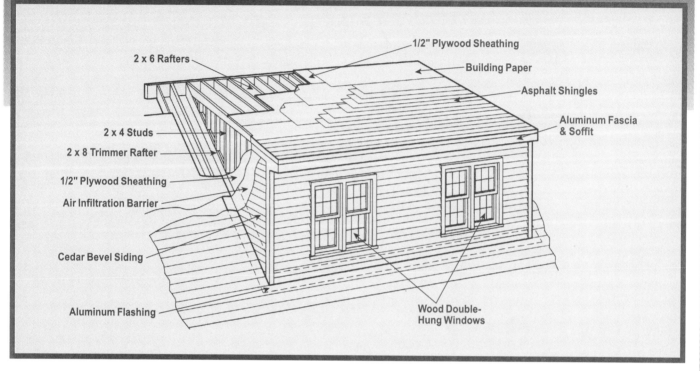

2 x 6 Rafters

2 x 4 Studs

2 x 8 Trimmer Rafter

1/2" Plywood Sheathing

Air Infiltration Barrier

Cedar Bevel Siding

Aluminum Flashing

1/2" Plywood Sheathing

Building Paper

Asphalt Shingles

Aluminum Fascia & Soffit

Wood Double-Hung Windows

If it fits in with the style of your home, adding a wide shed dormer to an upstairs room or attic can give you a huge amount of new living area. This is a valuable commodity for you and your family, but also for future buyers, should you sell your home. This type of dormer is usually placed on Cape-style houses and other structures with gable roof designs large enough to accommodate them.

Before you begin the planning, measure your attic to see that you have enough ridge clearance to accept the dormer and enough roof and floor area to make the project worthwhile. A height of roughly 9' from the bottom of the ridge board to the floor of the room below or, in the case of an attic improvement, to the top of the ceiling joists, is a reasonable minimum measurement. Once the minimum height requirement has been met, the width and length of the dormer can be determined according to the intended use of the facility, the rafter spacing in the existing roof, and the location of interior partitions.

Materials & Tools

This model project is a basic design for the exterior shell of a 9' x 20' shed dormer with no interior materials included, except insulation. Although basic, a substantial number of exterior materials are needed to construct this dormer. If you're doing the building on your own, compare prices from several suppliers before you place the order. Also, make sure that delivery is included in the final price, as the materials will have to be trucked.

The dormer requires different materials for the various components as the structure is built, including 2 x 4s and 2 x 6s for the support system and frame, plywood sheathing, and various exterior coverings and finish materials. If you do the work yourself, you'll need standard hand and power carpentry tools, safety glasses and a dust mask, gloves, and a ladder or rented equipment, such as scaffolding, for working at a height.

What's Involved

The most critical step in the project is determining the location and laying out the section of roof to be opened for the new dormer. In most cases, the new structure will be located in an area that begins on the lower part of the roof and runs to the ridge board. This width measurement will vary from house to house, depending on the size and pitch of the roof, aesthetic considerations, and the desired amount of new living space. The length of the roof area to be opened depends on such variables as the location of existing interior partitions, the design of the new living space, and the spacing of the existing roof rafters. The cost of this shed dormer project is based on a 9' x 20' structure placed on a roof with a steep pitch, and will vary somewhat if the dormer dimensions are modified.

The roof section can be opened by removing the roofing, sheathing, and rafters. After the opening has been made, the new work begins with the installation of trimmer rafters, the heading-off of the cut rafter ends, and, if conditions warrant, the placement of additional support. Again, each case will vary, and additional costs and time may be required to bring the roof and dormer support system to safe standards. The frame of the new dormer should not be placed until its support system has been correctly installed and fastened. Professional consultation or hiring a contractor to do the work will save you money in the long run if extensive roof support reconditioning is required.

The framing and closing in of the shed dormer proceeds in much the same order as that of other wood frame structures. The face of the dormer is framed with 2 x 4s and then braced temporarily while 2 x 6 ceiling joists and rafters are placed. Then the side walls are framed, and the entire enclosure sheathed with 1/2" plywood. An air-infiltration barrier is applied to the walls, and 15# felt to the roof. The roofing material (in this project, asphalt shingles) is then applied to the shed roof before the window units are installed, and the finish siding applied. Fiberglass batt insulation is installed between the rafters. The roofing, siding, and window units will vary with each installation, as the products that are selected for the dormer should match or approximate the materials already in place on the house.

Level of Difficulty

This project is a major home improvement that demands a considerable amount of carpentry skill. The roof location increases the difficulty and adds to the time and general inconvenience of the operations. You may need special equipment for roof and upper-story staging, roof jacks, sturdy extension ladders, and pump jacks, to name a few. Some of these devices can be rented. If you need them, allow for their rental cost and plan ahead to make sure they're available during the construction. Also, get instruction in their operation before you use them. Generally, beginners should not attempt this or other dormer projects, as the

necessary skills and location of the work are too demanding. Experts and intermediates should add 25% and 75%, respectively, to the labor-hour estimates and be willing to seek professional advice when necessary.

What to Watch Out For

If the materials you need differ from those listed in the estimate, the cost may be affected. For example, cedar roofing costs more than asphalt shingles, and quite a bit more than asphalt roll roofing. The windows included in this plan are quality thermopane double-hung units, but standard single-pane windows can be installed at less expense. Generally,

quality window units are worth the extra cost in the long run because of their return in energy savings and low maintenance.

Additional cost may also be incurred, and more installation time required, if a gutter and downspout are needed for the dormer. If the face of the dormer is far enough back from the edge of the existing roof, you probably won't need it, but if it is positioned close enough to the edge to cause spattering or icing problems from the roof run-off, then you should spend the extra money to install one.

See also:
Gable Dormers, 12' Shed Dormer, New Window.

Shed Dormer, 20' x 9'

Description	Quantity/ Unit	Labor-Hours	Material
Removal of roofing shingles, asphalt strip	300 S.F.	3.4	
Roof cutout and demolition, sheathing to 1" thick, per 5 S.F.	36 Ea.	48.0	
Framing, rafters, 2 x 6 x 12', 16" O.C.	192 L.F.	5.2	124.42
Trimmer rafters, 2 x 8 x 18'	36 L.F.	1.1	35.86
Studs and plates, 2 x 4 x 8', 16" O.C.	240 L.F.	4.2	97.92
Sub-fascia, 2 x 6	22 L.F.	0.6	14.26
Headers, double, 2 x 8 x 4' long	16 L.F.	0.8	15.94
Headers, double, 2 x 12 x 20' long	40 L.F.	2.1	76.80
Ceiling joists, 2 x 6 x 12'	192 L.F.	2.5	124.42
Sheathing, roof and walls, 1/2" thick, 4' x 8' sheets	384 S.F.	4.4	285.70
Housewrap, spun bonded polypropylene	130 S.F.	0.3	24.96
Building paper, 15 lb. felt	300 S.F.	0.6	10.80
Drip edge, aluminum, 0.016" thick, 5" girth	44 L.F.	0.9	13.20
Flashing, aluminum, 0.019" thick	40 S.F.	2.2	31.20
Shingles, asphalt, multi-layered, 285 lb. per square	3 Sq.	6.0	176.40
Siding, cedar, beveled, rough-sawn, stained	130 S.F.	4.3	469.56
Soffit and fascia, aluminum, vented	42 L.F.	6.1	108.86
Windows, double-hung, insulating glass, 3'-0" x 4'-0"	4 Ea.	3.6	926.40
Gypsum wallboard, walls, 1/2" thick, taped and finished	384 S.F.	6.4	115.20
Wall insulation, fiberglass, 3-1/2" thick, R-11, paper-backed	130 S.F.	0.7	42.12
Roof insulation, fiberglass, 6" thick, R-19	300 S.F.	2.1	165.60
Totals		105.5	$2,859.62

Contractor's Fee, Including Materials: **$9,326**

Exterior Doors and Windows

Any projects that involve opening up interior space to the outdoors deserve special pre-planning and extra care. Consider the following:

- Have the tools and materials assembled in advance to prevent delays in installing the new door or window once the old one has been removed.

- Pre-hung doors can be heavy and awkward, and windows fragile. Have someone available to help you with these projects.

- If you're planning to install a window or door where there wasn't one before, determine the wall thickness by measuring the jamb width of another window or exterior door. The manufacturer can make frame jambs to match your wall thickness.

- All wood doors in this book are priced pre-mortised for hinges and pre-drilled for cylindrical locksets. If you're planning to re-use hardware from an existing door, specify when ordering that you want a "blank" door, as opposed to a pre-drilled door ready to receive hardware.

- Many pre-hung doors come with average-quality hardware. Specialty knobs and hinges, such as solid brass or bronze, can add considerably to the appearance, and also the price. Locksets are purchased separately.

- Depending on the type and style of exterior wood door you choose, a steel substitute could cost about half the price. Both steel and fiberglass doors provide a better thermal value; steel requires less maintenance.

- New wood doors should be sealed (primed, or primed and painted) before they're installed according to the manufacturer's recommendation. Don't forget to cover all six sides, including the top and bottom edges.

- When deciding on a style of window for an older home, look at books that show homes from the same period for ideas on how to stay within the appropriate style. Major window manufacturers offer special lines of architecturally styled windows, and can provide illustrated catalogs to help you choose.

- When replacing windows, beware of cracked or broken glass. You may want to apply masking tape to the glass before removing the sash.

- Most quality windows must be ordered several weeks in advance. Don't remove existing windows before delivery of the new ones, since shipment could be delayed.

- Skylights should let in as much light as possible. The goal is to distribute the light evenly inside the room without creating a glare. Skylights should ideally open—for ventilation and to minimize heat gain in summer. They should provide thermal protection to inhibit heat loss in winter.

- If a ventilating skylight is too high to reach, consider motorized controls. When pricing skylights, also inquire about the cost of hardware extensions and handles to open and shut not only the windows, but also blinds and shades.

- Talk to your door or window dealer about glass types and options, such as tempered glass for strength, R-values for energy savings (4.0 is optimum), and low-E glass, which prevents sunlight from damaging interior furnishings.

Bow/Bay Window

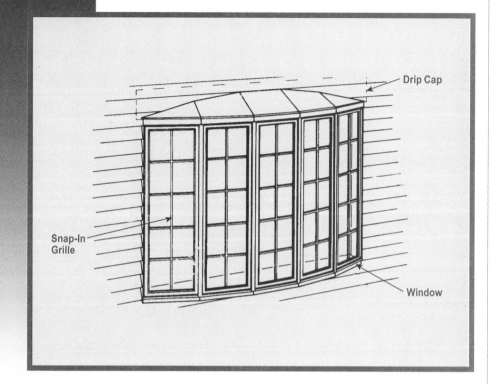

As homeowners and designers seek to integrate more natural light into homes, window manufacturers continue to design large and interesting shapes for individual windows and window groupings. In fact, one out of every five windows in a new home has a special shape, such as the bay window in this project.

On the inside, the effect of a bow or bay window is to make the room seem to expand outward, and the angles of the side-lites take in the view like a wide-angle camera lens. The bay also adds interest to the exterior façade. If your house style and budget are right for a bay window addition, the end results can be impressive.

Materials & Tools

Bay windows are available in four basic styles, based on the angles they form with the plane of the house wall: 90°, 45°, 30°, and bow (made up of a number of sashes in a curved configuration). Because a bay projects outward, forming in effect a separate, tiny "room," it requires some sort of roof, as well as a support system to prevent sagging from its own weight.

The most common bay roof, used when the main roof is high above the window, is a hip-style matching the slope of the main roof, and shingled, either with cedar or other materials that match the house. Copper roofs are also popular.

Most manufacturers offer complete window kits with roofing, or you can customize your own. If your home has a narrow soffit, and if the bay is located close under it, you may need to purchase and install an extension to the main roof to cover the bay so that it blends into the existing exterior features and doesn't look like a poorly designed afterthought.

This project will require the tools and materials necessary to cut and frame a rough opening (including the support structure necessary if it is a load-bearing wall), and roofing materials. A level, hammer and nails, circular saw, and framing square are needed for the framing. The window installation will require shingle shims.

Depending on the trim materials you choose, you may need a miter box, along with the finish nails and paint to match your existing finishes. Insulation, caulk, and vapor barriers are also on the materials list for this project, to make your window weathertight and avoid drafts and higher energy costs.

What's Involved

Begin this project by assessing the visual impact the new window will make on both the interior and exterior of your home. Factors to consider include size, light, view, structural constraints, interior appearance, insulation, exterior appearance, technical and functional features, and cost.

If your window is located in a load-bearing wall, temporary supports are needed to bear the load while the studs are being cut out, and the header installed. These are installed in the manner of a framed wall with vertical studs between horizontal plates. This temporary wall should, of course, go at right angles to the floor joists, above and below.

If the ceiling wallboard is attached to furring strips, be sure the top plate of the support wall is located on a strip (furring strips are usually 1 x 3s spaced 16" on center); otherwise all the overhead weight will be supported by about 1/2" of wallboard, which will collapse when the studs are cut, and the load transferred to the support wall.

Make sure to have temporary supports outside, 2 x 4s cut to length, and ground pads of wider 2x lumber for them to rest on. These will bear the weight of the window when it's set in the rough opening. Setting the unit in place for a trial fit will allow you to identify and correct any problems, and once the window is centered and plumbed, you can mark exactly where to cut the siding to get a clean, tight joint where it meets the trim.

An adequate support system is essential to ensure that the window's weight doesn't cause it to pull away from the house, or to sag and rack the frame out of square, thereby causing the moving side sashes to bind. The most obvious way to provide this support is to install angled members under the window that will transfer the load to the main frame of the house. These can be simply two or more short pieces of 2 x 4 with 45° cuts on each end.

Decorative brackets can also be used for support or designed and made on the job, but they should, in either case, match the style of the house. The number of brackets needed depends on the size and weight of the window.

The best support system calls for suspending the bay from overhead using steel cables. The cables, which have threaded bolts on their lower ends, are fed up through holes in the seat- and head-boards, and fastened to cleats anchored to the framing of the house. The bay can be leveled by adjusting the nuts on the cable bolts. The cables themselves are hidden by the window's inside corner trim. With this system, any exterior angled skirt, or brackets, are merely decorative, and do not offer any structural support. Be sure that any

lower trim work is built so as to allow access to the cable bolts if future adjustments become necessary.

The bay roof should be built with as much care and regard for proper flashing and sealing as would be given to any roofing job. You may want to consult a professional for this part of the project.

Level of Difficulty

Installing a bay window is a major undertaking. Experience is needed to cut and frame the rough opening. If care is taken during this phase to make the rough sill level, it will greatly facilitate the installation. At least two people are needed to put the window in place, and for a very large and heavy unit, more help will be required. Building a roof, whether from a kit or from scratch, is always tricky.

This is not a project for beginners or many intermediate do-it-yourselfers to attempt. Even competent and fairly experienced intermediates should have experienced advice in advance and help with the job. Intermediates should add at least 75% to the professional time estimates for any work they can do. Experts should figure on 25% more

time, and should have a helper for putting the window in place.

What to Watch Out For

Careful planning makes every job easier. Remember that you're cutting a hole in your house; have everything—tools, materials, and help—on hand so that it can be closed in and made weathertight as quickly as possible.

Bay windows can lose a lot of heat through the glass and be cold spots in the winter. You can minimize heat loss by purchasing double-glazed sashes, and by taking care to adequately insulate above, below, and around the window. A combination of rigid board insulation and standard fiberglass batts can be used to achieve the highest possible R value (R = resistance to heat flow).

Be sure to check with your local building inspector about code requirements for headers. There may be limitations on the height of the rough opening, and thus limits regarding the size of the window.

See also
New Window, New Opening in Wall*, Replacement Window.

*In *Interior Home Improvement Costs*

Bow/Bay Window, 8' x 5'

Description	Quantity/Unit		Labor-Hours	Material
Demolition, cut opening for window, per 5 S.F.	8	Ea.	2.7	
Blocking, misc. to wood construction, 2 x 4	60	L.F.	1.9	24.48
Headers over opening, 2 x 12	24	L.F.	1.3	46.08
Casement bow/bay window, wood, vinyl-clad, 8' x 5' insul. glass	1	Ea.	1.6	1,380.00
Trim, interior casing, stock pine, 11/16" x 2-1/2"	33	L.F.	0.1	1.02
Paint, int. trim, incl. putty, primer, oil base, brushwork	33	L.F.	0.1	0.02
Int. trim, 2 coats, oil base, brushwork	33	L.F.	0.1	0.04
Drip cap, vinyl	8	L.F.	0.1	0.32
Caulking	32	L.F.	0.1	0.18
Snap-in grilles	5	Ea.	1.3	204.00
Totals			9.3	$1,656.14

Contractor's Fee, Including Materials:	**$2,822**

New Window

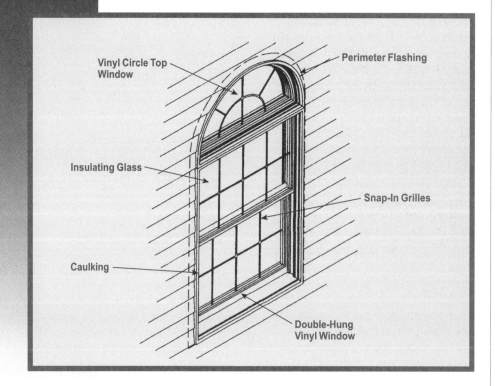

Vinyl Circle Top Window

Perimeter Flashing

Insulating Glass

Snap-In Grilles

Caulking

Double-Hung Vinyl Window

Adding a window is an exciting project, as it can bring new light, atmosphere, and fresh air into your home. With the wide variety of energy-saving, UV-filtering, low-maintenance, and decorative choices now available, it makes sense to invest some time checking out your options.

The new window should match or blend with your existing windows in terms of location, style, size, and shape. Consider the look of the window both inside and out. If the ideal interior location looks out of balance with architectural elements on the exterior, you may have to come up with a compromise location that works for both.

To help envision what it will look like, cut a piece of roofing felt (tar paper) to represent the new window and tape or staple it in place. Then view it at various distances, from different angles, focusing your attention on how the shapes of windows, doors, roof lines, and walls balance and blend. Take your time and get professional advice.

Windows, perhaps more than any other single element, affect a home's external character and beauty. Window shapes and styles include double-hung, casement, fixed pane, sliding pane, awning, half-round, and elliptical, among others. Manufacturers offer them with a variety of options such as primed wood or vinyl-clad casing, true divided lights with removable screen and storm panels, thermopane or double insulated glass with snap-in grilles; and crank or lever-type opening mechanisms.

The latest terminology includes *solar heat gain coefficient (SHGC), daylight transmittance, U-value and R-value,* and *low-E coatings.* The right combination will allow the desired amount of light, while enhancing your energy savings. A window/door/millwork shop or home center can provide you with information and descriptive brochures for the various brands and answer your questions.

Expect to pay for quality. Inexpensive windows look insubstantial, often perform poorly, and offer little in the way of guarantees. Remember, it costs as much to install a cheap window as a

high-grade one. Trying to save a few dollars on a feature of your house as important and visible as a window is false economy.

Materials & Tools

In addition to the window itself, you'll need framing lumber, usually 2 x 4 studs with a 2 x 6 header. In older houses, the dimensions of the framing lumber, inside finished wall, and outside sheathing can vary widely from what is standard today. Other materials include shims and trim, as well as caulk and paint. Tools include standard hand and power tools, plus a pry bar, painter's tools, saws (hacksaw, handsaw, reciprocating saw, circular saw), framing square, level, safety glasses, dust mask and gloves, and, if the window is higher than ground level, a stepladder or rented scaffolding.

What's Involved

Preparation and installation is similar for all types of windows. Start by drawing an outline of the new window on the inside wall. Next, cut the opening in the inside wall, making exploratory cuts with a wallboard saw to make sure that your line of cut is free of hazardous obstructions. If you encounter wires, they can usually be rerouted around the new window. Pipes, and particularly heating ducts, are harder to deal with and will require a plumber. Once the area is clear of wiring and pipes, you can use a circular saw. Then remove the wallboard, insulation, and any strapping. The framing members used to create an opening in a wall provide support for the unit being installed, though you may have to build a support frame if a framing member has to be removed to accommodate your window.

Once the proper supports are completed, the operation moves to the outside, where the siding will be cut out. Stripping and replacing the exterior siding may require a contractor if your house has aluminum or vinyl, and most certainly if it is brick or stone. Wood siding, such as clapboard or shingles, makes it easier. The pry bar is used to strip the siding back from the rough opening in a staggered pattern so that the vertical joints don't line up and result in leaks. Be sure the sheathing is covered with building paper properly overlapped.

When the window is in place, make sure it's level and plumb, as this affects not only the appearance of the window, but its operation. Water-shedding components, such as building paper, drip cap, flashing, siding, and caulking are among the last, but crucial steps. Before closing up the inside wall and trimming out, be sure any insulation you've removed is replaced, and stuff crevices around the new window with insulation. Painting and installation of hardware, including window locks, are the final task.

Level of Difficulty

Because of the skills involved, the cost of a quality window, and the potential for leaks, this project is not recommended for beginners. An intermediate could tackle an uncomplicated installation, but will need a helper when leveling the unit and nailing it in place—one person on the outside and the other inside. Any installation above the ground floor should be undertaken only by an expert with an experienced helper. The difficulties and hazards are challenges for even a seasoned professional. The risks for the inexperienced do-it-yourselfer outweigh the cost of hiring a skilled carpenter. Intermediates should add 75%, and experts 30% to the estimated time. For an upper-story installation, experts should add an extra 20%.

What to Watch Out For

Before deciding where you want to place the window and especially before cutting into the wall, check for electrical wiring, water pipes, and heating ducts. There is, literally, no more shocking surprise than encountering a live wire with the blade of a reciprocating saw.

Check building codes for header size requirements. A large picture window in a load-bearing wall calls for more header support than a small ventilation window in an attic gable. Under no circumstances should a window frame be used to support a vertical load. Follow manufacturers' recommendations for correct clearance and fastening sequence.

Schedule this project during warm weather months if you can, and be prepared to make the opening weathertight in the event of a storm or delay. Even if all goes smoothly, the simplest window project can take two or more days to complete. Handling tools at heights and in awkward positions will add time to the job.

Options

Adding a half-round or other transom window above a standard unit can give a room an appearance of greater height, and a lot of visual interest—inside and out. Such variations, and any complex window shapes, usually require the expertise of a professional carpenter.

See also:
Replacement Window, Bow/Bay Window, Skylights, Patio Doors.

New Window, 3' x 4'

Description	Quantity/ Unit	Labor-Hours	Material
Demolition, cut opening for window, per 5 S.F.	2.50 Ea.	0.8	
Blocking, misc. to wood construction, 2 x 4	40 L.F.	1.3	16.32
Headers over opening, 2 x 8	8 L.F.	0.4	11.33
Window, wood, vinyl-clad, premium, 3' x 4', insulated glass	1 Ea.	0.9	390.00
Trim, interior casing	15 L.F.	0.5	42.84
Paint, interior, trim, primer and 2 coats, enamel, brushwork	2 Side	2.7	5.06
Caulking	14 L.F.	0.5	2.52
Snap-in grilles	1 Set	0.2	139.20
Totals		7.3	$607.27

Contractor's Fee, Including Materials:	**$1,251**

Replacement Window

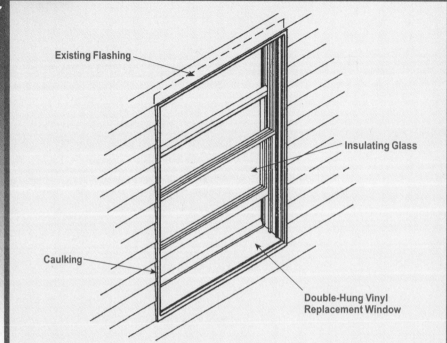

Existing Flashing

Insulating Glass

Caulking

Double-Hung Vinyl Replacement Window

Replacing an older window with a new energy-efficient unit is a simple way to improve your home's appearance while decreasing your heating and cooling costs. The new ENERGY STAR®-rated windows are 100% more efficient than those available only 10 years ago. Other advantages of replacement windows include protection from UV rays that can damage your furniture, floors, and artwork; less noise from outside; and increased comfort (by eliminating drafts).

In terms of return on investment, *Remodeling Magazine* indicated in a 2003 "Cost vs. Value Survey," that replacing existing standard, double-hung windows with mid-priced vinyl- or aluminum-clad, double-glazed, wood replacement windows (and matching trim) brings an 84.8% return when the home is sold.

Materials & Tools

The three most common materials used in replacement windows are vinyl, wood, and aluminum. New windows come with a frame that includes new balances and hardware. They are almost always custom made to fit a particular window opening. Windows should last a long time, so it's best not to skimp on quality just to save a few dollars. Today's vinyl replacement windows come with many options for color, style, and architectural appeal. Other features include energy-efficient low-E/argon glazing, tilt-in systems for cleaning, and grille systems in a variety of configurations. Care should be taken to choose a unit that best suits your budget and the look of your home. Shop carefully and make sure you understand the measuring procedure, measuring twice before ordering a replacement window. Once a window is made, it's yours. You may want to request that a sales representative measure the opening for you.

If you hire a contractor, he or she will measure and order the window for you. If you're doing the installation yourself, you'll need shims (and possibly trim) and caulk, foam insulation, and standard hand and power tools, including a pry bar, hacksaw or reciprocating saw, a level, safety glasses, gloves, and, if the window is higher than ground level, a stepladder or rented scaffolding.

In considering energy-efficiency, you'll come across several factors:

- The U-factor—rate of heat transfer through the window. (With a lower U-factor, less heat is transferred.)
- The solar heat gain coefficient—the amount of heat the house gains from the sun. (The lower the SHGC number, the less heat is gained.)
- Your region's specific energy needs (defined according to climate).
- ENERGY STAR®'s independent energy performance ratings.

What's Involved

Replacing just the window sash can be relatively easy and involve minor preparation of the frame to receive the new window. If you need to replace the entire window, including the frame, as in this model project, it's a much bigger job and can require careful planning, scheduling, and siding and trim removal and replacement. You'll start by removing the interior casing from the old wood window, then prying off the stop from the jamb using a pry bar. (If your old wood window has counterweights behind the side jambs, you can access and remove them through a panel on the jambs.) You can remove the nails that attach the window jambs to the house frame with a small hacksaw or reciprocating saw. Then, remove the window sash and parting bead so they won't be broken when you take out the window frame. Take the window frame out from the exterior of the building by removing any nails that are holding it in place and prying it out. If the old window is aluminum- or vinyl-clad, nails may have to be removed from a flange around the window perimeter.

To install the new window, first center it in the opening, leveling and shimming it from both sides as needed, then fastening it in place. For aluminum and vinyl-clad windows with flanges, you'll nail the flange to the sheathing and framing. For wood windows with a casing, nail through the casing into the house framing. Complete the nailing when the window is level and plumb. Finally, place insulation between the frame and the jamb and make sure the wall's vapor barrier extends to the jamb frame. Reinstall any siding that has been removed, and install the casing.

Level of Difficulty

Basic carpentry skills are involved in installing a replacement window. Although not recommended for beginners, this project is within the capabilities of an intermediate do-it-yourselfer, but a helper may be needed when positioning the replacement unit. Depending on the type of window unit and the amount of preparation required for the opening, experts and experienced intermediates should plan on 50% more time than that estimated here; beginners should plan on 100% more time.

What to Watch Out For

If your home is in a historic district or other area where there are restrictions on the types of home improvements you may undertake, be sure to check

Built by J.P. Gallagher, Hanover, MA

with your local building department before replacing any windows. Also, be sure to choose a style of window that will complement the style of your house, no matter what its age.

Installation above the ground floor should be taken on only with careful preparation. Installing upper-story windows involves working from a ladder or even scaffolding. Working at heights and in awkward positions can add time and difficulty to the job.

Because windows are susceptible to leaks, ensure that your existing window frames are clean and in good condition and seal them well with caulking when the new window is installed. Before replacing the window stops on the interior, be sure to fill any open areas with insulation to avoid heat loss.

This type of project is best done in times of mild weather to avoid problems related to poor weather conditions.

Be prepared to make the opening weathertight in the event your work is interrupted due to a storm or other unforeseen delay.

The final step in the installation of any replacement window requires some form of caulking. If you are doing the installation yourself, be sure to do a thorough job. All your efforts could be diminished by poor caulking.

Options

If you're not quite prepared for a window replacement project, you can still save energy by insulating your old windows. Hold a candle or piece of thread up to the window on a windy day, and find out if there are drafts and where exactly the air is coming in. Remove damaged caulk and weather-stripping, and replace them with self-stick foam or rolled rubber weather-stripping. Solid interior shutters, window shades, blinds, curtains, and lined draperies offer some insulation as well.

See also:
Bow/Bay Window, New Window.

Alternate Materials

Replacement Window
Cost per Each, Installed
Window, wood replacement,
3' x 5', insul. glass $360.00

Replacement Window, 3' x 5'

Description	Quantity/Unit	Labor-Hours	Material
Demolition, remove existing window sashes and interior trim	1 Ea.	0.4	
Window, vinyl replacement unit, 3' x 5', insulated glass	1 Ea.	1.0	696.00
Caulk exterior perimeter	16 L.F.	0.5	2.88
Insulate voids as necessary	16 S.F.	0.1	4.03
Install window trim	1 Opng.	0.6	23.88
Paint, interior, trim, primer and 2 coats, brushwork	16 L.F.	0.4	1.54
Totals		3.0	$728.33
Contractor's Fee, Including Materials:			**$1,203**

Fixed Skylight

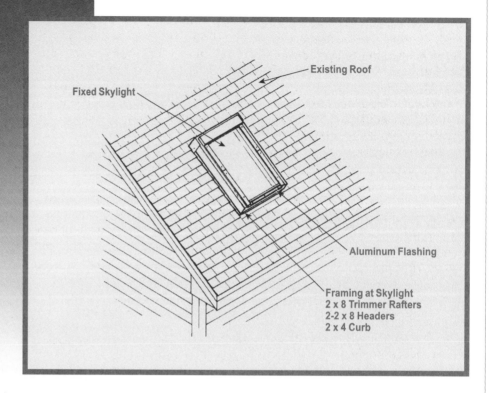

Fixed Skylight

Existing Roof

Aluminum Flashing

Framing at Skylight
2 x 8 Trimmer Rafters
2-2 x 8 Headers
2 x 4 Curb

Skylights are popular features in new homes, providing desirable natural light, while reducing energy costs. Skylights can also be a great addition to rooms that have no exterior wall, and therefore no possibility of a standard window. In an attic conversion or other space with a sloped roof, skylights can be a great window option if you want to avoid the more substantial construction required for a dormer.

Materials & Tools

Choose a skylight that will provide the right level of light, in a roof or ceiling style. If the skylight is intended for the kitchen or bath, a ventilating style might be a better choice. (See "Options" at the end of this project, and the Operable Skylight project.) The inexpensive fixed skylight used in this model project is great for adding light to an attic or dormer, where ventilation is not an essential feature. This medium-sized model is rectangular, with tinted insulated glass.

The other materials needed to complete this project are relatively inexpensive, unless there are unusual conditions involved in your installation. The trimmer rafters are constructed from 2 x 6s or framing lumber that matches the existing roof rafters, and a 2 x 4 curb is needed for the opening. Flashing is needed to seal the roof opening, and milled trim for the interior casing. All of these items can be purchased at most lumberyards and home centers. Skylights are factory-prepared units that include a flashing kit, basic fasteners, and complete installation instructions. Skylights can be purchased from door and window retailers and most home centers. Standard hand and power carpentry tools are needed, including a reciprocating or circular saw.

What's Involved

This project can be a challenge for do-it-yourselfers because it involves cutting through the existing roofing and installing flashing—both crucial and somewhat complex tasks. Beginners and less-skilled intermediates are encouraged to hire a professional for the exterior work and, in some cases, for the interior part of the installation. Aside from the actual placement of the skylight, the most important operation in the project is preparing the roof and ceiling to receive the unit. In some situations, this preliminary procedure is basic, but in others it can be quite difficult and costly.

This plan allows for the materials and installation time under basic conditions, as in an attic, loft, or winterized porch, where the finished ceiling follows the roof line and is attached to the rafters. If your conditions are not this accommodating, the interior work can be considerably more difficult and expensive. This is because the ceiling must be opened and the joists trimmed in a procedure similar to that done on the roof. Also, if there is an attic between the room where you plan to install the skylight and the roof, a light shaft between the two openings must be framed, sheathed, insulated, and finished. Because the dimensions and design features of the shaft will be different from project to project, the cost and time to install it will vary greatly. Remember that a longer shaft is required for a skylight located near the ridge than for one placed near the eaves.

In this project, the roof area affected by the skylight is first laid out, carefully squared, and stripped of its finish covering. The sheathing is then cut out with a reciprocating or circular saw and a section of the intermediate rafter is removed. The opening is trimmed with 2 x 6s and 2 x 8s, as required.

Although they will add slightly to the cost, joist hangers and angle brackets may facilitate the header installation and add strength to the reconditioned support system. If the rafters are spaced at 24", as in some older homes, rafter cutting may not be required. However, additional cost may be incurred in this case to bolster the support of the old roof. If extensive roof reconditioning is required, the increase in cost may be substantial, and a professional contractor may have to be consulted.

Level of Difficulty

Beginners and less-skilled intermediates will find much of this project too challenging, especially the roof work, and particularly when extra reconditioning is needed. Joist work and the building of a shaft, if needed, would also be difficult for inexperienced do-it-yourselfers. Beginners should consider hiring a qualified carpenter to complete this project. Experts and experienced intermediate do-it-yourselfers should add at least 25% and 75%, respectively, to all tasks, with slightly more time allowed for the careful handling and placing of the expensive skylight unit.

All do-it-yourselfers should take into consideration the general inconvenience and hazards of roof work before they tackle the project. Extra expense may be incurred if staging or other specialized equipment is required for high ceilings and steep roofs.

What to Watch Out For

Any project that involves cutting into the roof presents the risk of roof leaks and damage to wood structural supports and sheathing, if the work is not done properly. If you are cutting into a ceiling, determine whether electrical wiring might be in the area beforehand, and plan accordingly—either re-positioning your skylight, or arranging to have an electrician disconnect the power and move the wires and light fixtures, as needed.

Options

During the summer, skylights may contribute unwanted extra heat to the house. Tinting or shades (manual or motorized) can be used to control the heat, and insulated shades can also help keep the heat in, in winter. (Insulated thermal glass is another way to save energy.) Tinting and shades also help protect your furniture and furnishings from UV light.

A ventilating skylight is a better option for a kitchen or bathroom, where you'll want to be able to dispel humid air. A variety of opening and closing systems are available, including automatic temperature sensors, an electric wall switch, a remote control, or a motorized or hand crank.

Tubular skylights are a fairly new solution for small spaces that don't have room for standard skylights. These units have a width of 10"–14", and are often used in windowless bathrooms, hallways, and closets.

See also:
Operable Skylight.

Fixed Skylight, 24" x 48"

Description	Quantity/ Unit		Labor- Hours	Material
Roof cutout and demolition incl. layout, sheathing per 5 S.F.	8	S.F.	2.1	
Demolition, roofing shingles, asphalt strip	8	S.F.	0.1	
Demolition, framing rafters, ordinary 2 x 6	8	S.F.	0.2	
Trimmer rafters, 2 x 8 x 14'	28	L.F.	0.7	35.95
Headers, double, 2 x 8 stock	16	L.F.	0.8	15.94
Curb, 2 x 4 stock	12	L.F.	0.4	4.90
Sky window, fixed thermopane glass, metal-clad wood, 24" x 48"	1	Ea.	3.2	654.00
Flashing, aluminum, 0.013" thick	14	S.F.	0.8	5.88
Trim, interior casing, 9/16" x 4-1/2"	12	L.F.	0.5	23.47
Jamb, 1 x 8 clear pine	12	L.F.	0.4	22.46
Paint, trim, primer, oil base, brushwork	24	L.F.	0.3	0.58
Trim, 1 coat, oil base, brushwork	24	L.F.	0.3	0.86
Totals			9.8	$764.04

Contractor's Fee, Including Materials: $1,585

Operable Skylight

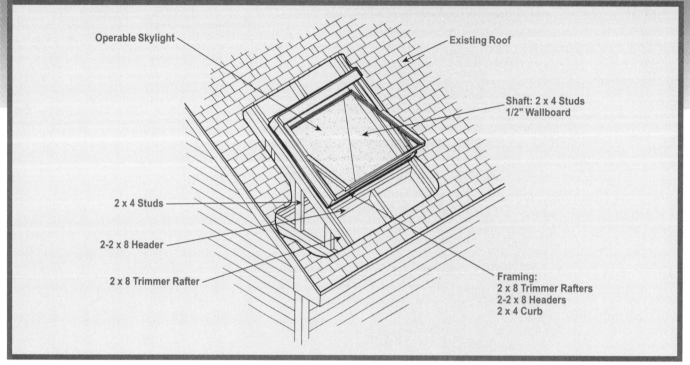

Operable Skylight

Existing Roof

Shaft: 2 x 4 Studs
1/2" Wallboard

2 x 4 Studs

2-2 x 8 Header

2 x 8 Trimmer Rafter

Framing:
2 x 8 Trimmer Rafters
2-2 x 8 Headers
2 x 4 Curb

Operable skylights offer the benefits of attractive, natural light and potentially lower electrical bills for daytime lighting, with the added advantage of ventilation. Kitchens and bathrooms are obvious locations for operable skylights because these rooms have humidity and odors to expel. Other rooms, such as converted attic spaces that don't have other open-able windows, would also benefit from the fresh air offered by this type of skylight.

Materials & Tools

This project features a 44" x 57" top-of-the-line skylight with an operable sash. Units like this are equipped with insulated tinted glass and factory-prepared curbing. A wide selection of accessories is also available to suit the particular needs of each installation. Only a small number of other materials are required, since the pre-fabricated skylight comes as a kit with fasteners and other components. These include framing lumber, flashing, and interior wood trim. The trimmer joist/rafters,

especially, are vital structural elements that will be used to restore ceiling and roof support after the opening has been cut.

If your situation requires a shaft to be built between the roof and the ceiling that houses the skylight, you'll need additional materials based on your shaft size and design. Standard hand and power carpentry tools, including a reciprocating or circular saw, will be needed to complete the project.

What's Involved

A considerable amount of preliminary roof work is necessary on the interior and exterior, before the skylight can be installed. First, take time to lay out the location of the window unit. Most of the popular brands of skylights are manufactured for easy placement between standard-sized rafter spacings. The frame of this 57" skylight unit can be located in the space opened when two 16" rafters are cut and partially removed. Whenever possible, locate the skylight between existing rafters, to save time and disturb the roof structure as little as possible. Be sure to square the layout and double-check all measurements before the hole is cut.

After the trimmers and headers are installed, a curb may have to be fastened to the perimeter of the opening, depending on the manufacturer's recommendations. The instructions will also provide specific information for proper fastening and flashing of the window frame. Use accepted roofing methods for replacing the disturbed area with finished materials. Reusing salvaged pieces (in good condition) of the old roofing will neatly blend the new work with the existing roof.

In many instances, skylights are placed in attic rooms, family rooms with vaulted ceilings, and other rooms with slanted ceilings. The installation process is easier in these cases because the roof rafters and ceiling joists are often one and the same. After the skylight unit has been secured in its opening and made weathertight on the outside, the interior finishing is a relatively easy undertaking. The inside of the opening is trimmed with appropriate material, usually wood, and matching trim is applied for the casing.

In some instances, the interior finishing may be more difficult and costly. If, as in this plan, the roof and ceiling are separated by unused attic space, and separate holes are required for the exterior and interior openings, you'll have to build a light shaft between the two. The cost and difficulty of this procedure will vary with each situation, depending on your shaft's dimensions, finish, and design features. Usually, a 2 x 4 frame, sheathing, and insulation make up the basic structure. Trim work and an appropriate covering are then applied to the shaft's interior to complete the procedure. If the bottom of the shaft is joisted on a horizontal ceiling, the opening could be flared to get the maximum benefits of light and ventilation. Remember to figure in some extra time and cost for cutting and trimming the ceiling opening and joists as part of the shaft installation.

Level of Difficulty

If the basic conditions are right, the installation of this skylight can be a manageable project for do-it-yourselfers experienced in the use of tools and roof work. Beginners and less-skilled intermediates, however, should seriously consider hiring a professional because of the importance and risks of roof work and the dangers and awkwardness of using power tools in difficult positions, such as from a roof ladder. Nearly 20 square feet of roof area will be opened to the elements, so fast and effective work is required to cut the hole, trim it, place and flash the unit, and restore the finished roofing to weathertight condition—ideally within a few hours' time.

Experts and skilled intermediates should be able to complete these tasks with 25% and 75%, respectively, added to the professional time. Inexperienced do-it-yourselfers should not attempt the exterior work and should double the estimated labor-hours for finishing the

interior. All do-it-yourselfers are reminded to allow extra time for handling this expensive and fragile skylight unit and for the general inconvenience of roof work.

What to Watch Out For

If a long light shaft is required for the installation of your skylight, you might want to put some extra effort into planning its design features. Although the length and width of the shaft's top opening are fixed, the sides, bottom, and length of its run can be creatively arranged to provide maximum results and aesthetic appeal. One of the best ways to increase ventilation and light penetration is to flare the shaft by increasing the width of the opening at the ceiling end. This design requires more time and considerable expertise to install because the rafters, joists, 2 x 4s for the frame, and sheathing must be

cut at precise angles. The additional cost for materials is relatively small.

Options

In addition to tinting, you may want insulated thermal glass in your skylight, to help keep heat out in summer, and in, in the winter. Shades (manually operated or motorized) allow you to control the amount of light at any time, maximizing it in winter, and reducing it in summer.

Consider the various opening and closing systems available with operable skylights. Among the options are automatic sensors, electric wall or remote control switches, or cranks (hand or motorized).

Tubular skylights are another option for: very small spaces. They are round, resembling ceiling light fixtures, and are typically 10"-14" across.

See also:
Fixed Skylight.

Operable Skylight, 44" x 57"

Description	Quantity/ Unit		Labor- Hours	Material
Roof cutout and demolition incl. layout, sheathing per 5 S.F.	8	S.F.	4.8	
Demolition, roofing shingles, asphalt strip	20	S.F.	0.2	
Demolition, framing rafters, ordinary 2 x 6	20	S.F.	0.4	
Trimmer rafters, 2 x 8 x 14'	28	L.F.	0.7	35.95
Headers, double 2 x 8	16	L.F.	0.8	15.94
Curb, 2 x 4 stock	14	L.F.	0.4	5.71
Sky window, operating, thermopane glass, 44" x 57"	1	Ea.	1.3	516.00
Flashing, aluminum, 0.019" thick	14	S.F.	0.8	10.92
Shaft construction, 2 x 6 x 12" trimmer joists	24	L.F.	0.6	30.82
Headers for joist, 2 x 6 stock	20	L.F.	0.9	12.96
Framing, 2 x 4 stock	64	L.F.	2.0	26.11
Gypsum wallboard, walls, 1/2" thick, taped and finished	64	S.F.	1.1	19.20
Paint, shaft walls, primer, oil base, brushwork	64	S.F.	0.4	3.84
Shaft walls, 1 coat, oil base, brushwork	64	S.F.	0.4	3.84
Insulation, foil-faced batts, 3-1/2" thick, R-11	64	S.F.	0.3	30.72
Trim, interior casing, 9/16" x 4-1/2"	18	L.F.	0.7	35.21
Paint, trim, primer, oil base, brushwork	18	L.F.	0.2	0.43
Trim, 1 coat, oil base, brushwork	18	L.F.	0.2	0.65
Totals			16.2	$748.30

Contractor's Fee, Including Materials: **$1,915**

Patio Doors

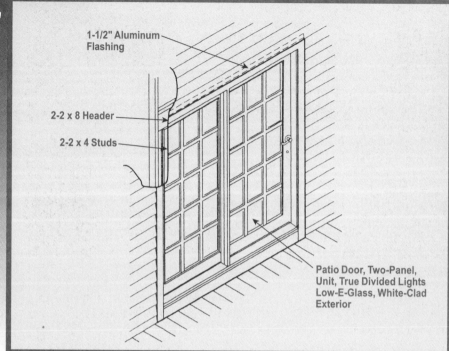

1-1/2" Aluminum Flashing

2-2 x 8 Header

2-2 x 4 Studs

Patio Door, Two-Panel, Unit, True Divided Lights Low-E-Glass, White-Clad Exterior

New patio doors can make a huge difference in the feeling of a room, providing natural light and a view. This project can dramatically enhance the look of your home—both inside and out, and allows you to take advantage of new technology in low-emissivity glass that can save on utility costs year-round.

Materials & Tools

Patio doors combine a light, airy, yet substantial look with highly efficient insulation and weatherproofing. Glass sliders continue to be a popular choice. Their aluminum construction makes them almost maintenance-free. They also admit a lot of light and take up no swing space.

Swinging French doors (or sliding doors with grilles meant to resemble French doors) also continue to be very popular. Your home's interior and exterior style will help you decide which style of door best suits your needs. As is the case with all building products and materials, it hardly ever makes sense to try to economize by purchasing cheap goods. Shopping around for the best price on a quality door unit is smart; choosing an inferior door based solely on price is not.

A patio door is made up of two glazed door units, or panels, in one wide frame. The operating door is usually hinged to the mullion between the panels, with the lockset located at the outside jamb. Manufacturers offer a wide range of decorative options in terms of size, shape, and finish, as well as more technical options that deal mainly with glazing, insulation, and protective coatings.

The doors used in this project are made of clear pine and are glazed with low-emissivity (low-E) glass, which reduces the radiative emission of heat from the glass surfaces, resulting in greater energy efficiency than triple-glazing without the extra weight and reduced light transmission. More insulation is provided by foam-filled weatherstripping that seals tightly and almost completely eliminates drafts.

The exterior wood is protected from the elements by a .055" silicone-polyester-coated extruded aluminum cladding with a baked-on white enamel finish that makes the door practically maintenance-free. The sill is made of a high-tech composition, trade-named Lexan™, which combines good

insulating properties with virtual indestructibility.

The solid brass patio door hardware is more expensive, but much more durable than lightweight, plated brass. Four heavy-duty hinges provide smooth movement and ensure that the door will never sag.

When you purchase your door, refer to the manufacturer's instructions for the tools and installation procedures. Most installations will need a level, a screwdriver or drill, shims, a hammer, and other basic carpentry tools. For finishing the interior and exterior trim after the door installation, you'll need a miter box, finish nails, and paint or stain. Sealant is used to seal the door from the elements and to prevent drafts.

If you're adding a patio door where there was none before, the process is of course more complicated and requires more tools and materials. A circular saw and other framing tools are needed for this extra phase. *(Refer to the Bow/Bay Window project for more on the tools needed to create an opening in an exterior wall.)*

Built by J.P. Gallagher, Hanover, MA

What's Involved

Any kind of door needs a well-framed rough opening, but the unique characteristics of a glass slider call for extra framing care. The bottom track will need to be perfectly plumb and level. Try to frame the opening to close tolerances (the rough opening for this unit is 96" x 80-3/4") to keep shimming of the aluminum frame to a minimum.

Cutting out and framing the opening for either unit is essentially the same as for any door. For a load-bearing wall installation, follow the procedure for erecting a temporary supporting wall.

Detailed instructions are generally provided with patio door units to guide you step by step through the installation process. Once the frame is secured in the rough opening, level, square, and true to plane, the fixed panel is put in place. Once the slider door is slipped on the track, the sliding panel can be installed, and the rollers adjusted. If the rough sill is level and square to the sides, the amount of roller adjustment should be minimal.

Level of Difficulty

If this is a same-size replacement project, it should be fairly easy and straightforward. When replacing a slider, the old unit is lifted out and, with casings removed, the aluminum frame is unscrewed and taken from the opening in reverse order of installation.

Cutting and framing a rough opening is a somewhat more serious operation, demanding experience in both planning and execution. A beginner could undertake a simple remove-and-replace project, but would need experienced assistance with a new installation. Intermediates and experts should encounter few difficulties. Transporting the heavy glass panels and lifting them into place is easier and safer with at least two people.

Time estimates for both types of door should be increased by 100% for beginners. For the sliding glass door installation, add 25% for intermediates, and 10% for experts. For the patio door, intermediates should add 50% to the time estimates, and an expert, 20%.

What to Watch Out For

A common design flaw of decks is to make them virtually flush with the inside floor. This means that rain can bounce off the decking and drench the lower portions of the door, filling the track with water that can work its way under the panels. If you're installing a sliding door that will open onto a new deck or landing, make sure that there is a full step down from the door sill to the deck.

Local building codes should be consulted to determine header specifications and whether safety glass is required. (Safety glass is tempered to break into small, gravel-like bits, rather than sharp shards and spears.) The problem is, safety codes take precedence, so you may be forced to opt for triple-glazed or other types of panes in order to be in conformity. Check out all safety requirements before you begin shopping for a door.

See also:

Bow/Bay Window, Security Measures.

Alternate Materials

Patio Doors
Cost per Each, Installed
Sliding doors, 6'-0" x 6'-8" $1,300.0

Patio Doors, 6' x 6'-8"

Description	Quantity/Unit		Labor-Hours	Material
Wood French door, 6'-0" x 6'-8", w/ 1/2" insul. glass and grille	1	Ea.	2.3	1,380.00
Interior casing, raised molding	22	L.F.	0.7	62.83
Drip cap, aluminum	7	L.F.	0.3	2.52
Entrance lock, cylinder, grip handle	1	Ea.	0.9	135.60
Paint, trim, incl. puttying, primer, oil base, brushwork	22	L.F.	0.3	0.53
Trim, 2 coats, oil base, brushwork	22	L.F.	0.4	1.32
Door, 2 sides, incl. frame & trim, primer & 2 coats, brushwork	1	Ea.	2.7	4.98
Totals			7.6	$1,587.78

Contractor's Fee, Including Materials:	**$2,670**

Main Entry Door

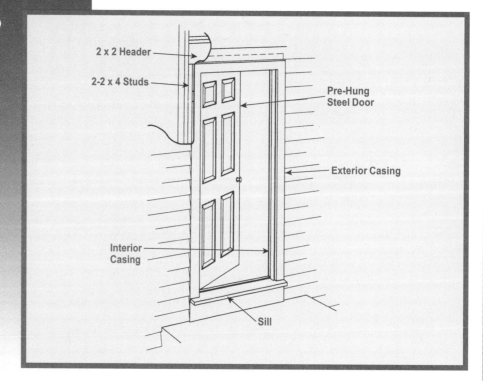

2 x 2 Header

2-2 x 4 Studs

Pre-Hung Steel Door

Exterior Casing

Interior Casing

Sill

Few components of a house get more use, or at least attention, than a main entry door. Replacing a worn one or upgrading to a newer style can greatly enhance your home's curb appeal, since it is one of the first things people see from the street. From the inside, the front door provides you with privacy, security, and protection from the elements, so the right style, durability, and energy efficiency are all important considerations.

Materials & Tools

Even with a screen, the main entry door is subject to weather and the potential warping and rotting that can go along with it. Careful selection of your door's material can help combat this. Also, choose a door that matches the house's overall look and style.

The variety and options available today are countless when it comes to door material and styles. Wood, steel, fiberglass, and vinyl-clad are common choices. The non-wood materials often have a wood grain, and offer significantly higher energy efficiency.

While wood doors typically don't have as high of an insulation level as steel or fiberglass, their elegance makes them a good choice for antique homes. Manufacturers are enhancing wood doors with foam insulation cores for increased energy efficiency. In general, a high quality wood door will cost you more than an insulated steel door.

In addition to the material itself, be sure to consider the arrangement and efficiency of glass panes, if any. Rectangular and oval windows, as well as custom shapes, allow in light and can add an attractive accent. To help you make your selection, visit building suppliers and home centers or seek the advice of designers and remodelers.

The entry door system in this model project is quite basic, but can give you a reference point when you consider other designs and decorative features. A door comes hinged in one of four ways: left or right in-swing, or left or right out-swing. Most exterior doors swing into the house to allow for an outside storm or screen door. A swing door, when open, takes up space in the entry or hallway, but most houses are designed to accommodate this.

A storm/screen door may be added. These are generally pre-hung, like most entry doors. Storm doors are available in standard and custom sizes. Most are set up to swing a particular way, but are usually reversible, so they can open on either side. An insulated steel door does not need a storm, but it's a nice feature for ventilation in good weather.

Tools for this project are standard hand and power tools, including a tape measure, level, hammer, and drill. In addition to the door itself, you'll need a nailset and nails, spackle, shims to level the door, and possibly siding to repair the perimeter if you're enlarging the opening for a bigger door.

What's Involved

The main entrance door of a residence is usually 6'-8" high, and either 2'-8" or 3' wide. Transoms and sidelights, fixed or operating, are optional and will add to the size of the rough opening. Another factor affecting dimensions is whether your house has wood or masonry siding.

Removing the existing door is easy and not too messy. Carefully remove the casings inside and out, take the door off the hinges, cut through the side jambs with a reciprocating saw, and pry the frame out of the opening. There's more work involved if you want to salvage the old door. You'll then have to cut the nails behind the jambs, push the frame out of the opening at the top, and carefully pry up the threshold.

Installing a pre-hung door is a fairly straightforward operation, especially if the rough opening does not need to be enlarged or reduced. Entry door replacement systems such as this one come with detailed instructions that need only be carefully followed. The same is true for mounting the hardware.

If your new door is exactly the same size as the old one, it should fit the existing rough opening. Center the unit in the opening with the casings removed and the door closed. Level the sill and shim between the side jambs at the bottom, tacking through both jambs and shims into the trimmer studs. Plumb the side jambs, shim at the top, and tack. Shim at several intermediate points along the side jambs and tack through the shims, making sure to maintain the proper joint between the door and jamb. Take care when shimming the jambs not to push the frame out of alignment. The door's smooth operation depends on the frame being level and plumb.

Check the operation of the door to see that it opens and closes without sticking or binding. Then, make any final adjustments, and complete the job by driving and setting all nails. Be careful not to leave any hammer "smiles"—those tell-tale marks of the unskilled hacker. Don't drive the nails all the way home with the hammer; leave the heads a tiny bit above the surface of the wood, and use a nailset to countersink them. Fill the nail holes with vinyl spackle after the wood has been primed.

Courtesy of Fenton Inc., Wellesley, MA

Level of Difficulty

A beginner should have experienced assistance during all phases of the main entry door project, and should add 100% to the professional time estimates. Both intermediates and experts should have little difficulty handling this job; they should add 20% and 10%, respectively. A storm door is a fairly simple add-on, and can be installed by a beginner with some carpentry knowledge and aptitude with tools.

What to Watch Out For

An entry door is an expensive, precision-made item of cabinetwork. Its finish appearance and proper functioning can be compromised by rough handling during transit and installation. A steel door is tough, but carelessness around it, for example with hammers swinging from tool belts, can cause scratches and dents that can be all but impossible to repair or hide. Store the door in a safe place and leave the protective plastic sleeve on while installing it.

Some manufacturers warn that dark-colored paint and/or storm doors should not be used with any steel door that has plastic trim. Sun exposure can cause excessive heat build-up, which may distort the trim. Read all the printed material that comes with your door for information on this and particular painting requirements.

See also:
Security Measures, Patio Doors.

Main Entry Door, 3' x 6'-8"

Description	Quantity/ Unit	Labor-Hours	Material
Remove existing door, 3'-0" x 6'-8"	1 Ea.	0.5	
Remove existing frame, including trim	1 Ea.	0.5	
Pre-hung door, insul, metal face, raised molding, 3'-0" x 6'-8"	1 Ea.	1.0	226.80
Door trim set, 1 head and 2 sides, 4-1/2" wide	1 Opng.	1.5	24.60
Entrance lock, cylinder, grip handle	1 Ea.	0.9	135.60
Dead bolt	1 Ea.	1.0	165.60
Paint, door and frame, primer	1 Ea.	1.3	2.18
Paint, door and frame, 2 coats	1 Ea.	2.7	7.08
Paint, trim, primer, including puttying	20 L.F.	0.2	0.48
Paint, trim, 2 coats, including puttying	20 L.F.	0.4	1.20
Totals		10.0	$563.54

Contractor's Fee, Including Materials: $1,343

Basement Bulkhead Door

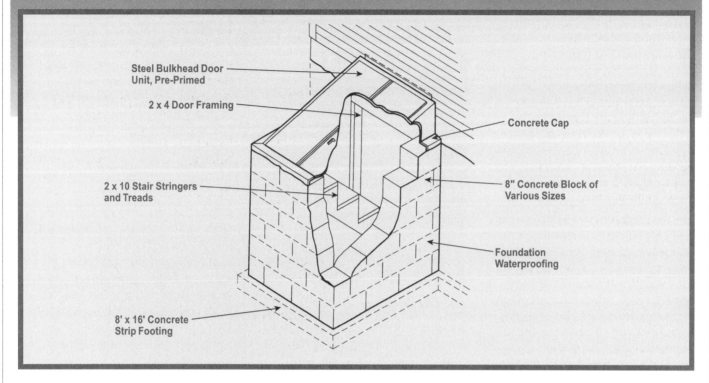

Steel Bulkhead Door Unit, Pre-Primed

2 x 4 Door Framing

2 x 10 Stair Stringers and Treads

8' x 16' Concrete Strip Footing

Concrete Cap

8" Concrete Block of Various Sizes

Foundation Waterproofing

Many houses with a full basement lack direct access to the outdoors, an inconvenience when having to move stored items, such as garden tools and outdoor furniture, through the living area of the house. When moving a large object such as a hot-water heater, there is the additional problem of having to negotiate at least two doorways, as well as typically narrow basement stairs. A bulkhead door may also be considered a good investment in the sense that it is a fairly basic necessity that prospective home-buyers may expect if and when you sell your home.

When deciding on the location of your new bulkhead door entrance, you must deal with interior and exterior constraints and considerations. On the inside, the opening must be free of obstructions and allow room for maneuvering large or long items that might need to be moved in or out. On the outside, the basement access is typically at the back or on the side of the house. Most homeowners prefer to have the bulkhead out of view from the front, and sometimes landscaping is

used as a way to screen the doors from view.

Installing a bulkhead requires excavation, and that means a backhoe has to be able to access the site. Also, the usefulness of this door is limited if it will be difficult to approach from the outside.

Materials, Tools, & Equipment

Materials for this project include the bulkhead door unit, the door and frame for the basement entry, concrete for the walls and foundation, masonry anchors and lag screws, metal enamel paint for the bulkhead door, a lockset and insulation for the basement door, and lumber to frame the door and build the concrete forms and stairs.

Equipment and tools include a backhoe and a concrete saw, as well as carpentry tools for installation of the steps and door, and concrete tools. A backhoe is the preferred equipment for the excavation, since a Bobcat (or skid steer loader) would require a backhoe

attachment anyway, and the cost to rent the Bobcat with this attachment would probably add up to the cost of the standard backhoe.

What's Involved

When you have decided on the best location, you'll need to determine the dimensions of the areaway opening, the size of the bulkhead door unit, and the length of the stair stringers. Door manufacturers can make these calculations, based on the height of grade (ground level) above the basement floor. The area to be excavated should be staked off, allowing at least 18" all around for the footing and working space. The hole should be dug to a depth equal to about 4" below the basement floor, with a perimeter trench to the depth of the existing foundation footing. The areaway opening is then cut through the foundation wall. This opening will later be framed with 2 x 8 lumber to provide the rough opening dimensions for the vertical door.

Once a form is constructed, the footing can be poured for the areaway floor. The next step is to coat the outside of the walls with mortar and, after letting it dry, to apply a heavy coat or two of foundation sealant. Once this is done, the excavated area surrounding the hole can be backfilled and tamped. The metal bulkhead door frame is assembled and mounted by bolting it to the house and anchoring it to the top of the block walls. Once the doors are attached to the frame, the bulkhead is weathertight, needing only two coats of metal enamel paint, matched to the color of the house, to be finished.

The next task is to set the stair stringers in place and secure them to the side walls with masonry nails. Treads are cut from standard 2 x 10 lumber and slid into the stringer slots. This type of stair system is not only economical in terms of materials and labor, but is also very functional, in that the treads can be removed at any time to allow the full areaway to be opened up for lowering large or heavy items into the basement.

The vertical door into the basement should be of exterior grade, either solid wood or insulated, and weatherstripped to seal it against heat loss. Once the door is in place, it can be cased off with appropriate lumber or molding. In an unfinished basement where looks don't matter much, simple, square-cut 1 x 6 #2 pine will do. Be sure to stuff the cracks and crevices around the door frame with insulation before nailing on the casing. Once the door is primed and painted, the lockset can be installed.

Level of Difficulty

A bulkhead door installation is a fairly involved remodeling project. Choosing a suitable location and determining the job specifications are easy. You merely have to take measurements, then look them up in manufacturers' tables, to know the dimensions of the areaway, the size of the bulkhead unit, the length of the stringers, and the number of stair treads.

Breaking through the foundation requires the use of a concrete saw. One can be rented, but it is a pretty formidable tool, heavy and loud, and it is not the most pleasant job for someone who is not accustomed to this kind of work. Most homeowners would be better off hiring a contractor for this.

Pouring the footing and building the block walls require masonry tools and skills. Putting up plumb, square, and level walls with masonry blocks is not an easy job and, therefore, not one to be undertaken by anyone less than an expert. The rest of the installation tasks—bulkhead, stairs, and vertical door—are not very difficult, though they can be time-consuming. Beginners and intermediates should hire professionals for the excavation, concrete wall breakthrough, and masonry work, and should add at least 150% and 75%, respectively, to the time estimates for all other tasks. Expert do-it-yourselfers should hire an excavator and a concrete-sawing contractor, and for all other tasks should add 50% to the times given.

What to Watch Out For

Scheduling is important during any remodeling job that requires the services of professional contractors. Once the hole is excavated, try to get the remaining work done as quickly as possible and keep the hole covered with a large plastic tarp to prevent rain from creating muddy and unstable soil conditions, and from running into the basement. Also, set up barriers around the hole to keep children and others from getting too close to the edge.

This is not a good cold-weather project. Not only do concrete and mortar take longer to cure, but it is difficult to completely seal the foundation opening against heat loss. Moreover, most backhoe operators will not subject the teeth of their buckets to the abuse of trying to break through frozen topsoil.

Before the cutting begins on the foundation wall, do everything you can to seal off the work area from the rest of the basement and the living space of the house.

Basement Bulkhead Door, 6' x 4'-7"

Description	Quantity/ Unit		Labor-Hours	Material
Rent backhoe-loader	1	Day		218.33
Demolition, cut opening for door, per 5 S.F.	5	Ea.	1.7	
Footing, 9" thick x 18" wide, 3000 psi concrete	5.33	C.Y.	14.9	646.00
Slab on grade, 4" thick, incl. textured finish, no reinforcing	23	S.F.	0.4	28.98
Vapor barrier, 6 mil polyethylene	420	S.F.	90.8	1,386.00
Block wall, 8" thick, conc. block 8" x 16" x 8", reinforced	96	S.F.	10.7	285.70
Dampproofing, bituminous coating, 1 coat	96	S.F.	1.2	8.06
Anchor bolts, 1/2" diameter, 8" long, 4' O.C.	8	Ea.	0.3	5.28
Bulkhead cellar door with sides	1	Ea.	1.9	384.00
Door framing, 2 x 8	18	L.F.	0.4	10.58
Residential stl. door, pre-hung, insul. ext., flush face 3' x 6'-8"	1	Ea.	1.0	230.40
Stair stringers and treads, 2 x 10	42	L.F.	5.2	59.47
Lockset, standard duty, cylindrical, keyed, single cylinder	1	Ea.	0.8	84.60
Door trim, 1 x 6 pine, #2	17	L.F.	0.5	24.28
Paint, door, incl. frame & trim, exterior, primer & 1 coat, latex	1	Ea.	2.3	19.92
Totals			132.1	$3,391.60

Contractor's Fee, Including Materials: $12,308

Garages and Carports

Garages and carports should be carefully planned based on their intended use (storage, workshop space, and/or vehicle shelter), the size and style of your house, the available and buildable (within requirements of the local building department) land, and the characteristics of the site. Following are some additional considerations.

- Design your garage—and its doors— to accommodate full-size vehicles even if your car is a subcompact. Not only will you appreciate having space to move around (on foot and in your car), but it won't limit your home's appeal to buyers, should you ever decide to move.

- Prior to finalizing plans, mark off the proposed garage on the ground with stakes and string. Get a feel for the size of the project.

- Garage floors cannot be flush with an adjacent, existing floor. A minimum difference is required for protection from gasoline spills. Check with your local fire and building departments for specific requirements.

- Unless you have a reason to call attention to your garage, the doors should be unobtrusive and harmonize with your house.

- If security is a concern, avoid garage doors with windows.

- When deciding between a carport and a garage, consider the value of full protection and security that a conventional garage offers, versus the lower expense and simpler-to-build carport project. Do-it-yourselfers can do much of the work to construct a carport on their own.

- Roll-up garage doors have more moving parts than any other door in your house and therefore require more maintenance to keep them working properly. Oil the roller bearings, pulleys, lock mechanism, and cables twice a year, and check track alignment once a year.

- In shopping for an automatic garage door opener, ask how loud the opener will be and request a demonstration, if possible. Generally, a 1/4 horsepower motor can handle a single door, and a 1/2 horsepower motor can operate a double door. Inquire about safety features as well, particularly if you have children.

- Check the local building code if you're building an attached garage. There are important requirements to ensure adequate fire separation that cannot be ignored.

- Garages require the same standards for seismic bracing as your building code requires for your home.

- If a project requires excavation (e.g., to install a foundation), be alert to the dangers of encountering and damaging underground electrical, gas, water, or septic lines. You or your contractor should contact your utility companies to verify the location of these services underground.

Attached Single Carport

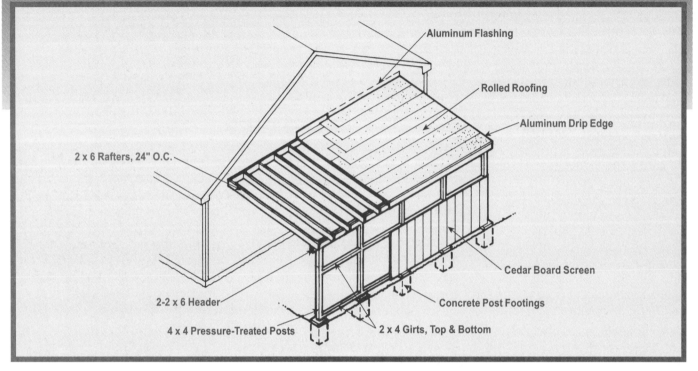

Aluminum Flashing

Rolled Roofing

Aluminum Drip Edge

2 x 6 Rafters, 24" O.C.

Cedar Board Screen

2-2 x 6 Header

Concrete Post Footings

4 x 4 Pressure-Treated Posts

2 x 4 Girts, Top & Bottom

In some areas of the country, particularly those subject to heavy snow or hailstorms, overhead protection for vehicles is a must. It can also increase your home's value and potential curb appeal if carefully designed and finished to match your house. Carports offer a simple and economical alternative to the more costly framing and finish work required for conventional garages. In addition to requiring less material, they're also easier to build than garages. Their uncomplicated design and structure offer the opportunity for do-it-yourselfers to do all or much of the work on their own.

Materials & Tools

The components of this carport project are standard construction materials that can be purchased at building supply outlets. Try to buy all of the materials at the same place, and be sure to make arrangements for curb delivery as part of the purchase price. Because the carport is attached to the house on one side and open at either end, materials are needed only for the side opposite the house and

the roof. Both of these sections are basic in their design and thus are fairly easy to build, if you lay them out correctly, follow good carpentry practices, and work slowly on tasks that are new to you.

If you plan to place the supporting posts in concrete, be sure to purchase pressure-treated lumber, as recommended in the plan. Standard grade 4 x 4s can also be used at reduced cost, but they should not be set in concrete and require fastening to anchors preset in the footings. This plan suggests rough-sawn 1 x 12 cedar siding boards, but many options are available to suit the style and decor of your home, including various wood sheet goods and fiberglass and metal panels. The cost of these products—and their installation time—varies widely.

As a driveway is most likely already in place, no consideration has been given in the plan for the cost of the floor surface within the carport. If a surface is needed, the cost of asphalt pavement, concrete, or gravel, and the time to

install it, will have to be added to the project estimate.

To build the carport, you'll need standard carpentry tools, such as a hammer, nails, screws, drill, saws, ladder, level, tape measure, stringline, and so forth. You'll also need a shovel or post hole digger to excavate post holes.

What's Involved

The side of the carport serves two purposes: to provide support for one end of the roof and for the girts to which the siding is fastened. Because the roof structure is large and heavy, the five 4 x 4 posts that support it must be set in concrete footings at least 3' deep. Precise placement of the posts is important, so take time to space them at even intervals, align them with a stringline, and plumb them with a level.

After the posts have been placed, the roof can be framed with 2 x 6s set 24" on center, and doubled headers. This roof framing task is the trickiest operation in the project, so get some help if you have not done it before.

Be sure to cut the joists at the correct angle on the house end and at the reverse angle on the other end. Also, prepare the house wall to receive the roof assembly by cutting back the siding to the sheathing and arranging for the roof flashing. The method of installing the flashing will depend on the type of siding, so costs may vary for this operation.

At the other end of the roof frame, temporary support should be provided for the first few joists before the headers are positioned and fastened. Be sure to square the frame before laying the 1/2" plywood roof sheathing, and stagger the termination seams of the plywood as you place it. Rolled roofing material should be used in place of asphalt shingles, because of the flat pitch of the shed roof. The installation of this type of roofing can go quickly with the right methods and equipment.

With a little advice and instruction beforehand, and the right tools, most intermediates can undertake the job, even if they have no prior experience with rolled roofing. Be sure to place the drip edge on all open sides of the roof before you apply the roofing material and install flashing on the house side. Once the roof is completed, the girts can be placed between the posts, and a 5'-high screen attached to them to make up the side of the carport.

Level of Difficulty

This carport plan is a manageable exterior project for most homeowners, including beginners who have the time to work slowly and are willing to learn as they go. Some of the tasks involved in the building process require knowledge that beginners may not have, but the required skills can be learned by trial and error and proper guidance from an experienced builder.

Two of the jobs in this project are critical: the precise setting and leveling of the support posts and accurate layout and placement of the roof frame. If

these two installations are correct, the rest of the job should go a lot more smoothly. Intermediates and experts should be able to handle all of the tasks required in the project, but may need to learn the details of rolled-roofing installation before they begin that part of the job. Generally, beginners should double the professional time for all tasks; intermediates should add 40%; and experts, about 10%.

Options

With the investment of a little more time and money, carports can be made to double as screened-in sitting and eating enclosures. Once the carport structure, as described in this plan, has been completed, screen panels can be made up to fit the open areas in the rear end and side of the enclosure. The front

end of the carport can be fitted with roll-down screening that can be raised when the facility is used to shelter the car. These modifications will increase the cost of the project, but they can be done at any time after the carport is finished.

See also:
Free-Standing Double Carport, Attached Single Garage.

Alternate Materials

Roofing Options
Cost per Square, Installed

Asphalt roll roofing	$46.50
Asphalt shingles	$91.50
Cedar shingles	$310.00
Clay tile	$805.00

Attached Single Carport, 11' x 21'

Description	Quantity/ Unit	Labor-Hours	Material
Excavate post holes, incl. layout	5 Ea.	5.0	
Concrete, field mix, 1 C.F. per bag, for posts	5 Bags		34.80
Post base, 4 x 4	5 Ea.	0.3	32.70
Post cap, 4 x 4	5 Ea.	0.3	16.02
Posts, pressure-treated, 4 x 4 x 10'	50 L.F.	1.6	46.20
Headers, 2 x 6, doubled	96 L.F.	3.1	88.70
Joists/rafters, 2 x 6 x 12', 24" O.C.	132 L.F.	4.2	121.97
Sheathing, 5/8" plywood, 4' x 8' sheets	288 S.F.	3.5	269.57
Roofing, rolled, 30 lb. asphalt-coated felt	3 Sq.	2.4	106.20
Drip edge, aluminum, 5"	48 L.F.	1.0	14.40
Load center, circuit breaker, 20A	1 Ea.	0.7	10.26
Non-metallic sheathed cable, #14, 2-wire	40 L.F.	1.3	8.16
Switch, w/ 20' #14/2 type NM cable	1 Ea.	0.5	9.72
Duplex receptacle, 20A, w/ 20' #12/2 type NM cable	1 Ea.	0.6	22.80
Lighting outlet box and wire	1 Ea.	0.3	11.28
Light fixture, canopy type, economy grade	1 Ea.	0.2	31.20
Girts, 2 x 4, for screening panel	40 L.F.	1.2	23.52
Siding boards, cedar, rough-sawn, 1 x 12, board on board	110 S.F.	3.4	266.64
Paint, all exposed wood, primer	625 S.F.	7.7	67.50
All exposed wood, 2 coats	625 S.F.	12.3	120.00
Totals		49.6	$1,301.64

Contractor's Fee, Including Materials: **$4,510**

Free-Standing Double Carport

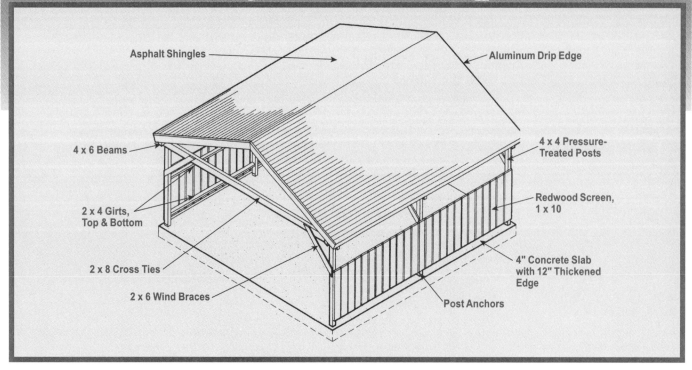

- Asphalt Shingles
- Aluminum Drip Edge
- 4 x 6 Beams
- 4 x 4 Pressure-Treated Posts
- 2 x 4 Girts, Top & Bottom
- Redwood Screen, 1 x 10
- 2 x 8 Cross Ties
- 4" Concrete Slab with 12" Thickened Edge
- 2 x 6 Wind Braces
- Post Anchors

Carports previously had a poor reputation for their unattractive appearance, when many were cheaply constructed with corrugated plastic roofing and stood out like the proverbial sore thumb. Over the years, and especially in the South, where intense sun and occasional hail require protection from overhead, but not necessarily walls all around, builders have increasingly upgraded carport materials and design. Carports are also being better tied into the style of the homes they serve.

This two-car model uses post-and-beam construction. Its simple design makes it a reasonable undertaking for do-it-yourselfers with intermediate-level skills, provided they're experienced in the use of tools and have a fundamental knowledge of carpentry. Because of the size of this structure, beginners should hire a professional to install the post and-beam support and roof framing, but could tackle other tasks in the project. Most skilled intermediates and experts can complete the entire project, with the exception of the concrete slab floor.

Materials & Tools

This project starts with the floor—a slab constructed of ready-mixed concrete, metal-rod or wire reinforcement, a sub-base of granular material, and a plastic vapor barrier. The post-and-beam frame is made of 4 x 6s and 4 x 8s (or doubled 2 x 8s). The roofing requires standard grade 2 x 6 rafters, plywood sheathing, and asphalt shingles. The siding in this project is cedar boards, but your choice will depend on what best coordinates with the style of your house. All these products, except the ready-mixed concrete, are available at lumberyards and building centers, so check around for the best prices. Arrange for curb delivery, since the materials are heavy and bulky. A good collection of standard hand and power carpentry tools will also be needed, along with some masonry tools.

What's Involved

The project begins with the excavation for, and placement of, the concrete floor. There are several important

aspects in this procedure that, when closely monitored and done right, make the rest of the project go more smoothly. The first is double-checking the square and level of the forms before the concrete is placed. The second is precisely positioning and aligning the post anchors for the 4 x 6s as they're embedded in the concrete. The wood structure will be supported by the slab and aligned according to the position of the anchors, so measure and square the layout carefully. If you're unfamiliar with this type of concrete placement, get help from an experienced person or subcontract this part of the project to a professional.

The post-and-beam construction of the roof support requires basic carpentry and some engineering skills. The three post-and-cross-tie assemblies can be laid out and fabricated on the slab before being raised and positioned on the anchors, and temporarily braced. Be sure that they're plumb, level, and square before the 4 x 6 plate beams are placed and fastened.

The roof frame is constructed of 2 x 6 rafters, set 24" on center, and a 2 x 8 ridge board. If you're unfamiliar with the process of cutting and setting rafters, get some instruction beforehand, and be sure to have helpers available. Since you'll be working 8' to 12' off the ground when placing the rafters, plan to rent or borrow staging, trestle ladders and planks, or other means of working safely and efficiently at the ridge board height. An alternative to installing the rafters is purchasing pre-manufactured 2 x 4 trusses and placing them 24" on center. The cost for materials will be higher, but the roof will go up faster.

The 1/2" plywood sheathing is placed on the rafters after the roof frame has been squared and fastened. The aluminum drip edge, installed before the roofing is laid, is an inexpensive but important item, as it provides run-off clearance for rainwater and creates a straight, neat, finished edge.

This plan includes 1 x 10 cedar siding installed on 2 x 4 girts as screening and then stained or painted. The post-and-beam supports, girts, rafters, underside of the roof deck, and the fascia board are covered with two coats of stain or paint.

Level of Difficulty

This carport plan calls for a moderate to advanced level of building expertise and carpentry skills. The size of the structure and the post-and-beam layout requirements involve complex planning that may also put parts of the project out of reach for beginners and many intermediates. After the supports and roof frame are in place, most intermediate do-it-yourselfers can finish the job, including installing the roof sheathing and shingles. Straight roofs are not that hard to install if you get started correctly.

Beginners should hire a professional to install the concrete slab. Intermediates and experts should allow extra time for careful layout of the forms and arrange to have helpers on hand when the concrete is placed. They should add

100% and 50%, respectively, to the professional time for the concrete slab installation, including building the formwork. They should add 40% and 10%, respectively, to the time for all other tasks. Beginners should not attempt the slab, post-and-beam support, or roof-framing jobs. They should double the professional time for all other tasks, and get ample help.

What to Watch Out For

The concrete floor must be correctly placed. If you're installing it on your own, set the forms accurately so that the edges of the slab are level and square. Be sure to crown the slab for surface-water drainage.

See also:
Driveway, Single Attached Carport.

Free-Standing Double Carport, 20' x 21'

Description	Quantity/Unit		Labor-Hours	Material
Grading, by hand, for slab area	53	S.Y.	1.8	
Edge forms for floor slab	86	L.F.	4.6	27.86
Vapor barrier, polyethylene, 6 mil	4.20	Sq.	0.9	13.86
Concrete floor slab, 4" thick, 3000 psi complete	420	S.F.	8.8	539.28
Post anchors, embedded in slab	6	Ea.	0.2	3.96
Posts, pressure-treated, 4 x 6 x 8'	48	L.F.	2.2	172.22
Crossties, 2 x 8 x 22'	132	L.F.	4.7	158.40
Plate beams, 4 x 6 x 12'	48	L.F.	2.2	172.22
Ridge board, 2 x 8 x 12'	24	L.F.	0.9	28.80
Fascia board, 2 x 8 x 12'	48	L.F.	1.7	57.60
Rafters, 2 x 6 x 12', 24" O.C.	288	L.F.	9.2	266.11
Sheathing, 1/2" plywood, 4' x 8' sheets	544	S.F.	6.2	404.74
Drip edge, aluminum, 5"	96	L.F.	1.9	28.80
Shingles, asphalt, 235 lb. per square	6	Sq.	8.7	223.20
Load center, circuit breaker, 20A	1	Ea.	0.7	10.26
Non-metallic sheathed cable, #14, 2-wire	60	L.F.	1.9	12.24
Switch, w/ 20' #14/2 type NM cable	1	Ea.	0.5	9.72
Duplex receptacle, 20A, w/ 20' #12/2 type NM cable	1	Ea.	0.6	22.80
Lighting outlet box and wire	1	Ea.	0.3	11.28
Light fixture, canopy type, economy grade	1	Ea.	0.2	31.20
Braces, 2 x 6 x 4'	40	L.F.	1.3	36.96
Girts, 2 x 4, for screening panels	84	L.F.	2.4	49.39
Screening, redwood channel siding, 1 x 10	210	S.F.	5.9	561.96
Paint, primer	1,100	S.F.	13.5	118.80
Paint, 2 coats	1,100	S.F.	21.7	211.20
Stain, 2 coats, brushwork on screen	420	S.F.	7.1	50.40
Totals			110.1	$3,223.26

Contractor's Fee, Including Materials: $10,531

Attached Single Garage

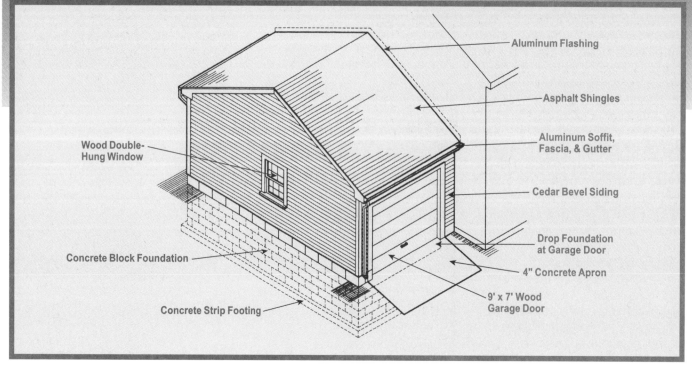

- Aluminum Flashing
- Asphalt Shingles
- Aluminum Soffit, Fascia, & Gutter
- Cedar Bevel Siding
- Drop Foundation at Garage Door
- 4" Concrete Apron
- 9' x 7' Wood Garage Door
- Wood Double-Hung Window
- Concrete Block Foundation
- Concrete Strip Footing

A garage can boost the value of your home, providing protection for vehicles, storage, and possibly a workshop for projects and hobbies. In areas of the country that have hail storms or heavy snow, garages may be considered a must by many home buyers. This model is a basic single-car garage attached to the house and built on level ground. Many do-it-yourselfers can accomplish this project with limited professional help, especially given the recent increase in "garage kits" on the market—complete with assembled roof components, pre-cut studs, and metal fasteners.

Materials & Tools

This garage plan calls for materials for the foundation and floor, three solid walls, and the roof, as well as some specialized products like gutters and a garage door unit. Because all of these items are usually readily available in home supply stores, prices are competitive, and it pays to shop around. You'll probably get a better package price if you buy all of the materials from the same retailer. You'll need framing lumber and sheathing, and siding and trim to match your house materials and design. Other materials include concrete block (for this model) for the foundation, roofing materials, gutters, a window, garage doors, a light fixture and electrical materials, and paint. Tools include the standard hand and power tools, plus sawhorses, a plumb bob, nail gun and staple gun, and hacksaw.

What's Involved

Before the foundation and floor are placed, the site for the new garage must be cleared and a perimeter trench excavated to a depth of about 4'. The site conditions will have an impact on cost. For example, a severe slope in the grade may require a higher foundation or backfill to raise the grade. Generally, if the terrain is level and free of trees and other obstructions, no extra cost will be incurred. The excavation of the 12' x 22' foundation trench should be dug to a level below the frost line and should be deep and wide enough to allow for an 8" x 16" footing and a 3' to 4' foundation wall.

It's probably best to hire a specialty contractor for the foundation work. In addition to a professional job, you'll get the benefit of having it done quickly and with much less aggravation. The concrete slab can be placed any time after the foundation has been completed. When excavating for the slab, be sure to allow for several inches of sub-base (4" of concrete), and at least one exposed course of block above the floor surface (except in the section that will include the door). If you're installing the floor on your own, be sure to allow a gradual slope toward the door opening for drainage.

Once the foundation has been placed and backfilled, the three solid walls can be framed and sheathed. Standard grade 2 x 4s placed 16" on center make up the walls, with a doubled plate on the top and a single shoe plate on the bottom. Be sure to fasten the shoe plate firmly to the 2 x 6 sill, which should be anchored to the top of the foundation wall. Remember to allow additional time to remove some or all of the siding from the house wall before framing. The other walls can then be framed, 2 x 6 joists can be placed, and the roof framing can be completed.

Framing the roof is more difficult than for the walls, but this model is still small enough for physically able do-it-yourselfers to tackle, provided they have help from someone with experience in placing a ridge board and rafter system. The keys to a square and level roof frame are consistency of rafter length and accuracy of angling and notching. Once the rafters have been laid out and cut from a template, placement and fastening usually move quickly.

After the frame is complete, you can install the sheathing, drip edge, roof, and shingles.

Level of Difficulty

Foundation, framing, and roofing operations are challenging tasks, regardless of the type of structure being built. Inexperienced do-it-yourselfers should assess their level of skills before taking it on. Because this structure is relatively small, with one side already in place, intermediates can improve their building skills, but they should seek assistance when needed. The concrete and masonry work may be out of the reach of many do-it-yourselfers. Electrical work should be done by a professional.

Ambitious beginners can complete most of the project once the foundation is in, but they should get instruction at the start and guidance along the way. They should double the labor-hours for the tasks they attempt. Intermediates and experts should add 40% and 10%, respectively, for the carpentry work, and more for specialty operations like the foundation, floor, and garage door installations.

What to Watch Out For

Get some assistance on the flashing installation if you haven't done it before. A poor job can cause water damage. Be sure to consult your local building department for permit and code requirements.

See also:
Free-Standing Garages and Carports, Skylights, New Window, Driveway.

Attached Single Garage, 12' x 22'

Description	Quantity/Unit	Labor-Hours	Material
Site clearing, layout, excavate for footing	1 Lot	7.0	
Footing, 8" thick x 16" wide x 44' long	2 C.Y.	5.6	242.40
Foundation, 6" concrete block, 3'-4" high	264 S.F.	23.2	529.06
Edge form at doorway	10 L.F.	0.5	3.24
Floor slab, 4" thick, 3000 psi, with sub-base, reinforcing, finished	265 S.F.	5.6	340.26
Wall framing, 2 x 4 x 10', 16" O.C.	38 L.F.	6.1	162.34
Headers over openings, 2 x 8	14 L.F.	0.7	13.94
Sheathing for walls, 1/2" thick, 4' x 8' sheets	384 S.F.	4.4	285.70
Ridge board, 1 x 8 x 12'	12 L.F.	0.3	17.57
Roof framing, rafters, 2 x 6 x 14', 16" O.C.	280 L.F.	4.5	181.44
Joists, 2 x 6 x 22', 16" O.C.	220 L.F.	2.8	142.56
Gable framing, 2 x 4	45 L.F.	1.3	18.36
Sheathing for roof, 1/2" thick, 4' x 8' sheets	352 S.F.	4.0	261.89
Sub-fascia, 2 x 8 x 12'	24 L.F.	1.7	23.90
Felt paper, 15 lb.	340 S.F.	0.7	12.24
Shingles, asphalt, standard strip	4 Sq.	6.4	194.40
Drip edge, aluminum, 5"	52 L.F.	1.0	15.60
Flashing, aluminum	21 S.F.	1.2	16.38
Fascia, aluminum	21 S.F.	1.2	53.93
Soffit, aluminum	38 S.F.	1.4	46.06
Gutters and downspouts, aluminum	44 L.F.	2.9	65.47
Load center, circuit breaker, 20A	1 Ea.	0.7	10.26
Non-metallic sheathed cable, #14, 2-wire	40 L.F.	1.3	8.16
Switch, w/ 20' #14/2 type NM cable	1 Ea.	0.5	9.72
Duplex receptacle, 20A, w/ 20' #12/2 type NM cable	1 Ea.	0.6	22.80
Lighting outlet box and wire	1 Ea.	0.3	11.28
Light fixture, canopy type, economy grade	1 Ea.	0.2	31.20
Wood, window, standard, 34" x 22"	1 Ea.	0.8	250.80
Garage door, wood, incl. hardware, 9' x 7'	1 Ea.	2.0	528.00
Siding, cedar, rough-sawn, stained	400 S.F.	13.3	1,444.80
Sheetrock, 5/8" fire resistant	484 S.F.	8.0	162.62
Painting, door and window trim, primer	85 S.F.	1.7	5.10
Painting, door and window trim, 2 coats	85 S.F.	1.7	5.10
Totals		113.6	$5,116.58

Contractor's Fee, Including Materials: **$13,758**

Free-Standing Double Garage

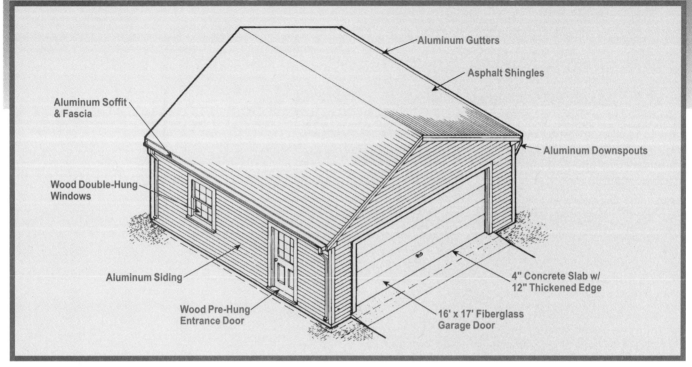

- Aluminum Gutters
- Asphalt Shingles
- Aluminum Soffit & Fascia
- Aluminum Downspouts
- Wood Double-Hung Windows
- Aluminum Siding
- Wood Pre-Hung Entrance Door
- 4" Concrete Slab w/ 12" Thickened Edge
- 16' x 17' Fiberglass Garage Door

Free-standing garages have a couple of advantages: they separate your workshop space (and noisy projects) from the rest of your house, and building one is less complicated than the attached garage in that you don't need to tie in to an existing structure.

This garage plan is basic in both design and materials, but it is still a major project—one that will challenge do-it-yourselfers of all ability levels. With patient and careful work, however, those with intermediate or advanced carpentry skills can save on labor costs by completing some or all of the project tasks on their own.

Materials & Tools

The basic components include commonly available building materials for wood-frame structures, as well as concrete for the foundation and finished floor. Standard 2 x 4 framing materials, plywood, and asphalt shingles will be needed for the roof. Several specialty products are also included: aluminum gutters, soffits, and fascias, and a double-width fiberglass garage door. An electric light, outlets, and wiring are additional items.

With an order this size, you have some leverage in getting the best possible price. Cost, quality, and convenience are factors to consider. Arrange for delivery. You'll need basic hand tools like a hammer, screwdrivers, a level, and squares, as well as a circular saw, sawhorses, electric drill, plumb bob, hacksaw, and nail and staple guns.

What's Involved

If the site for the new garage is level and dry, the suggested 4" reinforced concrete slab with a thickened perimeter is generally all that is required for a foundation. If you have poor soil conditions and frequent ground freeze, or if you're considering future attachment of the garage to another structure, you'll accrue extra costs for placing a footing and foundation wall. Site clearing and excavation will take some time because of the large surface area, so plan accordingly, particularly if you intend to clear and dig by hand.

It is recommended that you hire a contractor to install the foundation and slab.

Framing the walls and roof requires a command of carpentry skills. If you have not framed a four-sided structure before, you'll need assistance in laying out, erecting, squaring, and securing the partitions. After the wall framing is in place, 1/2" exterior plywood sheathing should be installed for support and as a base for the finish siding.

The roof framing is the most difficult task in the entire project, as the rafters and ridge board have to be positioned with accuracy. Sheathing and shingling can be handled by most do-it-yourselfers. Once the roof has been completed, the structure is ready to receive the windows, entrance door, siding, garage door, and other finish products. A specialty garage door contractor should also be consulted or hired before the installation.

unless you plan to insulate and heat the garage. Garage doors offer various architectural and functional features, in both wood and man-made materials.

See also:

Single and Three-Car Free-Standing Garages, Attached Garages, Carports, Driveway, Main Entry Door, Skylights, New Window.

Level of Difficulty

This project is a major undertaking, but if you're skilled at framing, you can save considerably on the labor costs. Two critical procedures, the slab installation and roof framing, can be costly in time and replacement materials if they're not done correctly. Seek professional assistance for these tasks if you lack experience.

Beginners should not attempt the concrete, framing, and specialty installations and should add 100% to the professional time for jobs they can handle. Intermediates and experts should add 40% and 10%, respectively, to the labor-hour estimates for the jobs they attempt. They should add more for the slab placement and roof framing. Intermediates and experts should consider hiring professionals for the specialty work, like the aluminum siding and garage door installation.

What to Watch Out For

If you plan to use the space above the ceiling joists for storage, remember that the joists in this plan are only 2 x 6s; 2 x 8s should be used if heavy materials are to be stored.

The windows and door should match the style of your home and your budget. The extra expense for insulating glass or deluxe airtight units is unnecessary

Free-Standing Double Garage, 22' x 22'

Description	Quantity/Unit	Labor-Hours	Material
Grading, by hand, for slab area	576 S.F.	19.8	
Edge forms for floor slab	88 L.F.	4.7	28.51
Concrete floor slab, 4" thick, 3000 psi, complete	484 S.F.	9.4	609.84
Anchor bolts, embedded in slab	20 Ea.	0.8	13.20
Wall framing, 2 x 4 x 8', 24" O.C., studs and plates	80 L.F.	10.2	227.52
Headers over openings, 2 x 12	28 L.F.	1.5	53.76
Sheathing for walls, 1/2" thick, plywood, 4' x 8' sheets	736 S.F.	10.7	733.06
Ridge board, 1 x 8 x 12'	24 L.F.	0.7	35.14
Roof framing, rafters, 2 x 6 x 14', 24' O.C.	336 L.F.	5.4	217.73
Joists, 2 x 6 x 22', 24" O.C.	264 L.F.	3.4	171.07
Gable framing, 2 x 4	75 L.F.	2.2	30.60
Sheathing for roof, 1/2" thick, plywood, 4' x 8' sheets	756 S.F.	8.6	562.46
Sub-fascia, 2 x 8 x 12'	48 L.F.	3.4	47.81
Felt paper, 15 lb.	7 Sq.	1.5	28.64
Shingles, asphalt, standard strip	7 Sq.	11.2	340.20
Drip edge, aluminum, 5"	100 L.F.	2.0	30.00
Fascia, aluminum	50 S.F.	2.8	128.40
Soffit, aluminum	70 S.F.	2.7	84.84
Gutters and downspouts, aluminum	64 L.F.	4.3	95.23
Load center, circuit breaker, 20A	1 Ea.	0.7	10.26
Non-metallic sheathed cable, #14, 2-wire	60 L.F.	1.9	12.24
Switch, w/ 20' #14/2 type NM cable	1 Ea.	0.5	9.72
Duplex receptacle, 20A, w/ 20' #12/2 type NM cable	1 Ea.	0.6	22.80
Lighting outlet box and wire	1 Ea.	0.3	11.28
Light fixture, canopy type, economy grade	1 Ea.	0.2	31.20
Windows, wood, standard, 2' x 3'	2 Ea.	1.6	501.60
Garage door, incl. hardware, hardboard, 16' x 7', standard	1 Ea.	2.7	840.00
Entrance door, wood, prehung, 2'-8" x 6'-8"	1 Ea.	1.0	326.40
Siding, aluminum, double, 4" pattern, 8" wide	680 S.F.	21.1	971.04
Painting, entrance door and window trim, primer	50 S.F.	0.6	1.20
Painting, entrance door and window trim, 2 coats	50 S.F.	1.0	3.00
Totals		137.5	$6,178.75

Contractor's Fee, Including Materials: **$16,399**

Free-Standing Three-Car Garage

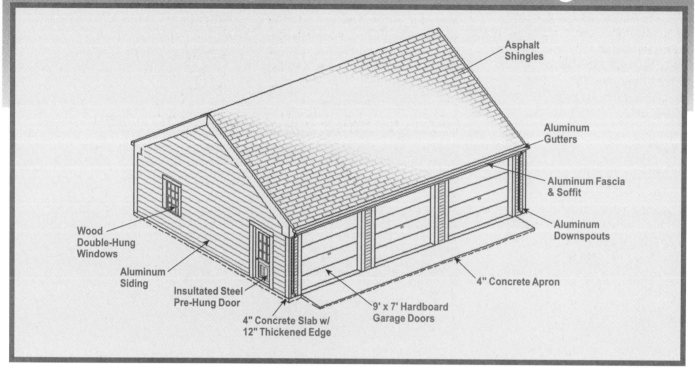

- Asphalt Shingles
- Aluminum Gutters
- Aluminum Fascia & Soffit
- Aluminum Downspouts
- 4" Concrete Apron
- 9' x 7' Hardboard Garage Doors
- 4" Concrete Slab w/ 12" Thickened Edge
- Insulated Steel Pre-Hung Door
- Aluminum Siding
- Wood Double-Hung Windows

If you need more space than the garages on the previous pages provide, whether for a car, workshop, or storing yard equipment, you might consider this improvement to your home. The plan is basic in both design and materials, but, like the other garages, this project is a big undertaking, made even more challenging by its size.

Fairly skilled do-it-yourselfers can accomplish much of this work. Do consider how much time you're willing to devote to it though—as well as the inconvenience of a large and ongoing construction site in your yard—when weighing whether and how much of the work to do yourself.

As with all significant-sized home improvement projects, be sure to consult your local building department early in the planning stages, not only for building requirements, but for restrictions in setbacks from property lines and other requirements.

Materials & Tools

This project requires the same materials used in the two-car garage on the previous pages. The foundation and floor consist of a 4" reinforced concrete slab with a thickened perimeter. The walls are 2 x 4s and plywood sheathing, and the roof is comprised of 2 x 6s, plywood, and asphalt shingles. Roofing felt can be added for an extra weatherproofing, but is not an essential. An electric light fixture, wiring, and outlets are also included.

Other materials in this project are aluminum fascias and soffits, aluminum siding, aluminum gutters and downspouts, and hardboard single-width garage doors. With an order this large, you are in a good position to get the best price, so shop around before you commit your order to a retailer. Arrange for local delivery in the final purchase agreement. Standard hand and power tools will also be needed.

What's Involved

Clearing and excavating the site is a major operation for the three-car garage since it's such a large area. Installing a concrete slab, as in this model project, will give you both the foundation and the floor for the garage, provided your site is level and dry. (You may need a footing and foundation wall if you have difficult soil conditions, live in an area where frost is an issue, or are considering attaching the garage to the house in the future.) For professional results, and considering the importance and size of this project's foundation, it might be best to hire a professional contractor for this part of the job, even if you have some home improvement experience. Remember, the floor should have a slight slope for drainage.

As mentioned in the two-car garage project, you'll need carpentry skill and experience to lay out, erect, square, and secure the walls for this structure. Nailing the exterior plywood sheathing to the frame, and attaching the shingles, can be accomplished by many do-it-yourselfers. Framing the roof is a demanding part of the job. It is more difficult to work at a height off the ground, and the framing must be accurate to avoid frustration later on, to say nothing of wasted time and materials.

After framing and sheathing the walls and roof, and shingling the roof, the windows, siding, entrance and garage doors, and trim can be installed. Specialty contractors are strongly recommended to install the aluminum siding and possibly the garage doors. Electrical wiring and placement and connection of the light fixture(s) and outlets should be done by a professional.

Level of Difficulty

Do-it-yourselfers who have experience in framing and exterior work should be able to complete much, if not all, of this garage. It's a substantial project because of its size, and it will take considerable time to properly plan and lay out, in addition to the work itself. On the other hand, you can save money by doing the tasks that you are qualified for. Hiring a professional is an excellent investment for any and all tasks for which you lack skill and experience.

Beginners should add at least 100% to the professional time estimates listed for the jobs they're able to undertake, such as nailing on the sheathing and roof shingles. They should leave the concrete, framing, and specialty installations to professionals. Intermediates and experts should add 40% and 10%, respectively, to the labor-hour estimates for the jobs within their capabilities. Add more for the slab placement and roof framing if you have not done this kind of work before. Professional contractors are a good way to go for the aluminum siding and garage door installations, and all electrical work should be done by a professional.

What to Watch Out For

As mentioned in the two-car garage project, be sure to use 2 x 8s instead of the model plan's 2 x 6s for the ceiling joists if you plan to take advantage of the space above for storage of heavy items. (If the peak of your garage is high enough, you have the potential of more

than 500 square feet of usable storage space in that area.)

Options

Take time to check out the options for windows and doors that match or complement the style of your house. Keep in mind the huge effect of the garage door on your home's curb appeal.

If you choose to add plywood over the garage ceiling joists to maximize storage space, you might also want to consider installing a pull-down, disappearing stairway.

See also:

Free-Standing Garages, Attached Garages, Carports, Driveway, New Window.

Free-Standing 3-Car Garage, 24' x 36'

Description	Quantity/ Unit	Labor- Hours	Material
Layout of building perimeter, 24' x 36'	120 L.F.	2.9	4.32
Strip topsoil, excavate for slab, 6" deep, by hand	16 C.Y.	16.0	
Load soil onto truck, by hand	16 C.Y.	10.7	
Gravel fill, 6" deep, compacted	864 S.F.	4.8	248.83
Fine grading of backfill, by hand	96 S.Y.	3.3	
Edge forms for concrete slab on grade	88 L.F.	4.7	28.51
Edge forms for concrete apron	40 L.F.	2.1	12.96
Anchor bolts emmbedded in slab	24 Ea.	2.1	27.07
Concrete slab on grade, 4" thick, complete	864 S.F.	18.2	1,109.38
Concrete apron, 4" thick, complete	128 S.F.	2.7	164.35
Wall framing, 2x4 16" OC, studs & plates, 8' high	93 L.F.	14.9	340.38
Wall framing, 2x4, gable studs & plates	24 L.F.	0.6	9.79
Headers over door & window openings, (2) 2x12	48 L.F.	1.5	184.32
Plywood sheathing for walls and gables, 1/2" CDX	896 S.F.	12.7	666.62
Roof ridge board, 2x8	36 L.F.	1.3	35.86
Ordinary roof rafters, 2x6 x 14' long, 16" OC	784 L.F.	12.5	508.03
Ceiling joists, 2x6 x 24' long, 16" OC	672 L.F.	8.6	435.46
Plywood sheathing for roof, 5/8" CDX	1,008 S.F.	12.4	943.49
Subfascia board, 2x8	72 L.F.	5.1	71.71
Fascia and soffit for aluminum siding	72 L.F.	10.5	186.62
Drip edge, 5" aluminum	128 L.F.	2.6	38.40
Felt paper, 15 lb.	10 Sq.	1.3	40.92
Shingles, asphalt, 235 lb. per square	10 Sq.	14.5	372.00
Windows, double hung, 2' x 3', incl screens	6 Ea.	4.8	1,317.60
Door, pre-hung, insul. steel, 2/8 x 6/8, 1/2 glass	1 Ea.	1.0	255.60
Lockset, keyed, single cylinder function	1 Ea.	0.8	84.60
Garage door, hardboard, 9' x 7', incl track & hardware	3 Ea.	6.0	1,332.00
Aluminum siding, double 4" clapboard pattern	896 S.F.	27.8	1,451.52
Gutter, enameled aluminum, 5" box	72 L.F.	4.8	94.18
Downspouts, enameled aluminum, 2" x 3"	52 L.F.	2.2	60.53
Paint windows, 12-lite, primer + 2 coats	6 Ea.	9.6	17.71
Paint entrance door, primer + 2 coats	1 Ea.	3.2	15.00
Paint garage doors, primer + 2 coats	189 S.F.	4.3	56.70
Totals		230.5	$10,114.46

Contractor's Fee, Including Materials:	**$26,864**

Landscaping

Much of the visual appeal of older, established neighborhoods comes from landscaping—mature, fully developed trees and shrubs, stone walls, and other features that increase property value. Whether you hire a landscape designer/contractor or work on your own, it helps to keep the following in mind.

- In planning landscape improvements, read through books, magazines, and catalogs, and consult your local nurseries. County agricultural extension offices can also provide up-to-date advice on which plant varieties thrive in your area, and which ones pose special problems.

- Large-scale landscape projects are often best left to contractors with properly equipped and experienced crews.

- When planning your project, try to create a three- to five-year plan to allow for growth and development of select shrubs and plantings. In particular, be sure to place shrubs a proper distance (at least 3') from the house to allow for their mature growth, and to put them beyond the roof's drip line. Make sure that foliage is always at least 6"-8" away from the house, in order to avoid mildew and other moisture damage, and to allow access to the exterior walls and windows for painting and maintenance. Leave plenty of room behind or between shrubs for access to oil propane tanks, meters, and so on.

- Create a sketch—a site plan showing the existing plants and features in your yard, and those you would like to change or add.

- Remember that professional landscapers tend to avoid harsh lines, adding natural curves that harmonize with the terrain.

- Provide lighting for safety and to accent your landscaping.

- Make sure paths are wide enough—at least 3' and preferably 4'—to accommodate two adults walking side by side. In choosing a pathway material, try to complement the textures and colors of the house.

- Select plants and shrubs that provide overall interest in all seasons, and are appropriately sized for your house and lot.

- Follow planting and transplanting directions provided by lawn and garden experts. Do not overdo fertilizers and maintain a regular maintenance program to protect your investment.

- Try to maximize your use of native plants that require the least watering, fertilizing, and protection from insects and disease.

Brick Walkway

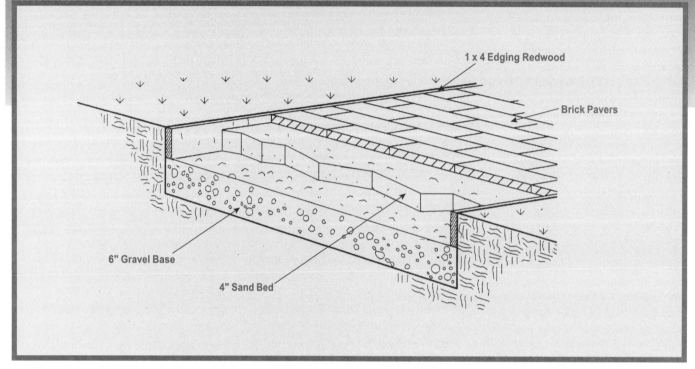

1 x 4 Edging Redwood

Brick Pavers

6" Gravel Base

4" Sand Bed

This project can bring instant curb appeal to your home and yard, and can be tackled by fairly ambitious do-it-yourselfers of all levels. A walkway is both attractive and practical: it provides a visual and physical link between elements in the landscape, directs visitors to your home's entrance and makes them welcome, offers dry and safe footing, and protects grass from foot traffic. If your walkway has deteriorated, is unattractive, or if one was never built, this project may be a good solution.

Materials & Tools

Brick, one of the oldest and most durable building materials, is compatible with most styles of architecture. It is also one of the most versatile and easily laid paving materials, and is widely available in home supply centers. A visit to a quarry or masonry supplier will reveal the surprising variety of colors, textures,

and shapes available. Individually, bricks are small and relatively lightweight, which makes them easy to handle, to arrange in interesting and attractive patterns, and to lay in a way that conforms with gradual changes of direction and ground level.

When you purchase the brick and the wood or plastic edging, you can also buy the sand or stone dust needed for the bed of the walkway. Unless your project is very small, it is wise to arrange for all of these materials to be delivered. You'll also need stakes and string, landscaping fabric, a patio broom, a rake, a rubber mallet, a pick and shovel, a hammer, a 3" masonry chisel, a wheelbarrow, and safety glasses and a dust mask. You may be able to rent a brick saw and plate compactor to make the work easier.

What's Involved

The simplest and most common way to build a walkway is to lay the paving material in a bed of sand or stone dust, without mortar. This is called *dry*

construction, and can be used with a variety of paving materials: bricks, concrete pavers, or flagstones. A walkway built this way drains well, and the process allows you to correct mistakes, replace broken pieces, or remove either a section or the whole of the walk at any time.

First, mark off the edges of the walkway area with string and stakes, and excavate to a depth of about 6". Dampen and tamp the earth in the walkway area, and put down a layer of synthetic landscaping fabric. This will help block the growth of weeds without inhibiting drainage. Cover the walkway with a layer of sand or stone dust to approximately 2" from grade level.

You'll need to decide on a pattern, or *bond,* for your walkway. There are many styles to choose from, the easiest and most economical being a *running bond*. Other bonds include jack-on-jack, herringbone, and basket-weave. Look at various walkways around your neighborhood and see which best suits your style and your budget.

Once you have chosen a bond, begin to lay the bricks by tamping down each one with a rubber mallet. As a general rule, allow five bricks per square foot of walkway, adding at least 10% for waste from brick cuts and defects. Add or remove sand to keep all the bricks in line and on the same plane. When all the bricks are laid, spread sand over them and sweep it into the joints, hosing down the walk to compact the sand. Repeat this procedure as many times as necessary to completely fill the joints. Laying the bricks tightly together keeps the joint spacing to a minimum; this helps the sand stay in place and makes the walkway almost maintenance-free.

Like highways, walkways should be crowned, or curved gradually towards both edges from a 3/4" to 1" high point in the center. This eliminates the possibility of ponding surface water by creating a natural drainage system. To crown the walkway, simply sculpture the base material to the desired shape. The bricks will follow this contour when placed.

Level of Difficulty

Most landscaping projects are physically demanding, and this walkway is no exception. If you're willing and able to do some hauling and bending, however, even a beginner can successfully complete this project. You can control the amount of labor involved in laying the bricks by the bond you choose. A running bond is easy to lay and requires few cuts. More complex patterns require more effort to arrange, and more cutting, which, of course, takes more time and creates more waste.

Beginners should get advice in choosing the most appropriate type of brick, and assistance in laying out the walkway, and should add 150% to the professional time estimate.

Intermediates can complete this job alone, adding about 50% to the time. Experts should add about 10%.

What to Watch Out For

Most walkway plans will require you to cut bricks to fit the pattern. This involves holding the chisel to the brick and hitting it sharply with a hammer. Your first few attempts may produce more rubble than usable brick, but you'll improve with practice. A short piece of wide 2" lumber makes a good cutting board.

One of the secrets to successful brick paving is properly constructing the border. It must be strong and lock the paving field in place. At steps or other places where the brick is terminated, you should lay the final course in mortar.

Removing an old walkway can add considerably to the preparation stage of this project, depending on its materials and construction. Concrete, for example, must be broken up with a jackhammer or sledgehammer, and then hauled away.

Options

Flagstones or concrete pavers can be used instead of bricks for your walkway, though remember that the style, cost, and ease of installation will vary depending on the material. Flagstones are cut from any type of a stone that splits into flat pieces, such as sandstone or slate, and are available in irregular or rectangular shapes, ranging in thickness from 1/2" to 2". Bluestone, granite, and marble are beautiful, but expensive, and look best in a formal setting.

Stones that are multiple-cut, cut to size, or of a special shape (e.g., round), are the easiest to arrange. Irregular stone pieces require time and care in laying, because they look best when fitted fairly close together, like a jigsaw puzzle.

Concrete pavers come in a huge variety of sizes, colors, shapes, and textures. They can be used alone or with other paving and border materials, such as brick, wood, or loose aggregates (e.g., wood chips or crushed stone). Some concrete pavers interlock, which helps them resist frost heaves. They're often used to create elegant-looking driveways.

See also:
Brick or Flagstone Patio, Driveway.

Alternate Materials

Material Options
Cost per Square Foot, Installed

Concrete pavers, 8" x 16"	$4.16
Flagstone, 1" thick	$14.45
Granite pavers, 4" x 8"	$56.50

Brick Walkway, 3' x 30'

Description	Quantity/Unit		Labor-Hours	Material
Gravel base, 6" deep	10	S.Y.	0.1	63.60
Sand, base fill, 4" deep	90	S.F.	1.7	77.76
Compaction in 6" lifts, hand tamp	1	C.Y.	0.4	
Hand grade, fine grading of base	3	C.Y.	2.1	75.60
Brick pavers, laid flat	90	S.F.	14.4	268.92
Edging, redwood, 1 x 4	60	L.F.	2.9	158.40
Totals			21.6	$644.28

Contractor's Fee, Including Materials: **$2,025**

Brick or Flagstone Patio

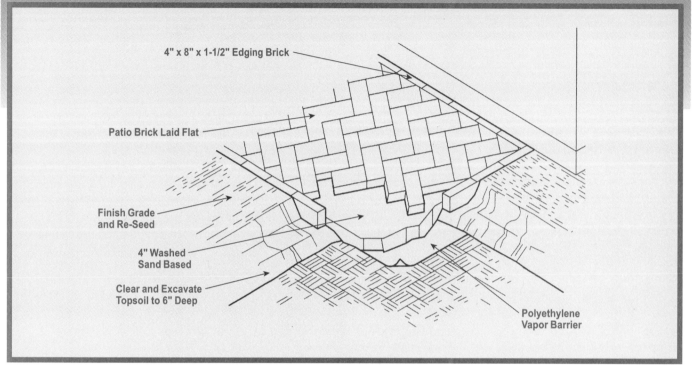

4" x 8" x 1-1/2" Edging Brick

Patio Brick Laid Flat

Finish Grade and Re-Seed

4" Washed Sand Based

Clear and Excavate Topsoil to 6" Deep

Polyethylene Vapor Barrier

Building a patio is an attractive way to link the indoors with the surrounding landscape. Brick or stone patios can be set in sand or mortar on a concrete slab—the latter being more difficult, but well within the reach of many do-it-yourselfers. If the slab is already in place, and the laying of the brick or stone is all that has to be done, the job is fairly easy. If you have to place the concrete slab first, the project is more involved and costly. Because of the specialized know-how required, inexperienced do-it-yourselfers should seek some assistance before they start.

Materials & Tools

This plan calls for brick or stone and mortar materials used for paving. Flagstones differ from brick in that they can be more expensive and often take longer to install because of their irregular sizes and shapes. While sand-set patios are less difficult to install than mortar construction, they are not as easy to maintain. Stones may need to be reset, and sand replenished periodically.

Additional cost may be incurred for a layer of gravel if the slab is being placed on poor subsoil. This will help to reduce cracking and heaving of the slab during freezes. Masonry sealer is worth the investment in time and money in the protection it provides for the finished patio. If you set your patio in sand, you'll need enough sand to cover the patio area to a depth of 4". You'll also need to place a polyethylene barrier beneath the sand bed to discourage vegetation from growing through the stones.

Several basic masonry tools, including a string line, a long mason's level, and a good trowel, are needed. Some of these tools may have to be purchased or rented, and the extra cost figured in.

What's Involved

Building a patio is really two installations in one. The first involves excavating and preparing the site and then placing the slab or sand bed; the second includes the placement of the flagstone or brick in the bed and the finish work. Before the slab or sand is placed, the patio area should be

carefully located, measured, squared, and then dug out by hand to a size slightly larger than that of the finished surface. If needed, you'll have to place and spread the gravel. For mortared patios, be aware that the edges may take some abuse during the slab placement. If you're fussy about appearance, you might consider placing the border after the concrete has set up. Be sure to square the form to its precise dimensions and then brace the corners and stake the sides to hold it firmly in place while the slab is poured. A slight pitch off level in one direction away from adjacent structures will help drain the surface.

Once the form has been placed and the reinforcing wire positioned, the slab is poured and finished. You must work quickly and efficiently, which can be challenging if you're new to this kind of work. Get some instruction beforehand and have plenty of help available when the ready-mix truck arrives. Be prepared and organized to minimize the delivery time. If the truck is tied up for too long, you'll be charged extra. After the concrete has been placed, it is leveled

and rough-finished in preparation for the placement of the bricks or stones.

Complete small, manageable sections at first, as you get the hang of the brick or flagstone placement. You'll pick up the pace as the project progresses. Flagstone requires more skill and installation time than brick because its irregular shape will necessitate cutting and fitting some pieces. Be sure to select "flags" of uniform thickness. Before laying a section of stones, place the pieces on the slab and position them by trial and error until you arrive at a combination that provides maximum coverage and joints that average about 3/4" in width.

Laying the bricks on the finished slab requires basic masonry knowledge, which you'll need to acquire before this project. Never mix more mortar than you'll use in an hour of brick-laying. If you need to cut bricks to fit your pattern, do it before mixing the mortar or starting the brick placement for that session. Laying bricks or stones in sand is more forgiving than working with mortar, which sets up quickly and leaves little room for re-working a mistake. When the bricks or stones are set, pack the spaces and cracks tightly with sand or grout to hold the materials in place and prevent damage over time. Make sure the patio is level and square as you proceed.

Level of Difficulty

Brick or stone patios set in sand are good projects for all levels of do-it-yourselfers.

Setting these materials in concrete is much more challenging, but manageable when taken in small steps. With proper preparation and patient work, even novices can complete the mortared patio with professional results.

The most demanding part of a mortared patio is placing the slab. Novices may consider hiring a contractor to form and pour the slab, then follow up with the brick work themselves. Experts and most intermediates should be able to complete either project completely on their own. They should add 10% and 40%, respectively, to the labor-hours estimates for the project. Beginners should add 75% to the professional time for all tasks.

What to Watch Out For

If bricks are set in concrete, you'll want to minimize the amount of mortar that gets on the brick or stones. Clean up any drops or smears with a damp cloth or sponge before they dry as you proceed. A final cleaning can be done with muriatic acid after the mortar has cured.

Options

Although special border features and materials usually cost more and take longer to install, they enhance the design and protect the edge of the surface. There are various options for rot-resistant wood edges, including redwood, cedar, locust, or pressure-treated pine. Patios that are recessed or terraced can be made more attractive with wood edging. A border of bricks set in soldier, sailor, or sawtooth arrangements is another option.

See also:

Brick Walkway, Brick Entryway Steps.

Alternate Materials

Walkway Options
Cost per Square Foot, Installed

Brick	$9.60
Redwood plank	$9.80
Granite pavers	$34.00
Bluestone	$14.40
Concrete	$4.22

Brick Patio, 12' x 18', Set in Sand

Description	Quantity/Unit	Labor-Hours	Material
Excavate for patio	4 C.Y.	8.0	
Hand tamp existing ground	4 C.Y.	1.6	
Wash sand delivered	0.67 C.Y.	0.1	21.31
Minimum delivery charge for small quantities	1 Ea.		63.00
Hand tamp sand 1" deep in place	0.67 C.Y.	0.3	
Place brick pavers 4" x 8" x 1 1/2"	216 S.F.	14.4	645.41
Place edging pavers in place	60 L.F.	7.1	194.40
Cleanup and remove debris	1 Ea.	2.0	
Totals		33.5	$924.12

Contractor's Fee, Including Materials: **$2,923**

Concrete Patio

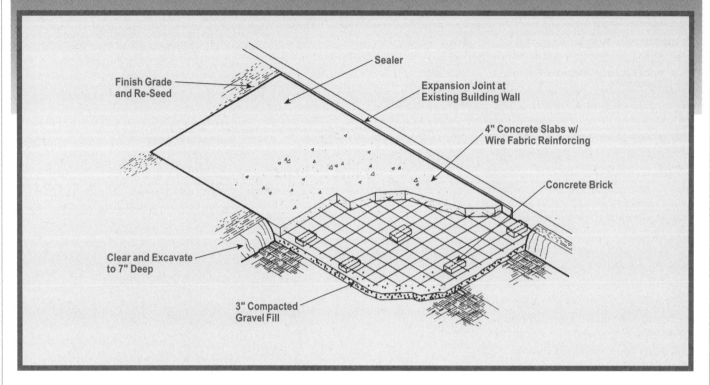

Finish Grade and Re-Seed

Sealer

Expansion Joint at Existing Building Wall

4" Concrete Slabs w/ Wire Fabric Reinforcing

Concrete Brick

Clear and Excavate to 7" Deep

3" Compacted Gravel Fill

The main advantage of using concrete for patios, walkways, or floors is its durability, but as an added benefit, it can also be shaped and molded into a variety of configurations and forms. Mixtures can also be added to create a color suitable to the surroundings. This model project is a standard 6' x 12' patio.

Materials, Tools, & Equipment

Before you begin work, you'll need some basic masonry hand tools: shovels, stiff rakes, a sturdy wheelbarrow, trowels, a float, a screed, and an edging or jointing tool. A string line, square, 3' or 4' level, 16' tape measure, and hammer are required for constructing the forms. Extra cost will be incurred if you have to buy the tools, but all are necessary to complete this project. Specialized concrete finishing tools, reinforcement materials, and pre-molded expansion joints can be purchased from a masonry

supplier. If you choose to mix the concrete yourself, you'll need to rent a concrete mixer.

What's Involved

All masonry work requires careful organization and preparation. As long as the grade on which the patio is to be built is level or gently sloped, and the subsoil easy to dig into, the project can be completed as estimated with no extra costs. The first step is to prepare the site, including marking the area, excavating it, and building the forms for the concrete.

The excavation work required for both patios is substantial, as you'll need to dig out the area with a shovel to a depth of at least 7" to allow for 3" of gravel fill for a sub-base under a 4" slab. The form should be pitched slightly away from any adjacent structures, squared, and braced firmly with stakes and corner ties to its final position.

Before you pour the concrete, you'll have to first install subsurface material (gravel), reinforcement wire, and an expansion joint. The amount of subsurface material will vary with the consistency of the soil, and the cost may increase or decrease proportionately. For example, if the soil is hard-packed and drains poorly, you'll need more gravel. If it's sandy and porous, less will be required. When in doubt, consult a professional.

The concrete can be prepared and poured using a rented automatic mixer, or delivered pre-mixed by truck to the site. If the site is accessible by a delivery truck, and the concrete poured directly from the chute, the job will go quickly and the most economically. If the concrete has to be transported from the street or driveway, then more time, labor, and, possibly, expense will be involved. Wheelbarrowing concrete is a physically demanding task that involves concentrated, rigorous work during the delivery period.

After the concrete has been placed, continue to keep some fresh concrete handy during the screeding process to add to low spots and to offset settling as the surface is finished. On warm days, you may have to work a little faster in the finishing process, but there should still be enough time to create the desired texture. If the surface sets up too quickly, you can use mist from a hose to dampen it and keep it workable. (Don't give it the full spray of a hose at this time, however, as this may expose the aggregate and damage the finish.)

Care should be taken when finishing the surface, as you don't want the surface to be too slick. A broom finish provides a safe, non-skid textured surface and is easiest to accomplish for do-it-yourselfers. Other finishes are also possible, but they may require more time. Imitation flagstone designs can be tooled into the surface, for example, or coarse ripples and swirls can be made by moistening the surface and reworking it with a sponge, float, or trowel. Don't overwork the surface during finishing, as too much troweling can cause the aggregate to settle and separate from the rest of the slab.

Control joints and rounded edges should be added in the surface after the finishing process. These help control the cracking that is part of all concrete installations. As a general rule, control joints should be placed at 4' to 6' intervals to a depth of about 3/4". When the surface has been finished and allowed to set up enough so that it will support your weight when distributed on a piece of plywood or lumber, use a jointing or edging tool to place the joint in a straight line across the slab. The three exterior edges of the slab should then be rounded or beveled with an edging tool before the finishing job is completed.

Level of Difficulty

Concrete patios are well within the abilities of expert do-it-yourselfers who have worked with concrete before.

Novices and intermediates can do much of the work, such as excavating, setting the forms, and placing the reinforcing, with help and guidance from someone experienced with concrete/masonry tasks. The most difficult part of the job is placing and finishing the concrete, which involves skill, as well as several hours of intense, physically demanding work, particularly for larger patios. Novices should add 75% to the professional time estimate for the tasks they can handle. Intermediates with experience in concrete work and experts should add 40% and 10%, respectively, to the labor-hours estimated.

What to Watch Out For

Carefully estimate the amount of concrete—delivered or bagged—that you'll need for the project. Concrete delivered for a small patio may cost more per cubic yard than it would for a complete foundation or cellar floor, because of the retailer's overhead for trucking costs. Be sure to work out all the delivery conditions with the dealer before the order is placed. The amount of time allowed for the delivery and the cost of additional truck tie-up time are important details that you should be aware of.

Good organization and advance preparation will keep the costs of the project down. Also keep in mind that concrete takes about three weeks to properly cure, so be sure to plan your activities accordingly—both for installation and use of the patio.

Options

If you already have a concrete patio, but a few stains or cracks have made you think about replacing it, consider re-surfacing it instead with a more natural material, such as flagstone, slate, or half-bricks. It's a great way to save money and get a new look. If your patio is level and in reasonably good shape, you can remove some of the hard edges and corners to create a more natural shape, then clean and re-mortar the remaining patio surface, applying the new flagstone or other material.

Concrete is also available in various molded paving units, some interlocking, which can add an interesting pattern to a patio or driveway.

See also:
Brick or Flagstone Patio.

Concrete Patio, 6' x 12'

Description	Quantity/ Unit	Labor-Hours	Material
Layout, clearing and excavation, by hand	1.33 C.Y.	2.7	
Compaction, vibratory, 8" lifts, common fill	1 C.Y.	0.1	
Gravel fill, 3" deep	0.66 C.Y.	0.1	17.82
Compaction, vibratory, 8" lifts, bank run gravel	1 C.Y.	0.1	
Edge forms, 4" high	24 L.F.	1.3	7.78
Expansion joint at existing foundation, premolded bit. fiber	12 L.F.	0.3	5.62
Concrete, ready mix, 3000 psi, for patio	1 C.Y.		97.20
Reinforcing, welded-wire fabric, 6 x 6 - W1.4 x W1.4	72 S.F.	0.3	16.42
Place concrete, slab on grade, 4" thick	1 C.Y.	0.4	
Concrete finishing, broom finish	72 S.F.	0.9	
Re-seed grass, hand push spreader, 4.5 lbs. per M.S.F.	0.04 M.S.F.	0.1	0.72
Sealer, silicone or stearate, sprayed on concrete, 1 coat	72 S.F.	0.1	26.78
Totals		6.4	$172.34

Contractor's Fee, Including Materials:	$541

Garden Arbor

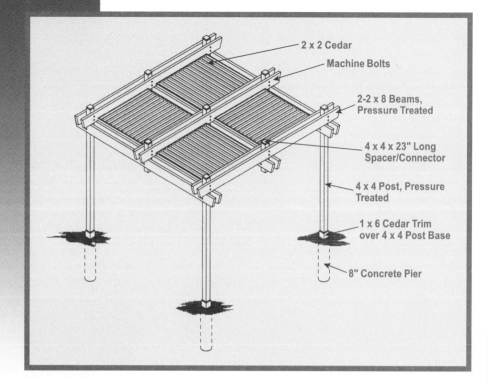

2 x 2 Cedar

Machine Bolts

2-2 x 8 Beams, Pressure Treated

4 x 4 x 23" Long Spacer/Connector

4 x 4 Post, Pressure Treated

1 x 6 Cedar Trim over 4 x 4 Post Base

8" Concrete Pier

The term "pergola" once referred to substantial masonry structures of the Renaissance era in Italy. The term has since taken on much wider meaning, to include wood garden arbors of various designs. Arbors or pergolas have become a popular and stylish way to add comfort and character to a home. They can dress up uninspired patios or soften the lines of an unattractive detached garage. They can be beautiful architectural features that make a home look elegant and more substantial, as well as a starting point for a whole new landscape design. A search of the Internet will turn up many inspirational photos, as well as arbor designs and kits.

Arbors can be created to suit just about every situation and style of architecture. Combined with a paved area or a deck, they expand your outdoor living area—and the value of your home—at a comparatively small cost. It is also a manageable project for homeowners who enjoy remodeling work, provided you have someone available to help lift and install the various wood pieces.

The arbor in this project can be either free-standing or attached to the house. This particular model is square, but the design can be customized to fit an existing patio simply by adding or eliminating segments. Larger segments may require deeper horizontal members.

Materials & Tools

A variety of woods can be used to build this structure. You can use cedar, redwood, or pressure-treated lumber, or you can mix different types of wood, using one for the structure, and the other for trim and accents. No matter what species of wood you select, the same principles will apply to the layout, fabrication, assembly, and finishing. Pressure-treated wood is used in this model project. If you choose another type of wood, take the time to apply at least one coat of wood preservative prior to assembling the pieces to ensure total coverage. The project also calls for concrete, purchased ready-mix in bags (and possibly tube forms) and fasteners (bolts).

Tools for this project include standard hand and power carpentry tools (including long drill bits), plus a combination or planer blade, as well as a dust mask, gloves, and eye protection.

What's Involved

The location of the post bases is critical to the success of this project. The framing members can be pre-cut from patterns that are laid out based on a plan. Using a pattern is recommended to ensure that all the components are consistent in size. Patterns also allow you to select the best pieces of lumber before you cut, thus minimizing waste and ensuring the project's appearance.

It may be necessary to remove sections of the existing patio to place the footings for the posts. If the soil is stable, the round fiber tube forms included in this project may not be necessary. You could simply place the concrete in the prepared holes. Before you place the anchor bolts in the concrete, check the dimensions and corners to be sure they are square. The post bases will allow you to make minor adjustments horizontally and vertically when you set the posts. Take care to make the tops of the footings level with one another. The bolts at the footings will need to be tightened after you've plumbed the posts one last time.

Once you've laid out the placement of the post bases, you'll be able to pre-cut the entire structure. The layout should be developed off of centerlines, to ensure the correct positioning of the members. Because this project is assembled with bolts at the main connection points, you will need to pre-drill the holes in the members. This is much easier to do on the ground than from a ladder. Once you've laid out and measured the wood for the first panel, you can use it as a pattern for the others.

When all the members have been cut, sand and pre-finish them before assembly. Temporary bracing of the vertical posts is recommended when you're assembling the arbor components.

Level of Difficulty

The arbor in this model project is not particularly difficult to build because it's square, and its segments are identical. The work requires good planning beforehand, and patience and a steady hand with the power saw once the job gets under way. Because all of the joints are exposed, good square cuts will make a big difference in the finished product. Using the cut members as patterns should reduce errors and eliminate waste.

Beginners should have little trouble with the basic structure, provided they have help with lifting and placing the members. Intermediates and experts might enjoy being more creative, shaping the ends of the structural members with a saber or scroll saw. The construction requires working off of

ladders or scaffolding, which might slow down inexperienced do-it-yourselfers.

Whatever your skill level, the job calls for two people working together, because the wood pieces are heavy. Intermediates and beginners should add 50% and 100%, respectively, and slightly more for creating fancy end shapes. Expert do-it-yourselfers should add 20% to the professional time for the total project.

What to Watch Out For

The key to building this structure is proper setting of the post bases. Be sure you have a tape long enough to check all dimensions. The easiest way to check the layout and the overhead structure for square is to tape the diagonals. If the opposite sides and the diagonals are equal, the structure is square.

A combination or planer blade is recommended for this project, because it makes a smoother cut. Be sure your drill bits are long enough to go through a minimum of 4" of wood. If you don't

intend to apply any finish to the structure, make all your marks lightly with pencil. Because all of the wood is visible in the finished project, you may want to select each piece of your lumber for straightness and pleasing appearance.

Options

Other materials for a garden arbor include composite lumber or foam-filled aluminum framing. Both are more expensive than the pressure-treated wood in this project, but they have a longer life than wood. Foam-filled aluminum framing is also lighter and therefore easier to handle.

Depending on your home's architectural style, another popular design includes classical round, white columns with a grid above.

All arbors gain shade and are enhanced by vines that are supported by their "roof" grids. Consult your nursery for suggestions of the right plants—flowering, deciduous or evergreen, fast- or slow-growing—and learn which best suits your arbor and its surroundings.

Garden Arbor, 12' x 12'

Description	Quantity/ Unit		Labor-Hours	Material
Pier form, round, fiber tube, 8" diameter	12	L.F.	2.5	21.74
Post base, 4 x 4	4	Ea.	0.2	26.16
Post, 4 x 4, pressure treated	48	L.F.	2.0	72.00
Lumber, 2 x 8, pressure treated	148	L.F.	2.5	147.41
Exterior trim, 1 x 2 cedar	74	L.F.	2.2	25.75
Exterior trim, 2 x 2 cedar	230	L.F.	8.0	138.00
Exterior trim, 1 x 6 cedar	8	L.F.	0.3	7.97
Machine bolts, 7-1/2" long	36	Ea.	2.2	27.22
Nails, 6d, galvanized	10	Lb.		17.28
Concrete, field mix, 1 C.F. per bag	2	Bag		14.64
Totals			19.9	$498.17

Contractor's Fee, Including Materials:	**$1,838**

Driveway

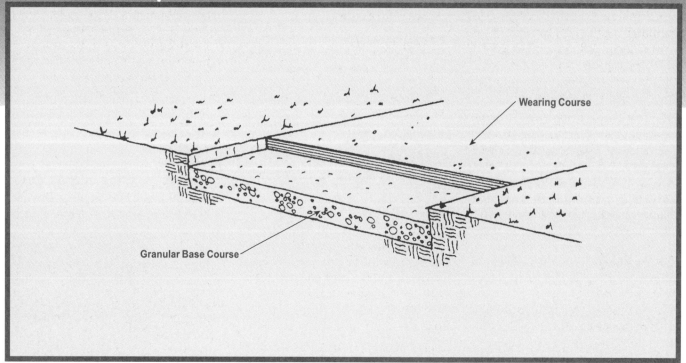

Wearing Course

Granular Base Course

A driveway is often one of the largest visual elements of your property. As such, it can make or break your home's curb appeal and can even impact resale value. Done right, a new driveway can make parking and other outside activities easier, as well as help tie together the landscaping of your yard.

Materials, Tools, & Equipment

Driveway materials for the most part are fairly traditional—asphalt, brick, concrete pavers, crushed stone, or gravel. Gravel, as recommended in this model project, is economical and often considered more attractive than other materials because of its natural look and the wide range of available colors and stone sizes.

You can dress up a simple gravel driveway—and help keep the stones in place—by installing brick or wood borders and edging, at some additional

cost. A natural redwood border can be striking and tie into the landscaping of the rest of the yard, but economical rigid plastic or metal edging can also be used, and has a low profile.

The tools you'll need to prepare the surface and install the new gravel driveway include a shovel, trowel, rake, carpenter's level, and wheelbarrow. Depending on the size and condition of your driveway, and whether this is a replacement or new installation, you may also need to rent a plate compactor and skid steer loader.

What's Involved

A gravel driveway is something most homeowners can install with some careful planning and instruction from an expert. Grading and base preparation are similar whether the driveway will be gravel, brick, concrete, or concrete pavers. Asphalt driveways require special equipment and experienced installers.

When deciding on a new driveway's layout, be sure to consult your local zoning/building codes, which will determine the required set-back distance from your property lines. Other considerations include the size of the space you need (such as to accommodate double or triple garages or children's play areas), whether there are utility lines or a septic system running under the proposed location, and how and where the new driveway will drain.

While this plan calls for a straight 13' x 21' driveway, you might consider adding slight curves to enhance the aesthetic effect or showcase particular lawn features. Be sure to plan for a way to tie the finished driveway into a path or entryway to the house as well.

Installing a new driveway where there has been none before requires excavation, usually done with a skid steer loader. You'll need to be able to maneuver this equipment easily in and out of the area. The excavation site should be clearly staked off.

If you're replacing an old asphalt driveway, usually that material will have to be thoroughly broken up and removed, at additional cost. For brand new driveways, excavate the topsoil to a minimum depth of 6", and place a base of bank run gravel to a minimum depth of 3". Hand-grade (with a shovel and rake) the base, and compact it using a vibratory plate compactor.

When the base is compacted, lay down a top layer of pea gravel and compact it in the same way. Finally, backfill any open areas around the driveway using the top soil that was excavated. Be sure the new driveway is level so that there won't be puddles or standing surface water after a rainstorm.

Level of Difficulty

A gravel driveway can be installed by most homeowners. The most difficult part of the installation is the excavation, which can be done by a professional excavator if you're not comfortable using the rented equipment. Most of the work involves the physical labor of spreading the gravel with a rake. A compactor can also be rented and used by a beginner.

Beginners and intermediate do-it-yourselfers could install a brick or interlocking paver driveway, but should consider complicating factors such as the weight and storage location of the materials, and requirements for cutting materials. (Refer to the patio projects for more on brick work.)

Experienced do-it-yourselfers may want to rent all of the aforementioned equipment and do the work on their own. They should add 100% to the estimated time. Beginners and intermediates should probably restrict themselves to the basic preparation and layout, and the final grading and landscaping. Seriously consider hiring a professional for the brick, concrete paver, or concrete driveways, and definitely for asphalt.

What to Watch Out For

Scheduling is important during any job that requires the services of a professional contractor. Be sure to arrange equipment rental in such a way as to avoid paying for it while it sits idle. Also plan for alternate parking while the work is under way, especially if street parking is not an option where you live.

In climates where snow is common, gravel can be easily displaced by shoveling and plowing. After several years, a gravel driveway may need to be regraded or improved with another layer of stone.

When considering paving, keep in mind that prices for asphalt and concrete are generally higher during the winter because of installation complications in cold weather.

Options

Interlocking concrete pavers come in a wide range of colors and designs, and can create an elegant look for a formal home. Some varieties have coordinating hollowed pavers that allow for a suitable ground cover to be grown in the center, alleviating the "parking lot" look of a big swath of paving, and adding more green—and curb appeal—to your property.

See also:

Attached Single Carport, Attached Single Garage, Brick or Flagstone Patio, Concrete Patio, Free-Standing Double Carport, Free-Standing Double Garage.

Alternate Materials

Surface Options
Cost per Square Foot, Installed

Asphalt	$0.79
Concrete	$3.40
Crushed stone	$0.61
Brick	$9.60
Paving stones	$14.40

Driveway, 13' x 21'

Description	Quantity/Unit		Labor-Hours	Material
Rent wheeled skid steer loader	1	Day		138.33
Rent vibratory plate compactor, 13" plate, gas	1	Day		22.00
Bank run gravel	273	S.F.	0.1	63.65
Pea gravel	2	C.Y.	1.7	58.80
Backfill perimeter of driveway	68	L.F.	1.1	
Totals			2.9	$282.78

Contractor's Fee, Including Materials: **$534**

Retaining Wall

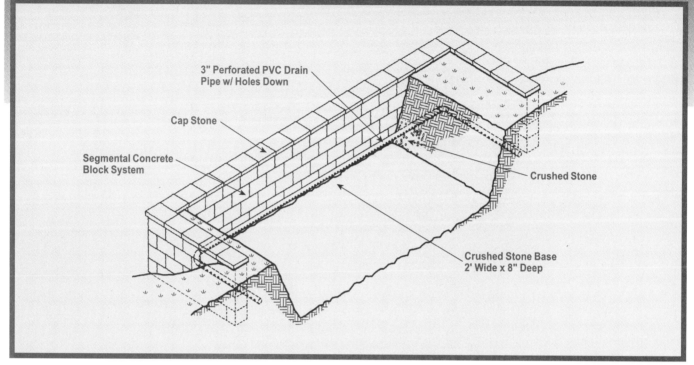

3" Perforated PVC Drain Pipe w/ Holes Down

Cap Stone

Segmental Concrete Block System

Crushed Stone

Crushed Stone Base 2' Wide x 8" Deep

Sloping areas and low hillsides are often difficult to landscape and maintain without major re-grading or construction, such as adding a retaining wall. This project is a great way to create a level-graded area for a lawn, shrubs, or flower beds, or to help prevent or correct drainage and erosion problems. A well-designed and constructed wall can also provide a substantial architectural element on your property, while adding visual interest and texture. Making your yard easier to maintain and landscape and enhancing your home's curb appeal are two important elements of this project that can add to the value of your property.

Materials & Tools

Harmonizing the terrain with the construction material is the challenge when planning and laying out a retaining wall. A dry-laid wall (without mortar) made of natural fieldstone (not cut and shaped) is one durable and attractive type of retainer that can be built to follow the most irregular grades and contours. Dry-stacked walls allow for additional drainage through the cracks between the stones themselves.

You may also choose to use the interlocking concrete block system featured in this project's estimate. These blocks are available in a variety of colors and textures, and offer the advantages of durability and structural integrity at reasonable cost. Installation is also easier—virtually error-free. This project also calls for a drainage pipe, cap stones, and crushed stone for materials, and a shovel, rake, wood stakes, and masonry twine for tools.

What's Involved

All walls begin with some sort of foundation. Once you have laid out the line of your wall with stakes and masonry twine, dig a trench about 2' wide and 8" deep, keeping the bottom as level as possible as a base for the first course of stones or blocks. Be sure to remove any grass, roots, or rocks. If the soil is hard-packed clay or otherwise impermeable, drainage will be better if you dig even deeper and increase the depth of the crushed stone to about 6". Butt the foundation blocks together and level them end to end. From front to back, they should be pitched back about 1°, or approximately 1/4", off level. The purpose of the pitch is to make the wall lean slightly back to resist the pressure of the soil and water behind it, without compromising its vertical stability.

The second course should be stepped back about 1/4" from the front edge of the first course, leaving small drainage gaps between the blocks. Be sure to stagger the vertical joints between the courses. For ties, spike the upper tier

Built by J.P. Gallagher, Hanover, MA

into the lower ones. Once the second course is in place, connect the PVC drainpipe and lay it, holes downward, on the crushed stone base so it is a few inches behind and about level with the upper ties. Then cover the pipe with the remaining crushed stone.

Hydraulic pressure in the soil can be quite powerful and potentially damaging to a wall such as this, especially in cold climates where winter frost may expand the soil and cause the wall to heave or become misaligned. A simple drainage system like this helps to relieve that pressure and to ensure the integrity of the wall. In warm, dry climates, or where soil conditions and terrain provide sufficient natural drainage, this system may not be needed. A local landscaper, stone dealer, or garden center should be able to advise you on drainage requirements.

Follow the concrete block manufacturer's recommendations about slope stabilization if your wall is greater than 2' high. After the wall is in place, you may want to cap it off with mortar or soil to be level with the upper grade of the yard. You may also choose to leave the stone exposed.

Level of Difficulty

Most landscaping work is physically demanding, but in good weather, can be invigorating and satisfying to complete. Block or stone walls are more labor-intensive than working with landscape ties because you have to place more units. If you're working with stones, they must often be individually selected and carefully fitted to ensure a strong and attractive wall. *(See the Dry Stone Wall project.)* It is slow work with hard, heavy material.

Considering the visual impact a retaining wall can have on a yard, beginners and intermediates would do well to get some advice regarding design and layout, especially where complex contours or other challenging landscape features are involved. Beginners should add 100% to the time estimates; intermediates, 50%; and experts, 20%.

What to Watch Out For

Keep in mind the most basic principle of all masonry: one over two; two over one. This is the way a brick wall is laid, and it ensures, in both brick and stone work, that there are no contiguous vertical joints to weaken the wall or threaten to collapse it. Backfilling can be done after the wall is complete, but it might be easier and produce better results if you backfill in layers as you add each course of concrete blocks or stones. That way, you can tamp each layer of soil to compact it and eliminate the looseness and voids that can cause settling and surface depressions later.

See also:
Dry Stone Wall.

Retaining Wall, 4' x 20'

Description	Quantity/Unit		Labor-Hours	Material
Segmental concrete retaining wall system, 8" x 8" x 12.5" blocks	128	S.F.	7.7	1,159.68
Cap stones	32	L.F.	2.6	407.04
4" PVC perforated drainage pipe	20	L.F.	1.3	46.08
Drain trench for pipe, crushed stone bedding	1	C.Y.	0.1	33.00
Crushed stone base, 8" deep x 2' wide, for wall	1	C.Y.	0.1	33.00
Totals			11.8	$1,678.80

Contractor's Fee, Including Materials:	**$3,027**

Dry Stone Wall

Dry Set Stone Wall

Dry stone walls have been built in many parts of the world for thousands of years, and the tremendous number still intact are outstanding examples of this form of construction. Stone walls serve as functional, attractive barriers between properties, along driveways, or along the street. They also provide important architectural accents, separating planting areas from walkways and patios, or creating visual interest by adding elevation to a relatively flat lot. Stone walls can also visually differentiate one level from another on a lot with minor elevation changes.

"Dry" stone walls, which are carefully fitted together without mortar, can also be used for retaining walls, but this application requires structural engineering, and is not covered in the project estimated here.

Materials, Tools, & Equipment

Because stone can be purchased in many colors, shapes, and sizes, it can accent your landscaping in different ways. Stone is sold by the ton, and prices range widely. To determine the quantities needed and the cost for your project, you will need to know the height, thickness, and length of the wall. A minimum of one foot is recommended for the wall's thickness, due to the irregularity of stone. You can use narrower wall widths, but this will mean breaking a lot of stones to obtain pieces with the proper dimensions. A narrower wall will also limit the height, because it offers less stability.

For those who are inexperienced in stone work, 4' is the recommended height limit. Professional stone masons build higher walls, but the stone has to be worked with chisels and other special tools to make it fit together snugly. This is a true art, and the North American economy has not produced many such skilled tradespeople. Higher stone walls are generally installed with mortar to bind them together.

Once you have determined the height, thickness, and length of your wall, you can calculate the cubic feet of stone. For example, a wall 2' high, 1' thick, and 45' long would require 90 cubic feet of stone. You can get a rough estimate of the stone quantity by using the conversion factor:

One ton of stone equals 15 cubic feet of wall.

Selecting the stone can take some time. Stone dealers typically offer a wide range of stone varieties, with prices averaging around $350 to $400 per ton for types that are appropriate for walls. This price range usually includes delivery if you are within about a 20-mile radius of the stone yard location. Small quantities will cost more, and large quantities will be less per ton. Because it is so heavy, most homeowners do not have the means to transport the stone themselves.

The tools and equipment for this project include a wheelbarrow or cart, especially if it is not possible for the delivery truck to get close to the installation site. Hand-carrying stone any distance is very fatiguing and time-consuming. If you have steep slopes to deal with, you may have to construct a working platform. The hand tools needed are a pick, a shovel, a string line, a folding ruler, a carpenter's level, a hammer, and a chisel for shaping some stone pieces.

What's Involved

You will start by laying out the wall using string lines. The next step is excavating the ground 1' wide and 3" deep for a wall less than 2' in height, and at least 6" deep for walls that are up to 4' in height. Use a string line to control the height of the wall. You can also use string lines to set the front edge of the wall. Due to stone's irregular shapes, you will not be able to maintain exact dimensions on both faces of the wall, so you'll need to decide which face should have the most finished appearance.

Use larger stones to begin the wall, to create a good footing. It is not necessary to compact the soil underneath the wall; the stone will do this by virtue of its weight. Place the larger stones first and use smaller ones to fill in the gaps and level the wall. When sorting through the stone that has been delivered, try to set aside some flat stones for the final cap course to help give the wall a level appearance.

Level of Difficulty

Because stone is so heavy, depending on the size of the pieces, there is a lot of physical effort involved in this project. Do-it-yourselfers who are not physically fit may find it difficult. It is possible for two people to build such a wall by lifting some of the larger stones together. Shorter walls are also easier

Built by J.P. Gallagher, Hanover, MA

because the stone does not have to be lifted as high. A short stepstool or bench can also help by breaking the lifting height into two shorter levels. Taking the stones from a wheelbarrow next to the wall also reduces the lifting height.

The technical level of difficulty should not deter the beginner, but do seek advice from an expert on the type of stone to select for a particular job and tips about fitting them together. The inexperienced beginner should increase the installation time indicated by 150%–200%. The intermediate do-it-yourselfer should add 100%–150%, and the expert, 75%.

What to Watch Out For

For safety, use protective eye goggles and gloves, and wear long pants and long sleeves when cutting and installing stone. Stone is very abrasive and can scratch and cut any exposed skin areas.

Special care should be taken when building walls higher than 2'. Do not continue to work when you are tired, because fatigue greatly increases the odds of dropping a stone and injuring yourself. For the upper courses on walls 4' high, it is recommended you get some help to prevent accidents, especially if the stone pieces are large. Be sure to wear leather boots with hard toes to offer some protection in case a stone is dropped. Soft-toed shoes, such as tennis shoes, are not a good idea. When working with stone, you should move it with a deliberate, smooth motion to prevent accidents. When lifting and setting stones down, keep your back straight, and use your leg muscles.

Options

While the character of stone is beautiful unto itself, a little greenery can add life and color to a wall. Creeping, drought-tolerant plants can be an attractive way to fill some of the spaces between stones. For example, in Zones 5-9, you could use Labrador Violet, Creeping Thyme, or Stonecrop Sedum, which provide the bonus of flowers or fragrance. There are many others to choose from. Consult your nursery for the best plants for your zone and for your shade, sun, and wind conditions. Plant them in potting soil within the wall spaces and water as necessary to get them well established.

Dry Stone Wall, 4' x 45'

Description	Quantity/ Unit		Labor-Hours	Material
Excavate base 6" deep	0.83	C.Y.	1.7	
Architectural stone delivered	12	Ton		5,328.00
Place dry stone wall	12	Ton	96.0	
Cleanup and remove debris	1	Ea.	2.0	
Totals			99.7	$5,328.00

Contractor's Fee, Including Materials:	$12,590

Wood Fence

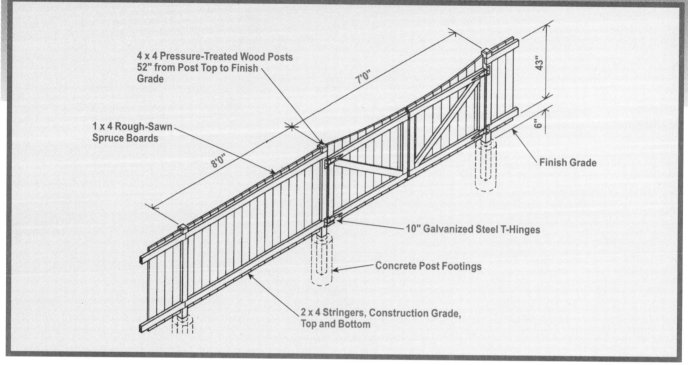

4 x 4 Pressure-Treated Wood Posts 52" from Post Top to Finish Grade

1 x 4 Rough-Sawn Spruce Boards

7'0"

8'0"

43"

6"

Finish Grade

10" Galvanized Steel T-Hinges

Concrete Post Footings

2 x 4 Stringers, Construction Grade, Top and Bottom

Fences are most often built to add privacy or a barrier between two homes, but they can also be an attractive enhancement to your landscaping. Depending on the fence's location, it can boost your property's curb appeal and potentially increase the value of your home. This project is designed as an economical way to enclose a yard, keeping children or pets in, and/or others out. As it is only 4' high, it is visually unobtrusive, yet provides enough vertical surface to serve as an attractive backdrop for shrubs and flowers.

Materials, Tools, & Equipment

The materials for your fence depend a great deal on the style you choose to match your house and landscape, the purpose the fence will serve, and, of course, your budget. The first step in choosing materials is deciding on a design. You'll need to determine exactly what style and purpose you're aiming for and where the fence will be located on your property. Answering these questions beforehand makes the design process a matter of narrowing choices by eliminating fencing styles and materials that are not appropriate to your needs. Using prefabricated panels will save you the labor involved in constructing them yourself and can reduce installation time, but these savings can be offset by the cost of the panels themselves.

The material components for this model project have been selected for their workability, practicality, and reasonable cost. The posts are pressure-treated 4 x 4s, rated for ground contact, which means they can be set directly in the ground and are guaranteed against rot for 30 years. The stringers are construction grade 2 x 4s. These are much easier to cut, screw, and nail than pressure-treated stock, and are about half the price. The same criteria govern the choice of 1 x 4 rough-sawn spruce for the fence boards, rather than pressure-treated lumber. The project also includes a pre-constructed gate, and concrete for setting the support posts.

You'll need a good quality preservative stain, which doesn't take long to apply, especially if you use a sprayer. The wide range of available colors allows you to choose one to complement your home or to help blend the fence into the surrounding landscape. Every few years, an application of the same stain or a clear coat preservative will keep the fence looking good and help protect it from the elements.

For digging the post holes, a long-handled shovel is an indispensable tool. You might also find a clamshell-type post hole digger useful. A long pry bar and an axe or hatchet may be needed to help remove rocks and tree roots. A rented power auger can make quick work of the hole-digging chore if the soil is fairly free of underground obstacles such as large rocks or tree roots. A two-man machine works best, but be sure you have available help and understand how to operate it safely before pulling the starter cord. You will also need a hammer, a saw, a carpenter's square, a mason's cord, a level, and protective gloves and safety glasses.

What's Involved

You could begin this project by staining or painting the wood fence posts and other components before digging in and anchoring the support posts. They can also be stained or painted in place when the fence has been installed. Otherwise, you'll start by laying out the line of the fence with stakes and a mason's cord. If the fence attaches to your house, you can work out from the house, using it as a fixed point of reference for determining measurements and angles. If the fence is free-standing, it's best to establish the position of the corner posts and lay out 90° angles to connect them.

Site building allows you to locate the posts where convenient, avoiding large, immovable roots and rocks. Eight-foot spacing between the posts is the general rule, but if necessary, shorter or longer spans can be used, the difference being hardly noticeable once the fence is complete. This is especially useful if the fence must follow the contours of a sloping or hilly yard.

The fence's support posts are set in concrete. The stringers can be joined to the posts in a number of different ways: with facenails or screws; toenails or screws; dadoes on the post sides; dadoes on the post fronts or backs; or with galvanized steel post brackets. The post tops can be left square-cut and plain, or beveled in various ways to dress them up and to help them shed water.

Once the posts are firmly in place and the stringers secured to them, the boards are nailed on. They should be butted tight together because some shrinkage is bound to occur as the boards dry out. The resulting gaps are generally less than 1/4" and should be narrow enough not to be noticeable, but they may affect the fence's overall appearance in the long run if not tight enough. Take the time before nailing the boards in place to step back and see how they are positioned.

The double gate in this model project is 7' wide, with enough room to allow access for vehicles and lawn equipment. It swings on heavy T-hinges and shuts with an external-mount latch. An easy and attractive way to set off the gate is to cut a curve across the top. Lay the gate sections side by side and scribe a large radius arc, using a string-and-pencil compass, a plywood template, or a flexible wood batten, and make the cut with a saber saw.

Level of Difficulty

A custom-designed, site-built fence is a challenging project. By its very nature, a fence is a public construction project in the sense that it greatly affects the appearance of your house and yard—for better or for worse. Beginners should not undertake the job without qualified help in design, layout, and construction. Intermediates should have no trouble completing the work independently.

The project calls for a few basic construction skills: measuring, cutting rough dadoes and angles, making sure the posts are plumb and level, and nailing or screwing the parts together. The work also involves the drudgery of digging a number of post holes between two and three feet deep. Local conditions can make this chore either fairly easy, as with loose, sandy soil, or a real pain in the neck, as with hard-packed clay full of roots and rocks.

Beginners should have guidance and assistance throughout all phases of this project, and, for any tasks undertaken, should add 150% to the estimated times.

Intermediates and experts could build this fence unaided, and should increase the estimates by 75% and 40%, respectively.

What to Watch Out For

Fences built on a hill will need to accommodate the slope or be stepped. In the first instance, the stringers are run parallel to the slope; in the other, the stringers remain level. In both cases, the posts and boards must be perfectly plumb. Prior to installing a fence, be sure you know your property boundary lines and are familiar with any building codes or restrictions that may impact your project.

Options

Composite wood products are increasingly popular as an alternative to wood. They don't warp or expand, are insect-resistant, and don't require sealing or painting, so maintenance is minimal. These materials are typically available in white, gray, and light tan, and look and feel like painted wood. Special versions of these components are made with a finish that withstands severe weather.

Alternate Materials

Material Options
Cost per Linear Foot, Installed

Vinyl fencing, 4' high	$28.00
Recycled fencing, 4' high	$30.00

Wood Fence, 4' x 320'

Description	Quantity/ Unit	Labor-Hours	Material
Board fence, 1 x 4 boards, 2 x 4 rails, 4 x 4 posts, treated	320 L.F.	56.9	2,515.20
Gate, general wood, 3'-6" wide, 4' high	2 Ea.	5.3	141.60
Excavation, hand pits, sandy soil for post anchors	1 C.Y.	1.0	
Concrete, field mix, 1 C.F. per bag, for posts	13 Bags		90.48
Placing concrete, footings, under 1 C.Y.	1 C.Y.	0.9	
Totals		64.1	$2,747.28

Contractor's Fee, Including Materials:	**$7,150**

Garden Pond

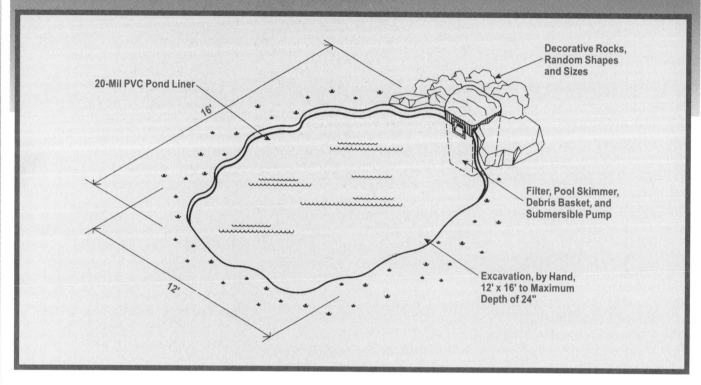

20-Mil PVC Pond Liner

16'

12'

Decorative Rocks, Random Shapes and Sizes

Filter, Pool Skimmer, Debris Basket, and Submersible Pump

Excavation, by Hand, 12' x 16' to Maximum Depth of 24"

Water gardening is one of the most popular trends in landscaping. Nothing fosters a greater feeling of peace and tranquility than a shaded garden nook beside a small pond or fountain. With the proper setting, a garden pond can be made to look like a natural feature of the landscape. If you're an avid gardener, a pond can open up a whole new realm of possibilities for incorporating aquatic and pond-side plants into your landscape.

Materials, Tools, & Equipment

Installing a garden pond has been made easier in recent years with the influx of kits available at home and garden centers. Some come complete with the liner, pipe, filter, pump, and everything else you'll need—aside from fish, plants, and of course, water.

Water lilies and cattails are popular choices for plants that bloom above the water. Submergent plans add oxygen to the water. For fish, consider species that can tolerate water temperature fluctuation, and scavengers to eat algae. There are also numerous plants, including some iris varieties, that thrive with moist roots along the edge of a pond.

Your backyard doesn't have to be very large to accommodate a garden pond, but the layout is important. The pond should look like a natural part of the landscape, and in the wrong setting, it will look out of place. If your yard is not suited for the type or size of pond treatment described in this project, you might consider a small, fixed-shape pool or fountain. Some suppliers offer as many as 60 different styles and shapes of fiberglass pools that can be sunk in the

ground and surrounded by bricks, tiles, or stone, along with suitable shrubs and flowers.

This project requires very few tools, aside from a shovel or backhoe to excavate the area, depending on your property and soil conditions.

What's Involved

Before undertaking this project, there is a great deal of careful planning and research to do. This may include consulting a landscape designer or water gardening specialist for ideas about integrating the pond into your particular yard, climate, and local water conditions. The ideal setting is a low area with a natural backdrop. In nature, water flows to and settles in holes and depressions to form ponds, and your aim is to replicate what nature does. Hillsides, ledges, trees, and shrubs all serve the purpose of a backdrop, which is to block unsightly or distracting views and focus attention on the pond.

The lay of the land and the size of the pond will determine the amount of excavation needed, and whether you want to do the work by hand or rent equipment. The pond does not need to be very large or deep; an irregular bed, sloping to a maximum depth of 18" to 20", is sufficient. If you plan to introduce a variety of fish, excavate deeper, from 24" to 36".

Line the bottom of the hole with 20-mil PVC, which is thin enough to conform to any contour, but thick enough to resist punctures and natural degradation. By adding a 1/2"-thick layer of damp newspapers under the liner, you can create a protective cushion, as the paper will eventually decompose to form an almost solid and virtually impermeable substratum. Rocks and crushed stone spread randomly over the liner will hold it in place and give the look of a natural pond bottom.

To prevent the pond from becoming a stagnant hatchery for mosquitoes, a filtered circulating system is required to skim off debris and aerate the water. This system allows you to create a natural ecosystem by growing aquatic plants, introducing fish, and attracting birds and dragonflies that feed on insects and their larvae.

A submersible pump draws the pond water through a swimming pool skimmer unit and pushes it out through a flexible plastic pipe that returns it to the pond, ideally rippling over rocks in a small stream. If the skimmer and pump, electrical hook-up, and plastic liner are all artfully concealed and surrounded with appropriate plantings, a year's seasoning will give the pond the appearance of always having been there.

Courtesy of Holway House Designs

Level of Difficulty

Excavation is really the only labor-intensive task in this project. The most challenging aspect of the project is probably coming up with the initial design. If your yard does not provide an obvious setting for a pond, you might do well to hire a landscape architect to create one. This is a major undertaking, and if the outcome is to be a natural-looking small pond, as opposed to an artificial-looking large puddle, paying for the services of a designer can be money well spent.

Beginners should not attempt this project without advice regarding design, layout, and construction, and should add 100% to the time given for any physical tasks performed. Intermediate-level do-it-yourselfers may also need assistance in formulating an attractive and workable design, and should add at least 50% to the time. Experts should add about 25% for all tasks. All do-it-yourselfers should leave the wiring for the electric pump and GFI receptacle to an electrician.

What to Watch Out For

After your pond is set up, let the pump run for a few days before you add plants or fish, so that any chlorine in the water will have a chance to evaporate. As shallow as this pond is, it still represents a potential drowning hazard to very young children. Take whatever precautions are necessary or are mandated by law. Also, if you live in a cold climate, be sure to shut down and drain the pond's pump system before the first freeze.

Garden Pond, 12' x 16'

Description	Quantity/ Unit	Labor-Hours	Material
Excavation, by hand, 12' x 16' to a max. 24" depth	7 C.Y.	14.0	
Rock, medium round stone, to line bottom	3 C.Y.	0.1	90.00
Liner, 20-mil PVC	200 S.F.	5.6	52.80
Filter and pump	1 Ea.	8.9	1,350.00
1-1/4" pipe with fittings	40 L.F.	7.6	65.28
Wire, type MC	100 L.F.	1.3	26.52
Outdoor receptacle, GFI, 15 amp	1 Ea.	0.7	57.00
Totals		38.2	$1,641.60

Contractor's Fee, Including Materials: $4,410

Outdoor Fountain

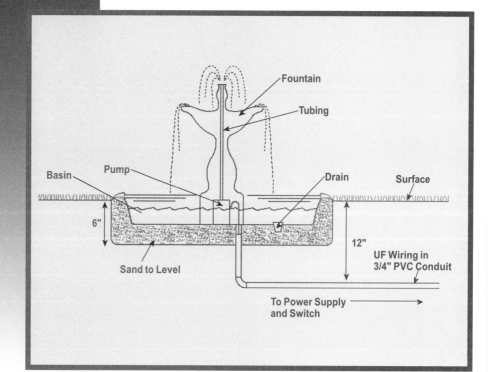

An outdoor fountain can take many forms—from a simple ceramic pot with stones to a formal sculpted piece of art or series of connected basins with plants and other aquatic life. The goal is an aesthetically pleasing focal point that provides soothing sound in your personal outdoor space.

Selecting the location for a fountain involves consideration of a few factors. It should be near an entryway, window, or in a courtyard or outdoor sitting area where it can be easily and often viewed and enjoyed. The location must also be level and allow for a trench and PVC piping that will carry the wire for the pump. Since the pump re-circulates the water, there is no need for piping. The fountain can be filled initially and then topped off periodically using a hose or bucket. Another issue to consider is the amount of sunlight the area will receive—as this will affect the types of plants that may surround or be placed in the water feature.

Materials & Tools

This model project includes a modest-size fountain unit, like those sold in many garden centers, large home centers, and some furniture stores. The basic materials include the basin, sand for leveling the fountain, PVC pipe, and the pump and associated wiring. You might also purchase a narrow board or bright-colored nylon tape to lay in over the buried wire under the topsoil, to call attention to the installation in the event of future excavations. A small trenching machine can be rented to make the excavation portion of the project less strenuous.

What's Involved

The first step is preparing the ground in the area where the fountain will be placed. The ground should be excavated 6" down, and horizontally to a size at least 1" larger than the outside dimensions of the fountain. The excavation should be filled with sand, compacted, and leveled to form a firm base for weight distribution and to allow for drainage. Hand-tamping the fill in the trench and under the feature should be accomplished in 2" layers. Larger fountains or series of ponds will require more excavation and area preparation.

The electrical service may be run from your house panel on an existing, under-used circuit. While the power requirements for your pump will be small, an electrician should assess the proposed circuit to make sure it can handle the additional load. The service is run to the outside of the house, enclosed in 3/4" PVC conduit, and a switch is wired inside the house to turn the pump on and off. From the point where the service leaves the house, the wiring must be contained in a service ell with a watertight cover plate to allow future access. The PVC continues down into a straight trench 12" deep where it will travel to the fountain.

The wire, which must be Type UF for underground feed, is brought to the surface again at the pump, where it will be encased in PVC conduit. (If you have large trees with unavoidable root systems, you may also want to encase the wire in pipe conduit for the entire run underground, as well as above-ground in order to prevent root damage.) PVC conduit may be cut with a wood saw or a special saw designed to cut this material. The pipe is joined using PVC cement for gray pipe.

When backfilling after you lay the wire in the trench, add 6" or so of clean fill (no stones), then compact it and place a narrow board or bright-colored nylon tape in the trench. Should someone excavate here at a later date, they will be alerted to the installation and less likely to damage the buried cable. If you have chosen to use direct-burial cable, include adapters on the open ends of the conduit so as not to damage the cable when it rests on the PVC conduit.

Level of Difficulty

Many aspects of this project, such as trenching and backfilling, and the final landscaping, are within reach of the novice do-it-yourselfer. An expert or a licensed electrician should undertake the wiring portion of this project. Accessing the circuit and tapping the service panel are not for amateurs, as the wrong procedure will endanger you and your home. If your yard has mature trees and bushes, trenching will be more difficult, and the path of the trench may have to angle around them. Novice do-it-yourselfers should add an extra 75% to the time given in the estimate for professional installation. Intermediates should allow an extra 25%, and experts 10%.

What to Watch Out For

Wear protective goggles for safety when cutting pipe, using PVC cement, and any other tasks where particles or fumes may enter your eyes. Gloves are recommended for much of the work. Only expert do-it-yourselfers experienced in electrical work should attempt the wiring and should use a circuit tester to be absolutely certain that the power is off in the affected circuit. Being in contact with the earth while doing the wiring creates an electrical ground which could prove fatal. (Electricians sometimes place lock-

out tags, and possibly a lock, on the circuitry to prevent a third party from energizing the circuit when it is being worked on.) Be sure to use gray conduit cement on the PVC piping, as water pipe cement will degrade the cable.

Large basins, urns, or fountains can be very heavy. Lining up an assistant or two is an excellent idea to avoid breaking an expensive item or injuring yourself. Also, if you are drilling a hole (usually 1") in an urn or similar receptacle to create a fountain, be careful not to apply too much pressure, as this could break the pottery. Also, keep in mind that in many parts of the country, fountains must be drained in winter to avoid cracking during a freeze.

Options

Among popular fountain designs are millstones (real, at several hundred pounds, and lightweight replicas at 10-15 pounds), a variety of stone types, and wall-mounted fountains for urban, small-space gardens. Cascading pools can be created to take advantage of a sloping yard. Professionally designed and installed fountains can range from $2,000 substantially upward, but capable do-it-yourselfers can achieve simple installations with a weekend's work for less than $500.

The colors in many types of stone or stone-like materials are deeper and more striking when wet. One option to maintain this look when the fountain is turned off is to paint masonry sealer on the parts of the fountain and surrounding stones that would get wet when the fountain is running. Two coats of the sealer should be applied when the fountain is in place, and before filling with water. Follow label instructions for drying time.

Larger fountains with pools can incorporate water plants that actually make the pond easier to maintain by limiting algae and providing oxygen. Water lily roots typically rest 2'-4' below the surface, while other water plants float on the surface or are completely beneath the water line. Dwarf ornamental trees can work well as part of the landscape surrounding a fountain, as they will remain in scale over time.

See also:

Garden Pond, Outdoor Living Area, Planting Trees and Shrubs, Patio Projects.

Outdoor Fountain

Description	Quantity/ Unit	Labor- Hours	Material
Conduit in trench	8 L.F.	0.3	5.57
Trench by hand	1.10 C.Y.	1.1	
Backfill	1.10 C.Y.	0.6	
Compation add to above	1.10 C.Y.	0.4	
Switch device	1 Ea.	0.5	9.72
Backfill	1 Ea.	0.3	1.87
GFI receptacle	1 Ea.	0.6	42.00
Wire interior	15 L.F.	0.1	0.72
Wire direct burial	30 L.F.	0.6	7.74
Sand	0.07 C.Y.	0.1	2.35
Fountain, 48" high with bowl and figures	1 Ea.	8.0	190.80
Totals		12.6	$260.77

Contractor's Fee, Including Materials: **$948**

Drip Irrigation System

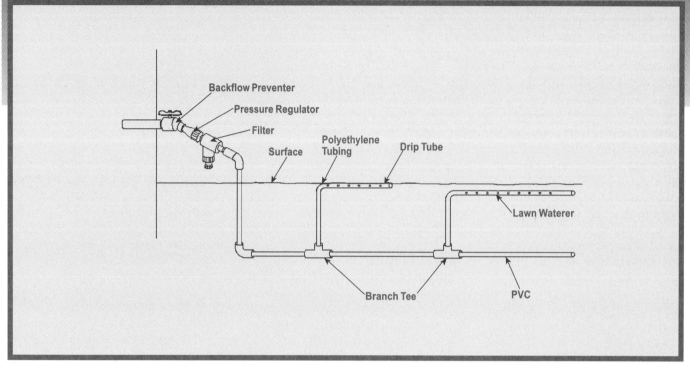

Landscape watering systems are recommended for their ability to conserve water while delivering moisture to plants' roots. They also avoid the evaporation waste that occurs with sprinkler systems and some fungal plant diseases, such as powdery mildew and black spot, by minimizing water accumulation on plants' leaves. Another obvious benefit is convenience—no more hand-watering.

Using a timer with these systems gives you the further convenience of meeting local water use restrictions and watering at the ideal time of day, early morning. (Also refer to the sprinkler system project on the previous pages. For some yards, sprinkler systems offer advantages in installation and cost. The sprinkler project also contains more information about the installation and materials used for the supply piping.)

Drip irrigation systems are more challenging and expensive to install than sprinkler systems, but their water-saving characteristics make them ideal in dry climates, and in areas with high water utility rates and restricted water

usage. These systems not only save water by eliminating evaporation and run-off, but in many cases improve the health of plants by providing the right amount of water without washing away the nutrients or drowning the root system. Avoiding run-off is especially important in areas with easily compacted soil (such as hard clay), a common condition in the southern United States.

Materials & Tools

Like the sprinkler system, this project covers a 20' x 40' lawn area. It requires underground piping to deliver water to the drip distribution tubing. PVC is an easy material to work with; the pipe is easy to cut and connect using simple tools (hand saw, emery cloth, pipe primer, and glue). If wiring is required to control valves, it's best to hire a professional unless you're qualified to do electrical work. If so, you'll probably already have the needed wire cutters, wire strippers, and underground wire cable or conduit.

Drip systems use controlled emitters with drip tubing appropriately sized to water the lawn or plants. Emitters and tubing should be duplicated for each location to make sure that the plants are not deprived of water if there is a clog. Metal stakes are needed to secure underground tubing for drip systems. Drip emitters for a lawn should be 1' apart in parallel rows 18" apart. (Since a lot more piping is required for a drip system than for sprinkler heads spaced 10' apart in a grid, another option might be to use a sprinkler head system for the lawn, and a drip system for shrubs, flowers, and trees.) Each drip emitter will water all the plants in a 3' radius. For flower beds and bushes, emitters should be at least 2' apart.

The mains can be PVC, but the supply from the emitters is usually low-pressure polyethylene tubing, as well as drip tubing. Because drip irrigation system emitters and tubing are sensitive to both pressure and clogging (due to their small openings), they must be protected by a pressure regulation device that prevents water pressures from exceeding 40 PSI (pounds per square inch). They should also have a cleanable filter with at least a 150 mesh size. (A 200 mesh size filter is highly recommended.) The filter should be at least as large as the supply valve. As with sprinkler systems, you can add an electric timer for automatic watering, and remote electric valves to water separate zones of the lawn.

What's Involved

The first tasks are laying out the system design, then digging trenches for the supply water lines with a shovel or rented equipment. (See the sprinkler system project for information on drain valves in colder parts of the country.) Next, the piping is connected and placed at the proper depth to protect it from damage.

Drip irrigation system emitters are supplied by tubing from the main line and installed about 4"-6" below the surface with the drip tubing itself at the surface (to be covered later with wood chips or another light material.) The exception is drip tubing for lawns, which is installed about 4"-6" inches *below* the surface. Drip emitters for mature trees should be about 12" below the ground to accommodate their more mature root systems. The tubing supply piping for drip systems should be staked at least every 3' to prevent it from moving in the ground. (This movement is caused by changes in air temperature inside of the small tubing creating a lifting effect like a hot air balloon.)

Level of Difficulty

PVC piping is fairly easy to install. If you need to use metal piping, it's more difficult to work with. (See the sprinkler project for more on depth and pipe materials.) Do-it-yourselfers who have never worked with metal pipe will need guidance on pipe cutting and installation. Novices can dig the trenches, though it's a big job, especially with heavy clay or rocky soils. This task can be made easier with rented power

equipment. Intermediates and experts can probably handle all but the electrical work, for which they should hire a professional contractor. All non-professionals should research the system they plan to install and get professional advice on selecting the drip emitters and other parts, as well as on installation. For the tasks they can handle, beginners should add 100% to the time given in the estimate. Intermediates should add 75%, and experts 50%.

What to Watch Out For

Wear protective shoes and clothing, as well as gloves and safety glasses for excavating and pipe-cutting. Sand the cut edges of PVC pipes and prime them before gluing to prevent leaks. As with the sprinkler system, you'll need to flush the installed system with clean water before you install the emitters to get rid of any soil. The emitters should be 4" below the surface to avoid damage from tilling in cultivated areas.

Options

Using native plants, known as *xeriscaping*, is an ideal approach that minimizes the need for watering, and also cuts down on your use of chemical

fertilizers and pest control. Local nurseries can advise you on the best of these plants for your area and for your yard's particular conditions.

Another way to save water is to collect rainwater in a cistern or barrel, or to use "gray water," recycled from showers, baths, and laundry (provided no phosphates or bleach are used). If you're thinking of using a rainwater collection system, do some research on your monthly rainfall and calculate your total roof area, as these are factors to consider. If a gray water system is used, it must be clearly labeled with no cross-connects to potable (drinking) water, and the water should not be used on plants that will be eaten, for seedlings, or for potted or acid-loving plants.

See also:

In-Ground Sprinkler System, Planting Trees and Shrubs, Seed or Sod, Outdoor Fountain.

Alternate Materials

Equipment Options
Cost per Each, Installed

Automatic timer	$108.00
Rain sensor	$145.00
Automatic drain valve	$20.00

Drip Irrigation System, 20' x 40' Lawn Area

Description	Quantity/Unit	Labor-Hours	Material
Excavate using chain trencher 4" x 18" deep mains	50 L.F.	0.5	
Excavate using chain trencher 4" x 4" deep drip lines	560 L.F.	9.0	
Install supply manifold (stop valve, strainer, PRV, adapters)	1 Ea.	2.0	47.40
Install 3/4" supply line and PVC risers to 4" drip level	75 L.F.	2.0	16.20
Install fittings (16 tees, 16 elbows, 14 adapters, 1 endcap)	1 Ea.	12.5	43.80
Install drip lines (14 Ea. 40' long)	560 L.F.	3.7	67.20
Backfill trenches	3.17 C.Y.	2.3	
Cleanup area	1 Ea.	2.0	
Totals		34.0	$174.60

Contractor's Fee, Including Materials: $2,088

In-Ground Sprinkler System

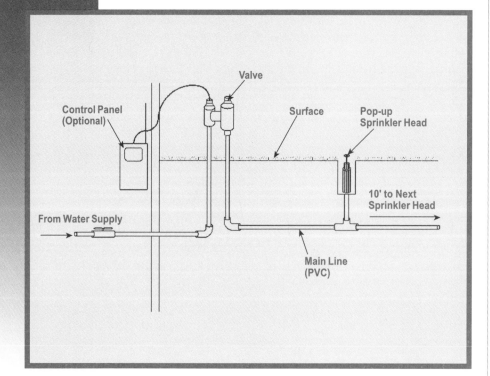

The task of hand-watering your yard with a hose, especially at certain times and days to meet local water-use restrictions, can make an efficient, automated sprinkler system an attractive alternative. These systems save you the chore of hauling hoses and moving sprinklers, and with timers added, can give your plants the water they need—even when you're away from home. In-ground sprinklers not only water lawns, but can also be set to water shrubs, flowers, trees, and even large container plants. (Check the Drip Irrigation System project, as it may be more suitable for some situations. You can also combine these two systems to take advantage of their unique characteristics.)

Sprinkler irrigation systems have lower installation costs than drip irrigation systems, and are usually easy to repair if damaged. Sprinkler systems are best where evaporation rates are not too high (Northern U.S. and Canada) or where water is abundant and rates are low, although they are popular throughout the country. Sprinkler systems tend to use less water than manual watering with a hose.

Materials & Tools

This project consists of a sprinkler irrigation system in a 20' x 40' lawn area. An underground piping system, typically PVC plastic with glued fittings, is required to deliver water to the sprinkler heads. PVC is easy to work with and requires only simple tools (hand saw, emery cloth, pipe primer, and glue). If your system has electric-controlled valves, and wiring is involved—and you are qualified to perform this part of the work—you'll need wire cutters, wire strippers, and underground wire cable, or conduit to contain the wire.

The number of sprinkler heads and amount of piping you'll need is determined based on the pattern and spacing. Lawn sprinklers are usually about 10' apart in a grid pattern, although spacing can be wider if your sprinkler heads provide wider coverage. *(Note: wider spacing may require larger supply pipes if adequate water pressure is not available to supply the larger irrigation heads.)*

The last major items are the water supply and controls. Sprinkler irrigation

systems operate at different pressures, and do not normally require pressure regulation devices. Nor do they require filter systems since their large spray heads don't clog easily.

An electric timer can be added to control watering automatically, and remote electric valves can be installed to water separate zones, or portions of the lawn, using the timer. (This is useful where the water pressure is not adequate to serve all areas of the yard at once.)

What's Involved

You'll need to lay out the system, then excavate for the supply water lines. You can dig the trenches by hand using a shovel or rent a small gasoline-powered chain digger. If you live in an area subject to freezing, you'll need to have drain valves installed at the ends of the main lines so the system can be thoroughly drained at the end of the season. *(Note: The best way to protect the system is to pump compressed air into the piping at the end of the season to completely evacuate the water. There are companies that perform this service, or you can rent an air compressor and do it yourself.)*

The main lines should be buried at least 24"-30". Some contractors install systems at a very shallow depth of 6"-8", but the risk of damage to the system piping is increased as the depth decreases. If you need to install the system at an unusually shallow depth due to rock or other soil problems, use metal pipe such as PVC-coated pipe or copper pipe or accept the possibility of future repairs to your system. Metal pipes can take more abuse without being broken, but you'll still need to provide a drainage valve and flush the system with air at the end of the season.

Some systems have remote-controlled valves that allow you to water different areas using the same supply water irrigation mains. If you're using electric-controlled valves, install the wiring below the piping for better protection. If you're not having the entire system professionally installed, at least seek professional assistance when installing the timing devices and the power supply. Improperly wiring these devices can seriously damage the equipment, leading to expensive replacements. If you're not experienced and knowledgeable in electrical work, get a professional to do the wiring.

In installing the sprinkler heads, follow your planned layout. Set each sprinker head's radius, overlapping the spray 10%-20% to ensure that the plants or lawn are adequately watered. Sprinkler heads are often raised to a height between 1'-4' above the plants to spray over flowers and shrubs, or at least positioned so that they don't damage delicate plants.

Level of Difficulty

If you're installing a plastic (PVC) piping system, the skills needed are minimal. Metal piping, on the other hand, requires some training in cutting and installation. Excavation is less strenuous with power equipment. If you have rocky or hard clay soil, a power chain excavator is highly recommended.

Seek advice from your home improvement supplier on the best sprinkler heads for your application. Novice do-it-yourselfers should add 100% to the professional contractor labor hours given in the estimate. Intermediates should add 75%, and experts 50%.

What to Watch Out For

When excavating, wear long pants and protective leather-soled shoes, as well as gloves and safety glasses. When gluing the plastic joints, sand the raw edges to remove any "burrs" and use primer to prepare the pipe for the glue. This will prevent leaks in the water system. Flush the system with clean water before installing the sprinkler heads to remove any trapped soil that can lead to clogging. The sprinkler heads should be encased in a plastic or metal box and recede when not in use to 2" below the ground to prevent damage from lawnmowers and rakes.

Make sure your sprinkler heads are directed so that they don't spray your house's foundation or painted surfaces. Also ensure that water is not being directed to sidewalks or driveways. Some cities and towns will fine homeowners for such waste, as well as for broken heads that allow water to "geyser" in non-landscaped areas. If a professional

installs your system, be sure to point out any problems with the direction and coverage of your sprinkler heads right away. It's a good idea to purchase a few extra sprinkler heads at the time of the installation so you'll be prepared if you need a quick replacement.

Options

Controls and timers are easy to justify if water use is restricted, and where water is expensive. Home centers stock not only sprinkler system replacement heads, but also diverters and tubing that allow you to divide the water output from one sprinkler head and deliver it to several different plants. This method can be used to water individual potted plants that might be grouped by your entranceway or on your patio.

See also:

Drip Irrigation System, Seed or Sod, Planting Trees and Shrubs, Outdoor Fountain.

Alternate Materials

Equipment Options
Cost per Each, Installed

Pressure gage	$24.00
Strainer	$36.00
Pump (1-1/2 HP)	$430.00
Automatic timer (4 zone)	$108.00
Automatic drain valve	$20.00
Shrub sprinkler	$23.00

In-Ground Sprinkler System, 20' x 40' Lawn

Description	Quantity/Unit		Labor-Hours	Material
Excavate using chain trencher 4" x 18" deep mains	94	L.F.	1.0	
Install supply manifold (stop valve, adapters)	1	Ea.	2.0	17.94
Install 3/4" supply line and PVC risers to sprinkler level	100	L.F.	2.7	21.60
Install fittings (7 tees, 5 elbows, 4 adapters, 4 risers)	1	Ea.	5.3	22.98
Install heads 15' radius plastic semi-circle pattern	4	Ea.	1.1	14.40
Backfill trenches	1.72	C.Y.	1.3	
Cleanup area	1	Ea.	2.0	
Totals			15.4	$76.92

Contractor's Fee, Including Materials:	**$934**

Post Light and Outlet

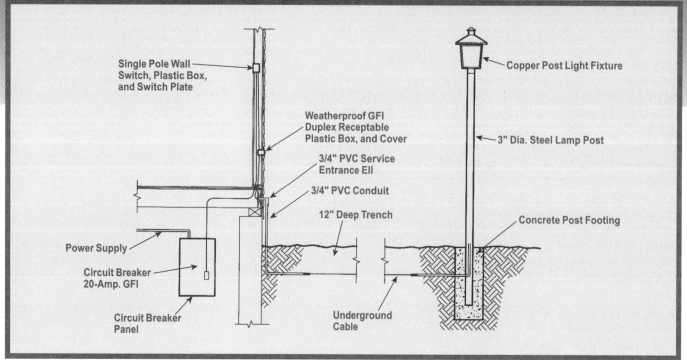

- Single Pole Wall Switch, Plastic Box, and Switch Plate
- Weatherproof GFI Duplex Receptable Plastic Box, and Cover
- 3/4" PVC Service Entrance Ell
- 3/4" PVC Conduit
- 12" Deep Trench
- Power Supply
- Circuit Breaker 20-Amp. GFI
- Circuit Breaker Panel
- Underground Cable
- Copper Post Light Fixture
- 3" Dia. Steel Lamp Post
- Concrete Post Footing

If you don't already have it, adding electrical power on the outside of your house is a very convenient improvement. An outside receptacle provides power for tools and appliances. Well-planned outdoor lighting adds night-time safety to steps and walkways and security to hidden corners and entrances. If wiring can be run from a new 20-amp GFI breaker in your home's main service panel, this project is fairly straightforward. All electrical work should be done by a professional, in order to protect both you and your home—but the investment will enhance your outdoor activities and help make your yard a more enjoyable living space.

Materials & Tools

This basic underground wiring for a 120-volt receptacle and post light can easily be adapted for different types of fixtures, or expanded to include more outlets to service other fixtures or

equipment, such as on walkways, or for security or spot lighting. You can purchase the lamp post and fixtures separately or as a kit, from the huge assortment of styles available at home centers and lighting stores. Corrosion-resistant aluminum lamps are popular and moderately priced, as are similar rust-proof alloys. The most durable (and to some the most attractive) exterior lamps are made of brass and copper. They are also among the most expensive. Beware of plastic fixtures; except for their inexpensive price, they have little to recommend them.

To install the post in the ground, you'll need a post-hole digger and concrete anchor. A trench-digging machine, though not essential, can help, especially with a long trench that will contain the wiring from the house to the fixtures. Tools and other necessary items include a drill, screwdriver, pliers, shovel, gray conduit cement, 3/4" PVC conduit, and Type UF (underground feed) cable.

What's Involved

It may be possible to get at your home's electrical system by tapping into an underused circuit. The post lamp will probably contain a 75- or 100-watt bulb, and most tools and appliances that you would be likely to plug into the receptacle don't draw much power, so the demand on the circuit's current should not be very great. Don't make this assumption, however; have an electrician evaluate the proposed circuit and determine if the circuit can handle the extra load.

If you don't have a circuit to tap into, or if you're planning more extensive lighting or additional outlets, a new circuit will have to be added, either at the main service panel or at a sub-panel. In many cases, this approach is actually quicker and easier than tapping an existing circuit, and has the advantage of allowing for future expansion without the need to upgrade its capacity.

From the new breaker, the wire is run through a hole drilled in the sole plate to the switch location, which is usually near an outside entry door. A wire from the switch is fed to the exterior receptacle box, after cutting a hole in the outside wall 18" or so above the foundation in a convenient, but unobtrusive spot. From there, the wire is snaked through another hole in the sole plate to exit the house via 3/4" PVC conduit through a hole in the rim joist.

PVC male adaptor fittings are used when the wire enters or exits the conduit to prevent it from being damaged by the cut ends of the conduit. Also, gray conduit cement should be used, rather than water pipe cement. Since the cement can degrade the outside coating of the wire, it's important to complete the joints before the wire is pulled through.

The service entrance ell has a watertight cover plate that can be removed to allow the necessary length of wire to be pulled to the outside. The wire continues down through the conduit to an elbow, where it enters the bottom of a 12"-deep trench to reach the lamp post. Type UF cable has a solid plastic, impermeable covering. For residential branch circuits, the *National Electric Code* requires UF cable to be buried a minimum of 12" deep. It must, however, be protected by rigid conduit wherever it is above ground. Your electrician will be aware of specific local building code requirements.

Next, a post hole must be dug about 2' deep at the selected location. A hole is drilled in the post at the bottom of the trench, and a PVC elbow inserted through the hole. Concrete is poured to anchor both the post and the conduit. Then the wire can be fed to the top of the post and connected to the lamp. After the cable has been run, the trench can be backfilled to the first 6" or so of its depth with clean fill (no stones). A 1' x 4' pressure-treated board or a bright colored nylon tape is then laid in the trench before you fill it in. This will eliminate the future possibility of someone digging and accidentally damaging the cable.

Level of Difficulty

Electrical work should be undertaken only by an expert or a professional contractor. Knowledge, experience, and skill are needed both to assess the suitability of an existing circuit for this extension, and to tap into it efficiently so as to minimize the mess and the amount of patching needed. A tie-in to the main service panel is equally demanding. When it comes to electrical matters, ignorance is dangerous and puts both you and your home in jeopardy. Hiring a licensed electrician eliminates the risk of hazardous errors and assures you that the correct components are properly installed. Most electricians will gladly adjust their price downward if you agree to dig the trench and post hole, mix and pour the concrete, and do the backfilling.

Both beginners and intermediates should hire professional help for the wiring, and should increase the time estimates for any tasks within their competence by 75% and 25%, respectively. An expert should add 15% to the time.

What to Watch Out For

Even if you're qualified to do the electrical work yourself, you must always remember that outdoors you're in contact with the ground and, therefore, any shock could be fatal. Never perform any work on wire or equipment unless you're certain that they're completely disconnected from the main service panel. Check wires with a circuit-tester to be doubly sure they're dead before you handle them.

In addition to the post light, this project also includes a ground fault interrupter (GFI) outlet, as required by the *National Electrical Code* for all areas that could get wet—both indoors or out. A GFI senses a leak of current or a false ground, such as a wet hand, and immediately cuts the power to that circuit. A GFI breaker can be installed in the service panel in the same way as an ordinary breaker.

See also:
Low-Voltage Landscape Lighting.

Post Light and Outlet

Description	Quantity/ Unit		Labor- Hours	Material
Conduit in trench	8	L.F.	0.3	5.57
Outlet boxes, plastic, square, with mounting nails	1	Ea.	0.3	4.13
Switch box, 1 gang	1	Ea.	0.3	1.87
Switch device	1	Ea.	0.5	9.72
GFI receptacle	1	Ea.	0.6	42.00
Waterproof cover	1	Ea.	0.3	5.28
Outdoor post lamp incl post, fixture, 35' #14/R-type MNC cable	1	Ea.	2.3	220.80
Concrete for post footing	1	C.Y.		93.00
Placing concrete, footings, spread, under 1 C.Y.	1	C.Y.	0.9	
Totals			5.5	$382.37

Contractor's Fee, Including Materials:	$874

Low-Voltage Landscape Lighting

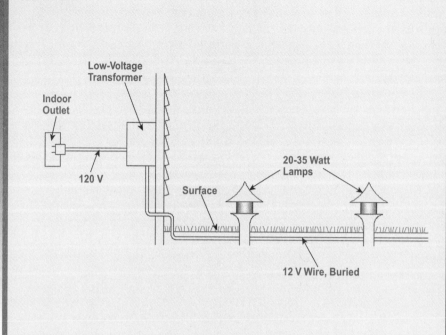

Outdoor lighting can add safety at steps and pathways, provide security by depriving prowlers of dark hiding places at windows and entrances, enhance your home's landscaping and special architectural features, and add to your outdoor living space. Tapping into your home's electrical system may require the services of an electrician, but it is an investment that will pay dividends if the project is well-planned.

Twelve-volt, low-voltage, or solar-powered landscape lighting systems are inexpensive and can be installed by non-professionals. The wire can be placed on the ground or buried just below the surface, which means it can be easily reconfigured to allow for future changes.

To plan your landscape lighting, look at your home in two ways: through the windows from inside to the outside to decide which features of your landscape you'd like to highlight. Then walk along the front perimeter of your yard, at the street, and the side and back yards, and consider which areas need lighting—for security, entertaining, and to enhance

the look of your landscape at night. To get an idea of what you want, you can plug a portable light with a long, outdoor-rated cord into an exterior outlet and use it to light up different areas at night. This will help you see where the light would provide the best visual effect, and the most important areas to light for safety. You can then draw up a layout plan, which will give you an idea of how many fixtures, the wattages, and length of wiring you will need.

Low-voltage light fixtures are available to suit different applications. For example, low-profile ground-level light fixtures can shed light on shrubs or shine upward to light up a garden area or an architectural feature. (Experts recommend placing accent lights no farther than 12' from the feature you want to highlight.) Spotlights can be installed in trees, or as decorative staked fixtures along pathways. Other types of fixtures can be inserted into steps and decks to provide safety and ambiance for entertaining.

Materials

Low-voltage lighting is available in kits that include the light fixtures, the transformer, and the wire (cable) all in one box. Or, you can buy everything individually at a lighting or home center and customize your project. Additional features include an outdoor timer with a photocell that will turn the lights on at dusk, or a motion sensor to turn on the lights when a moving heat source is detected.

The wire for this project can be sized based on the combined wattage of your lamps. The wire sizes are: #14 AWG/2 conductors UF (underground feeder) in copper with ground wire, or #16 AWG/3 conductor flexible cord, #12 AWG/2 conductors, and #10AWG/2 conductors direct-type burial cable—all made of special UV and weather-resistant material. Consult a lighting specialist or electrical supply store for guidance on the appropriate type of wire for your particular project.

A low-voltage transformer will also be needed. There are several economical types, including direct burial, fiberglass, stainless steel, and powder-coated. Some have timers built-in. The transformer for your project will need to have adequate capacity to match the number and wattage of light fixtures you plan to use.

Outdoor light fixtures are available in aluminum and plastic, brass, and copper. Decorative fixtures—for pathways and post and wall lanterns—are available at home centers and lighting stores in a variety of styles, such as Mission and tulip shapes in copper, brass, black, green, and gold tones. Prices range from about $20-$40 for individual path lights on spikes, to $100 and up for wall lanterns. Accent light fixtures for use in trees and shrubs should be unobtrusive (preferably black or green).

What's Involved

This project starts with identifying a power source. It may be possible to add the low-voltage lighting on an unused or under-used circuit. The low-voltage lamps will probably be between 20-35 watts, so, depending on the size of your yard and the number of light fixtures you plan to use, the demand on the circuit's current should not be too high. In most houses, branch circuits to rooms other than utility rooms, baths, and kitchens tend to be under-used, and offer expansion possibilities. An electrician should evaluate the proposed circuit to find out if it can handle the extra load. If you don't have an under-used circuit, or if you plan to add more outdoor lighting in the future, a new circuit will be needed.

Installation begins with the transformer, a weatherproof box that takes an incoming 120-volt household current and transforms it into an outgoing 12-volt current. The transformer is not difficult to install, provided you have an excellent understanding of electrical work, and completely disconnect the electricity at the main circuit breaker of the load center (panelboard) before you begin the installation.

The low-voltage wire is the next item to be installed. Once the wires have been properly connected at each end, you'll need to cover them with enough soil (or hide them behind shrubs or under a layer of mulch), (120-volt wire would require much deeper burial.)

Then the lights are installed, based on your earlier experiments with the portable light and the plan you have come up with. You will probably need to make some adjustments once you have the actual light fixtures with the correct wattages and can see their effects.

Level of Difficulty

Tapping into the service panel or sub-panel and connecting to a transformer should be undertaken only by a licensed electrician or an expert in electrical work. Knowledge and experience are needed to assess circuits to see if they can be used for this additional load, to make the connections, and to install a new circuit if one needs to be added. You can save money by creating your own lighting plan, laying out the wiring and mounting the fixtures, as well as burying the wire. You will want to have someone help you position the lights so that one person can hold them in place, while the other can see the effect from the desired location.

Both beginners and intermediates should increase the time estimates for any tasks within their competence by 50% and 20% respectively. An expert in electrical work should add 10%.

What to Watch Out For

Try to conceal the light fixtures—unless they are decorative and add to the effect. Try to keep any lights mounted in trees at a 10' height or lower so changing bulbs won't be difficult. Keep in mind that light attracts insects—aim fixtures so that they provide ample light, but don't shine directly on the sitting/entertaining area. And don't overdo the light or direct it at your neighbors.

Some professionals recommend PVC conduit when burying wires to protect them from damage by water, lawn mowers, or shovels. Steel conduit is a good choice if the terrain is rocky, and you don't have soil to bury the wire.

Options

An inexpensive, easy option is a kit that uses a solar cell to collect power during the day and to light the fixtures at night. Solar lighting tends to be dimmer than normal fixtures, and are only that effective if the solar cells on the fixtures get good sunlight exposure during the day.

See also:
Post Light and Outlet.

Low-Voltage Landscape Lighting

Description	Quantity/Unit	Labor-Hours	Material
Landscape lights, solar powered, 6″ dia x 12″ h, one piece w/ stake	8 Ea.	0.8	240.00
UF ugnd fdr cable, CU w/ grnd, #12, 2 conductor	1.50 C.L.F.	3.4	54.90
Outlet boxes, plastic, 4″ diameter	1 Ea.	0.3	2.74
Transformers low-voltage	1 Ea.	1.0	70.20
Landscaping lighting, bronze, pathlight	4 Ea.	4.0	480.00
Totals		9.5	$847.84

Contractor's Fee, Including Materials:	**$1,793**

Seed or Sod

Whether your house is newly built and the landscaping is not yet complete, or you have an older home surrounded by terminally ill grass, putting in a new lawn can dramatically improve your home's appearance and enhance its value. If you're sprucing up with the idea of selling your home, this project is probably one of the best improvements you can make in your home's all-important curb appeal. Creating a lawn is something just about any homeowner can successfully complete, given a little advice from a local garden center, proper materials, and a willingness to work up a sweat and a few blisters.

Materials, Tools, & Equipment

Both seeding and laying sod can produce a beautiful lawn, if you follow the correct procedures. Preparing the site for either project requires similar items. If you need to remove an existing lawn, rent a rototiller and spear the broken-up lawn pieces with a pitchfork. Additives such as nutrients, fertilizer, and lime need to be purchased and mixed into the soil.

Seeding is generally less expensive, especially if you do the work yourself, and gives you flexibility to choose different grass varieties. Sod, on the other hand, satisfies the more impatient among us. This "instant lawn" is sold in 18" x 6' strips, and can be laid from spring through fall. Sod does not

generate erosion problems, as seed can, and is less likely to fail.

If you decide to seed, you'll need enough to cover the entire area evenly. A general rule of thumb for determining how much seed to use is to spread two pounds for every 1,000 square feet. You'll also need a spreader, a spring-tooth rake, a roller, and enough fine mulch (sawdust) to cover the seed with a 1/8" layer for protection.

What's Involved

Site preparation is virtually identical for both seeding and sodding. Assuming the area has a sufficient depth of topsoil, spread 2"-3" of peat moss, some starter fertilizer, and, if your soil is particularly acidic, a dusting of ground limestone over the area. Dig these additives into the top 6" of soil so that all ingredients are thoroughly mixed. Rake the site to clear it of stones and twigs, then roll it smooth with a roller. Fill depressions, knock down high spots, and continue rolling until you've achieved a reasonably smooth and even soil surface. Consult with your local nursery or garden center to learn how best to prepare your soil for grass. A soil testing kit will help you determine the acidity and what needs to be added for proper balance.

Seeding is best done in the fall or spring. Your local supplier can recommend the best mix of seed for your lawn, as well as the proper rate of application. The seed should be sown on dampened soil and

scratched in lightly and carefully with a rake. Then roll the area lightly and spread a mulch-like plain sawdust over the lawn. Use a spreader to produce an even application that doesn't bury the seed too deep. Water daily unless it rains, until most of the lawn has sprouted, then let the surface dry out between waterings so the seedlings don't develop fungal diseases. When the young lawn needs mowing, use a sharp reel mower for the first several cuttings to avoid tearing out the new plants.

Sod strips have about 3/4" of soil attached to the roots, so prepare the surface that much lower than you want the level of the finished lawn. Lay the strips side-by-side with ends tightly butted together and with the joints staggered, as in brickwork. If the yard slopes, lay the strips horizontally across the slope to minimize erosion along the seams. Roll the sod thoroughly to ensure good contact with the soil, and water it daily until it's established—about three weeks in warm weather. Mow the grass when it needs it, but discourage other traffic. While sod may be called "instant lawn," it is really just loose dirt and grass until it takes root.

Level of Difficulty

If you need to remove what remains of an old lawn, the work increases considerably. Spreading the additives and mixing them into the soil, and rolling the area smooth is also heavy work, but it goes better if you work at a moderate, steady pace. The actual seeding is light work. Sodding, on the other hand, involves carrying (or wheelbarrowing) rolls of heavy, damp sod to the lawn, laying them out, maneuvering them into position, and possibly trimming them to fit around shrubs or other obstacles—all on the same day it's delivered. For a large lawn, be sure to arrange plenty of able-bodied assistance.

A beginner, using either method, should plan on at least 150% more time than listed to accomplish the job. An intermediate do-it-yourselfer should add 75% more to the time for sod, and 50% more for seed; an expert should add 20% and 10% respectively. All should line up enough help to complete the sod-laying in one day.

What to Watch Out For

When seeking advice at your local garden center, give them all the specifics. Factors can vary considerably even within a fairly restricted area. For example, the soil in your town may be somewhat acidic, but if your house is surrounded by pine trees, your soil will be even more so, and will require more lime to neutralize. A lawn in constant sunlight will call for a different type of grass seed or sod than one in constant shade. Slopes can present water run-off problems, which could lead to erosion, and low areas with slow percolation rates may require a drainage system before any type of lawn is attempted.

One way to prevent seed on a slope from being washed away by water and rain is to cover it with tobacco netting—a lightweight gauze available at most garden centers. It can be held down with staples of bent wire. The grass will sprout through the netting, which,

along with the wire, will eventually decompose into the soil.

If you're not careful to remove every last trace of vetch, crabgrass, and weeds when preparing the site, they'll invade your new lawn. Use the special fertilizers that are available for fast lawn startups. Within two to three weeks, a lawn may look lush and established, but the new lawn still needs your care and watering to survive.

The chemical lawn care service industry is under scrutiny by environmental organizations. Questions have been raised concerning the use of herbicides and pesticides, their methods of application, and the health hazards they pose to humans and animals. It's prudent and reasonable to be aware of them, and judge for yourself the pros and cons. Some lawn care companies use environmentally friendly products.

Options

Keep in mind that large lawns require more water, chemicals, and time to maintain than other types of landscape. Alternative, low-maintenance, low-water-requirement plantings include ornamental grasses and plants that are native to your area.

See also:
Drip Irrigation System, In-Ground Sprinkler System, Planting Trees and Shrubs.

Seed or Sod, 1,000 S.F.

Description	Quantity/Unit	Labor-Hours	Material
Scarify subsoil, residential, skid steer loader	1,000 S.F.	0.3	
Root raking and loading, residential, no boulders	1,000 S.F.	0.5	
Spread topsoil, skid loader	18.50 C.Y.	1.6	466.20
Spread limestone	110 S.Y.	0.1	10.56
Fertilizer, 0.2 lb./S.Y., push spreader	110 S.Y.	0.1	9.24
Till topsoil, 26″ rototiller	110 S.F.	0.9	
Rake topsoil, screened loam	1,000 S.F.	1.0	
Roll topsoil, hand-push roller	18.50 C.Y.	0.1	
Seeding, turf mix, 4 lb./M.S.F., push spreader	1,000 S.F.	1.0	7.68
Mulch, oat straw, 1″ deep, hand spread	110 S.Y.	1.9	42.24
Add to contractor's fee for sodding....$495			
Totals		7.5	$535.92

Contractor's Fee, Including Materials: **$1,134**

Planting Trees and Shrubs

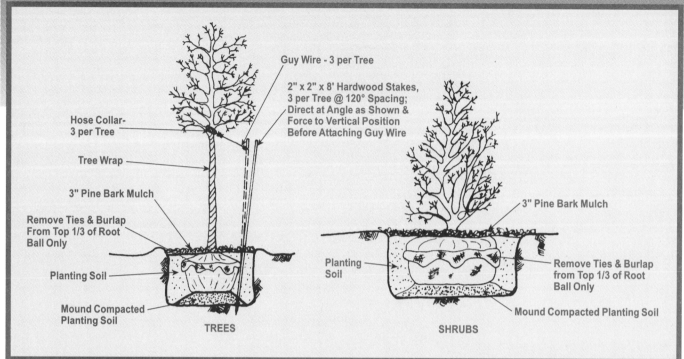

Guy Wire - 3 per Tree

2" x 2" x 8' Hardwood Stakes, 3 per Tree @ 120° Spacing; Direct at Angle as Shown & Force to Vertical Position Before Attaching Guy Wire

Hose Collar- 3 per Tree

Tree Wrap

3" Pine Bark Mulch

Remove Ties & Burlap From Top 1/3 of Root Ball Only

Planting Soil

Mound Compacted Planting Soil

TREES

3" Pine Bark Mulch

Planting Soil

Remove Ties & Burlap from Top 1/3 of Root Ball Only

Mound Compacted Planting Soil

SHRUBS

Curb appeal is defined by a home's architecture and its landscaping. Whether you're landscaping from scratch around a newly built house, or improving the established yard in an older home, new trees and shrubs can enhance your property's beauty and value. They can also provide privacy and lower your energy costs with shade in summer and wind breaks in winter, not to mention absorbing run-off and cleaning the air.

Materials, Tools, & Equipment

Trees and shrubs come packaged in one of four ways: balled and burlapped (B & B), potted, container-grown, or bare root. You'll find the first three types at your local nursery or home center. Bare root trees and shrubs are generally sold through mail-order catalogues. B & B-packaged plants usually offer the widest selection in size. In addition, you'll need peat moss, top soil, mulch, and nutrient additives recommended by your nursery or landscape designer/contractor.

Tools include a shovel and possibly a wheelbarrow. (For large trees, your best bet may be to hire a landscape contractor with specialized equipment for transporting and planting.) If you live in an area subject to high winds, you may want to purchase stakes and guy wire to hold young plants in place.

What's Involved

If your property needs completely new landscaping, a landscape architect or contractor/designer can create a master plan. This won't be cheap, but can ensure a successful result and even save you money in the long run by avoiding mistakes.

If your project is on a smaller scale, there's less risk, and you can save money by laying the plan out yourself, with advice from your nursery. Sketch your plan on paper including correct distances between shrubs and trees and the house.

There's an old saying: "Don't dig a $5 hole for a $10 plant." Even the healthiest tree or shrub will struggle to survive a poor job of planting. Dig the hole at least 12" wider and 6" deeper than the size of the root ball.

Throw some top soil into the hole, mixed with adequate peat moss, as recommended by the nursery. Water, then position the plant, making sure the top of the roots are level with the top of the hole. Cut and remove a third of the burlap without disturbing the root ball. Backfill the hole 3/4-full with topsoil, tamp it down, and fill with water.

After the water has soaked in, finish backfilling to grade level, and tamp the soil firmly around the plant to eliminate air pockets. Form a water-well by building up a low dike of soil around the hole's perimeter; fill this with water to soak the roots. Follow your nursery's recommendations for regular watering.

Level of Difficulty

Landscaping your entire property can be an enormous undertaking, possibly involving heavy equipment and large amounts of material for site preparation and drainage. If you're determined to do it yourself, attack a small section at a time, rent machinery and tools if needed, and be prepared to spend a number of weekends on this project. Most homeowners who can afford it would probably do well to hire a licensed landscape contractor to come in with a crew and get the whole thing, or at least the major construction and large tree planting, done in a few days.

A more modest project, such as planting a few shrubs and trees, can be done by average homeowners with advice from local gardening professionals. Digging holes and muscling heavy root balls into place calls for energy, a strong back, and a willingness to get dirty, offset by the rewards of an immediate improvement in the looks of the yard.

Beginners should add about 100% to the professional's time; intermediate do-it-yourselfers should add 50%, and experts, 20%. All should consult nursery professionals on the best varieties of trees and shrubs and their planting and maintenance.

What to Watch Out For

Purchase healthy plants without insects or injuries. Water frequently during the first two months or so, as new plants become established. Watch for pests such as spider mites and aphids, and contact your garden center for advice on treatment if necessary. Spread organic mulch around the base of the plant, and hold off fertilizing until the plant is well established, which may take a couple of years. Again, your local nursery expert is the best source of advice on how and when to fertilize.

Improperly installed landscaping can cause moisture problems around the house. Trees and shrubs require large quantities of water initially, which may contribute to dampness in basements. Before planting shrubs and trees or creating flower beds near the house, check your foundation for any cracks or signs of weakness that could leak. Most drainage problems can be solved by grading away from the foundation— easily accomplished by creating a low spot or swale some distance from the foundation.

Some additional guidelines for foundation plantings:

- Test the soil and, if necessary, improve it.
- Make your foundation beds wide enough—at least a third of the house's height.
- Don't use too many shrubs and trees, and leave enough space around them so they can develop without looking crowded. Position them so that when fully grown, there will be a minimum of 3' between them and the house.

- Pick the right plant for the site, considering light, moisture, and soil requirements.
- Select plants based on foliage as well as blooms. Look for subtle or striking contrasts that make the landscape more interesting.
- Don't plant too many species. Planting more than three or four types in a limited area creates a confused and jumbled effect.

Options

To reduce water, fertilizer, and pesticide requirements, consider indigenous trees and shrubs. They'll also attract native birds and butterflies. Plant trees with pinecones or berries if you want to encourage birds. Containers, arbors, benches, and other accessories can personalize and enhance entryway steps, the driveway, or other spots where plants in the ground have not grown well.

See also:

Seed or Sod, Retaining Wall, Garden Arbor, Garden Pond.

Planting Trees and Shrubs

Description	Quantity/Unit	Labor-Hours	Material
Stake out tree and plant locations	6 Ea.	0.4	
Tree pit, excavate planting pit	0.26 C.Y.	0.3	
Mix planting soil, by hand, including loam, peat, and manure	0.30 C.Y.	0.1	13.32
Backfill planting pit, by hand, prepared mix	0.30 C.Y.	0.3	
Bark mulch, hand spread 3" thick	0.35 C.Y.		0.85
Cornus Florida (white flowering dogwood) B & B, 5'-6'	1 Ea.		94.80
Picea Abies (Norway Spruce) B & B, 4'-5'	1 Ea.		67.80
Thuja occidentalis (American arborvitae) B & B, 4'-5'	1 Ea.		44.40
Syringa vulgaris (common lilac), 5-gal.	1 Ea.		23.70
Forsythia ovuta robusta (Korean forsythia) B & B, 3'-4'	1 Ea.		20.04
Rhododendron catawbeense hybrid, 5-gal.	1 Ea.		30.60
Totals		1.1	$295.51

Contractor's Fee, Including Materials: **$465**

Outdoor Living Areas and Structures

The time you spend at home, and the useable area of your home, can be greatly enhanced by adding convenience and enjoyment in the form of outdoor structures. Following are some tips for a successful project.

- Most cities and towns have setback requirements that limit construction to within a certain distance from property lines—front, side, and back. If the site chosen for your outbuilding doesn't meet these requirements, you must file an appeal for a variance.

- Gazebos can be challenging to build because of the carpentry skills required and the number of odd angle saw cuts and fastening techniques. Nevertheless, these projects can serve as a training ground for intermediate do-it-yourselfers interested in improving their carpentry skills. An accurate 2' or 4' level, calipers, and an adjustable square will help make the layouts and angle cutting easier and more precise.

- When fabricating recreational equipment, take care to make the cleanest cuts possible. Your lumber dealer should be able to recommend the correct blades for sawing. Carbide blades don't become dull as quickly as typical saw blades. Working on the ground exposes blades to dirt and grit that can easily dull a blade. Pressure-treated lumber dulls blades faster than ordinary lumber. When doing any sawing, always wear appropriate eye protection.

- Be sure to lay out the posts that support the play structure platform perfectly square and plumb, using braces whenever necessary. If even one post is out of line, it will throw off almost every measurement.

- If you're contemplating building an outdoor storage shed, plan carefully what items you will want to keep there, now and in the future. Be sure the door is wide enough and properly placed to allow you to move these items in and out easily. Ride-on mowers may require a ramp.

- There are many types of composite materials, such as those made of recycled plastics or sawdust. They are available in a variety of colors, and offer other advantages such as nontoxic, non-splintering surfaces. Consult your building material dealer for advice about these materials—particularly where children will be in close contact with them.

- Be sure to provide loose-fill materials below and around any children's swingset or play structure to cushion their falls. Wood chips, shredded bark mulch, or sand should be about 6"-7" deep. Consult play equipment manufacturers for more details.

- If you or your contractor are building projects that involve excavation, it's important to contact your utility companies to identify the location of—and avoid damage to—underground electrical, gas, water, and septic lines.

Outdoor Living Area

One of the biggest current trends in remodeling is creating outdoor "rooms" to expand a family's living area. These spaces often serve as outdoor kitchens or family rooms with a grill, comfortable seating areas, refrigeration, storage, and a music/media center.

The floor of the living area can be at ground level or on an elevated deck. Flooring might be wood, brick, stone, stamped concrete, or pavers. Most outdoor kitchens and living areas have some protection overhead, whether a roof, pergola, awning, or gazebo. You might also want to consider installing a trellis wall or decorative fence, depending on your view, wind exposure, and other factors. Possible features of these spaces are appliances, such as grills, small refrigerators, and outdoor fireplaces, as well as sinks or countertops for food and drink preparation and service.

Materials & Tools

This model project begins with a deck floor already in place, and includes installation of a small refrigerator, a gas grill, a sink, a weatherproof counter and cabinets, and a self-supporting awning to cover the seating area. The project also requires installation of electrical wiring and receptacles for the appliances, and piping and fittings for the sink and refrigerator. Two waterproof GFI (ground fault interrupter) duplex receptacles will provide electrical service on the deck. A gas grill for cooking may be a stand-alone or built-in unit, with a refillable tank or connected by piping to natural gas (if your house already has a natural gas connection). Some grills come with side burners (for cooking or warming in

pots), a built-in "countertop" area, and cabinet storage. Some elaborate units include pizza cookers, woks, and other specialty items. If you choose to add a sink, make sure you're able to drain it properly according to local codes.

What's Involved

The project begins with measuring your available space and deciding how you want to use your new living area. Check out home magazines and product manufacturers' Web sites to get an idea of the look and features you want. Gather information about the models of grills and refrigerators that suit your needs and budget—and the style of your home. Get details on appliance dimensions and power requirements. Draw a plan that incorporates not only the appliances, cabinets, and countertop, but adequate space for seating and other features, such as potted plants and patio furniture. Consider lighting changes or additions, and extras such as speakers and patio heaters. If you choose to do the project on your own, take your plan and any notes on appliances and features to a home center or retailer that specializes in grills and patio items. Find out exactly what's available, along with the prices, and get some advice that may improve your plan and layout.

As with interior construction projects, the rough plumbing and electrical work must be accomplished first. The electrical service will need to be run from your house panel to power the appliances and any additional lighting. A separate circuit is recommended. The power requirements for your project may not be large, depending on the appliances and lighting you include, but the available circuits should be

evaluated by a licensed electrician. Make no assumptions. The service will be run to the outside of the house in an armored, plastic-coated cable. (The actual cable/wire is dictated by local building codes.)

The plumbing work for the sink and refrigerator will depend on the specifics of your site. Drainage is an important consideration that must be evaluated by a plumber or plumbing expert based on local building code requirements. In colder climates, the piping should include the capability for drainage as part of the procedures for cold-weather shut-down. After the rough electrical and plumbing are completed, the cabinets, appliances, and sink can be installed.

Level of Difficulty

An expert or a professional electrician should undertake the wiring portion of this project. Knowledge, experience, and skill are required to assess the circuit and tap the service panel. The wrong procedures can endanger both you and your home. Someone with ample plumbing expertise should determine how the waste from the sink will flow, and whether a pump will be required. Installation of a countertop and cabinetry, as well as the awning, can be accomplished by an experienced do-it-yourselfer. A novice should hire professionals for the electrical and plumbing work and add 100% to the professional time listed for other tasks. Intermediates should add 75%, and experts 40%.

What to Watch Out For

For safety, wear goggles when using a saw or performing other work that may expose your eyes to flying particles or fumes. Wear gloves when handling rough or sharp-edged materials. Experts who are qualified to do the electrical work should use a circuit tester and know with certainty that the power is off when working on any aspect of the electrical installation. Exposure to electricity while working outdoors (in contact with the earth) establishes an electrical ground that could prove fatal.

Consider cleaning and maintenance requirements in selecting counter and other materials. Choose durable surfaces, such as tile or concrete, which will enable you to scrub and hose down the area.

Options

Though not in this project, outdoor fireplaces and patio heaters—gas, electric, propane, alcohol-gel, wood- and coal-burning—are another popular option for outdoor living areas. Some gas models, which can also be used in screened porches, gazebos, and lanais, vent through a front grill, so there's no need for a chimney. Portable campfire-style fire pits are available for cooking—in redwood, teak, metals, and pottery, among other materials. Specialty retailers that sell fireplace, grill, and patio products are a good source of information. Many home improvement centers also sell these items.

Consider upgrading the landscape around your outdoor living area. Features such as fountains and new plantings can add to the ambiance and provide shade and fragrance.

Insect control is another consideration, for which several options are available. Bug zappers use modest amounts of electricity, but can be noisy and kill beneficial insects as well as mosquitoes. Mosquito traps emit carbon dioxide or an attractant odor, or use a fan to blow a

Designed and built by Kooda Exteriors

specific temperature of air (similar to breath). Some of these devices have an electric eye that turns them off automatically at daylight. Mosquito repellers include timed misting systems installed around your property's perimeter, and another type that emits high-frequency sound to keep insects at bay.

Natural insect control methods include attracting bats and insect-eating birds (eastern or western bluebirds and purple martins, for example) by providing houses specifically designed for them, and offering a haven for toads in the form of a couple of upside-down flower pots. Certain flowering plants and herbs, such as lavender and pennyroyal, can also help control insects and rodents. Always remove standing water after a rain or sprinkler use to reduce mosquito breeding opportunities.

See also:

Outdoor Fountain, Planting Trees and Shrubs, Patios, Decks, Gazebo, Garden Arbor.

Outdoor Living Area

Description	Quantity/ Unit	Labor- Hours	Material
Wire	20 L.F.	0.1	0.96
Armored cable	40 C.L.F	1.5	44.64
Duplex receptacle	2 Ea.	0.4	2.74
Duplex waterproof box	2 Ea.	1.1	32.16
GFI	1 Ea.	0.6	42.00
Counter top	6 L.F.	1.8	11.23
Cabinet, base, 36"	1 Ea.	0.8	2,160.00
Cabinet, sink/range, 36"	2 Ea.	1.6	3,000.00
Cabinet, Filler strips, 2"	3 Ea.	0.3	115.20
Refrigerator	1 Ea.	1.1	342.00
Awning patio 16' x 14'	1 Ea.	13.3	2,370.00
Fireplace	1 Ea.	6.2	1,380.00
Sink	1 Ea.	2.9	302.40
Piping for sink	75 LF	22.6	206.10
Totals		54.3	$10,009.43

Contractor's Fee, Including Materials: $17,359

Gazebo

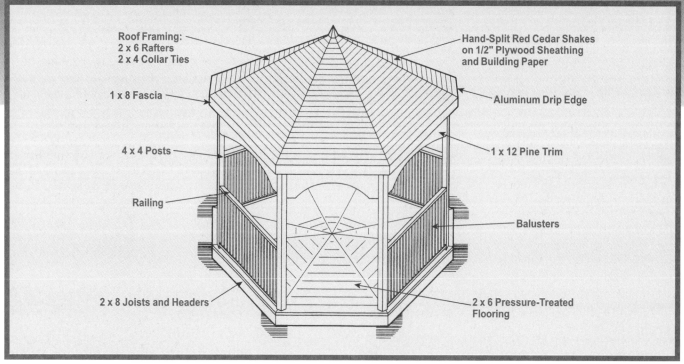

Roof Framing:
2 x 6 Rafters
2 x 4 Collar Ties

Hand-Split Red Cedar Shakes
on 1/2" Plywood Sheathing
and Building Paper

1 x 8 Fascia

Aluminum Drip Edge

4 x 4 Posts

1 x 12 Pine Trim

Railing

Balusters

2 x 8 Joists and Headers

2 x 6 Pressure-Treated
Flooring

A major trend in home improvements is the addition of well-defined outdoor living areas. If your yard has the space, a gazebo is one way to create an attractive spot for entertaining and relaxing, as well as adding potential curb appeal. Gazebos can be adapted to fit in just about any surroundings, and range in style from plain and rustic to elaborately decorative. The process of building a gazebo is similar to framing a small house, as it consists of a foundation, framed walls, and a roof. With that in mind, it's a good idea to hire a carpenter if you're a novice do-it-yourselfer.

Materials & Tools

Gazebos have become so popular that you can most likely purchase all the materials you'll need in a kit at your local home supply store. Lumber, roofing materials, nails, concrete, and tube forms for the supports are among the items you need. The order will be large, so arrange for delivery.

A good set of carpentry tools is essential for most home improvement projects, and the gazebo is no exception. Having

the right hand and power tools makes many challenging carpentry operations less difficult. Several different nail sets of various sizes will also help in toe-nailing angled pieces and nailing the floor, balusters, fascia, and other trim pieces.

What's Involved

The first step is to decide where you want your gazebo to be located, which could greatly impact how often you use it, and for what purposes. Consider placing it close to your house, situated so that you can see the most scenic elements of your yard from it. You'll also want to choose a spot where the gazebo itself looks most attractive from your windows inside the house, and from the street. Once you determine the location, make sure the terrain will be suitable for installing concrete piers that are 3' deep. These will serve as the basic supports for the structure. Because of the octagonal shape and the number of piers required, the operation is a little tricky. Careful planning and workmanship are crucial if the rest of the project is to go smoothly. Be sure to double-check all measurements before placing the

concrete. Tube forms will add to the cost, but they will make it easier to place and level the piers accurately.

The next phase of the project involves placing the platform and support posts. There are a few options available, and the cost and level of difficulty vary slightly among them. One method is to build the frame in place and then mortise it into the posts so that short legs extend below the frame onto the tops of the piers. An alternative is to place the frame directly on the piers and set the posts beside the perimeter of the frame at each point. A third method is to plant the legs in the concrete at rough height, level them by sawing them after the concrete has cured, and then build the frame of the platform around them. The roof supports should transmit the weight of the roof directly to the piers.

The roof rafters are difficult to install, as they must be precisely angled on both ends to fit correctly. Ceiling joists are not called for in this plan, but can be used in place of the collar ties if the new gazebo will have a finished ceiling. Sheathing and finished roofing are applied after the roof frame is complete, as well as the railing, the steps, and the finish trim.

Level of Difficulty

Gazebos are challenging projects even for expert do-it-yourselfers, though many can complete the project with patience, research, and guidance along the way. Beginners and intermediates should consider hiring a carpenter to complete the difficult initial layout and roof. The railing and steps of the gazebo also require skillful carpentry.

Generally, beginners and intermediates should add 100% and 50%, respectively, to the estimated time for all tasks other than the roofing, which should only be attempted by experts. Slightly more time is needed for installing any fancy trim work that involves careful work with expensive materials. Experts should add 20% to the professional time listed in the estimate for all jobs, and more for all aspects of the roof work.

What to Watch Out For

This gazebo is octagonal in shape, but five- and six-sided configurations are also considered authentic Victorian designs. Eight-sided structures provide more platform space, but they also require more materials and added construction time. Hexagonal and pentagonal types are slightly easier to build, because fewer piers are required, but roof construction is equally complicated. Also, be aware that considerable waste is expected on roofing materials, no matter what the size. The key to choosing a style that meets your needs is to think ahead to your intended use of the gazebo—whether, for instance, you plan to have a table and chairs for meals, or simply benches or a chaise lounge.

Options

There are unlimited possibilities for the style and materials for this type of structure. You could really get creative

Courtesy of Holway House Designs

with decorative floorboard patterns and angled wainscoting, as well as an ornate cap with trim instead of standard railings. Lattice, screens, and various trim styles can be used for the opening above the rail and fascia. Also, many stairway designs are possible, from basic wood steps to decorative concrete, brick, and stone. You have a few options for roofing materials, including standard asphalt shingles or more expensive red cedar shingles. One recent popular trend, though pricey, is copper roofing, used in place of the more traditional shingled style. With any added options, keep in mind that ornate designs tend to be more expensive; a compromise plan with a simpler design can keep the project within budget.

Gazebo, 4' Sides, 10'-6" x 10'-6"

Description	Quantity/ Unit	Labor-Hours	Material
Excavate post holes, by hand, incl. layout	0.50 C.Y.	0.5	
Forms, round fiber tube, 8" diameter, for piers	24 L.F.	5.0	43.49
Concrete, field mix, 1 C.F. per bag, for piers	8 Bags		55.68
Placing concrete, footings, under 1 C.Y.	0.50 C.Y.	0.4	
Framing materials, pressure-treated lumber posts, 4 x 4 x 12'	96 L.F.	3.4	155.52
Headers, 2 x 8 x 12'	36 L.F.	1.7	35.86
Joists, 2 x 8 x 10'	70 L.F.	2.5	84.00
Flooring, 2 x 6 x 10'	200 L.F.	6.4	184.80
Railing with balusters	32 L.F.	11.6	543.36
Trim, 1 x 12 pine	32 L.F.	1.4	67.97
Roofing material, headers, 2 x 6 stock	36 L.F.	1.6	23.33
Rafters, 2 x 6 x 8'	128 L.F.	2.0	82.94
Collar ties, 2 x 4 stock	16 L.F.	0.3	6.53
Fascia board, 1 x 8 stock	36 L.F.	1.3	67.39
Sheathing, plywood, 1/2", 4' x 8' sheets	256 S.F.	2.9	190.46
Building paper, asphalt felt paper, 15 lb.	140 S.F.	0.3	5.04
Drip edge, aluminum, 5" girth	32 L.F.	0.6	9.60
Shakes, hand-split red cedar, 18" long, 8-1/2" exposure	2 Sq.	8.0	234.00
Stair material, stringers, 2 x 10	16 L.F.	0.6	28.61
Treads, 2 x 4 x 3'-6", 3 per tread	24 L.F.	0.8	22.18
Paint, seal and varnish, latex, sprayed	700 S.F.	11.8	58.80
Totals		63.1	$1,899.56

Contractor's Fee, Including Materials:	$6,227

Utility Shed

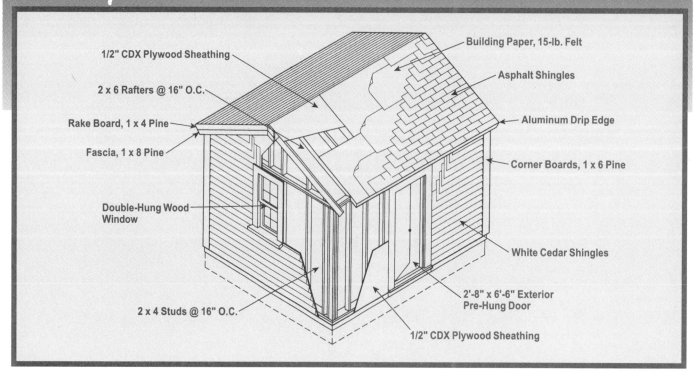

1/2" CDX Plywood Sheathing

2 x 6 Rafters @ 16" O.C.

Rake Board, 1 x 4 Pine

Fascia, 1 x 8 Pine

Double-Hung Wood Window

2 x 4 Studs @ 16" O.C.

Building Paper, 15-lb. Felt

Asphalt Shingles

Aluminum Drip Edge

Corner Boards, 1 x 6 Pine

White Cedar Shingles

2'-8" x 6'-6" Exterior Pre-Hung Door

1/2" CDX Plywood Sheathing

Finding enough storage space to house various tools and equipment, gardening supplies, bicycles, and other items can be a challenge for any homeowner. Even if you're lucky enough to have a garage, attic, basement, or back porch, you might want to save that space for other uses by building a simple, backyard storage shed. The 10' x 12' utility shed in this model project provides 120 square feet of storage space, and can, with a few finishing touches, be both useful and attractive.

Materials, Tools, & Equipment

This utility shed is a simple pitched roof design set on a slab floor/foundation made of concrete. The walls and roof are sheathed with 1/2" CDX plywood. Pressure-treated lumber is needed for the sill and anchor bolts to set the sill to the concrete. Roofing, siding, and trim materials can be selected to match your house or other existing outbuildings. The door can be simple and homemade, or you can purchase an exterior door equipped with hardware.

Only one small window is planned in this model, to allow for some light and ventilation, while leaving most of the interior wall area free for storage. You can also install shelving, pegboards, and racks, which can be purchased at any home improvement center or built on site from plywood or other materials.

Tools needed for this project include those found in a typical homeowner's tool collection, as well as a saw to cut the framing materials to size, sawhorses, and a step ladder to reach the roof. A level is also important to guarantee that the shed is plumb, level, and structurally sound.

What's Involved

First, you'll need to choose a location for the shed, keeping in mind it should blend in with your property and look as if it has always been there, rather than a poorly planned afterthought. Be sure it is far from enough from the house so as not to interfere with any possible additions to your home that you or future owners might consider building later. Also, the location shouldn't be

offensive to neighbors or impinge on their property. Finally, try to site the shed where it will be accessible—both while you're building it and when you're using it for storage.

Before starting construction, take your plans to your local building department and apply for a building permit. A contractor should be hired to excavate, grade, build a form, and level the concrete for the foundation. If the site is fairly level to begin with, this work could be accomplished by an experienced intermediate. You can rent an automatic mixer to prepare the concrete on site, or have a load of mixed concrete delivered from a supply house or contractor. A poured slab is at once a solid foundation and a durable floor. It should be set slightly above grade.

Wiring for lights, switches, and GFI outlets should be run as soon as the framing is complete. A narrow trench (one foot deep) will probably be required for an underground wire to tie into the house's electrical system. Depending on the nature and extent of the wiring, a subpanel in the shed may be recommended. Consult an electrical contractor for all electrical work.

Once the roof and walls are sheathed, the structure must be made weathertight by attaching the fascia and rake boards, and by shingling the roof. This protects the shed from the elements and reduces the need to make repairs in the near future that could otherwise be prevented.

Level of Difficulty

This project is an excellent opportunity for a beginner to learn construction skills. All the main phases of house carpentry are present here: site planning, excavating, laying a foundation, framing, sheathing, roofing, siding, installing a window and door, and applying finish trim. A true beginner would be advised to work under the guidance of a more experienced hand.

An intermediate could tackle this project as an opportunity to gain experience and thereby develop the confidence to undertake more ambitious tasks in the future. This little shed will forgive most mistakes, which can be learning experiences for larger projects. Beginners should add at least 100% to the times required for completion; intermediates, 50%. A person experienced in all phases of construction might come close to matching the professional's time, but it would be wise to add 10%-20% to be on the safe side.

What to Watch Out For

While getting ready to prepare the site for excavation, consider which, if any, sections of lawn have to be driven over by heavy equipment (for delivery of the shed kit components or concrete), and be sure that there is no septic system or dry well that could collapse under the weight of a large vehicle.

As framing a roof is the most difficult task involved in rough carpentry, a professional should also be consulted for this phase of the project. If necessary, loft the roof frame on a

suitable large and clean surface. This simply means drawing a full-size diagram in cross section on the entire frame so you can check the accuracy of your calculations. Finally, make test cuts on scrap lumber of all rafter cuts—ridge, bird's mouth, tail—before going to work on the actual (expensive) rafter lumber.

See also:
Re-Roofing with Asphalt Shingles, Re-Siding with Wood Clapboard, New Window, Trash Can Shed with Recycling Bins.

Utility Shed, 10' x 12'

Description	Quantity/ Unit		Labor-Hours	Material
Excavation, by hand, for slab area	1.50	C.Y.	3.0	
Fine grading, for slab on grade, confined area	13	S.Y.	0.4	
Concrete, forms in place, slab on grade, edge forms, to 6" high	88	L.F.	4.7	28.51
Welded wire fabric rolls, 6 x 6, for slab on grade reinforcing	120	S.F.	0.5	27.36
Slab, 4" thick with vapor barrier	120	S.F.	2.3	151.20
1/2" anchor bolts, 6" long	10	Ea.	0.9	11.28
Sills, 2 x 6 x 8', pressure-treated	44	L.F.	1.4	40.66
Plates, top and bottom, 2 x 4 x 8'	132	L.F.	2.6	53.86
Framing, studs, 2 x 4 x 8'	416	L.F.	6.1	169.73
Gable framing, 2 x 4 x 8'	128	L.F.	1.9	52.22
Ridgeboard, 2 x 8 x 6'	12	L.F.	0.4	11.95
Rafters, 2 x 6 x 10', to 4 in 12 pitch	220	L.F.	3.5	142.56
Wall sheathing, 1/2" CDX plywood, 4' x 8'	400	S.F.	4.6	297.60
Roof sheathing, 1/2" CDX plywood, 4' x 8'	230	S.F.	3.3	171.12
Rakeboard, 1 x 4, pine	60	L.F.	1.1	43.20
Fascia, 1 x 8, pine	60	L.F.	1.3	87.84
Aluminum gable vent, louvers, with screen, 8" x 8"	2	Ea.	0.4	22.44
Roofing, asphalt felt sheathing paper, 15 lb.	615	S.F.	1.3	22.14
Aluminum drip edge, 0.016" x 5", mill finish	24	L.F.	0.5	7.20
Drip edge, galvanized	5	L.F.	0.1	1.32
Roof shingles, asphalt strip	2.30	Sq.	5.3	150.42
Cornerboards, 1 x 6 pine	56	L.F.	1.1	61.15
Siding, white cedar shingles	4	Sq.	13.3	585.60
Double-hung wood window, 2'-0" x 3'-0"	1	Ea.	0.8	219.60
Pre-hung exterior wood door, 2'-8" x 6'-9"	1	Ea.	1.1	319.20
Door lockset, standard	1	Ea.	0.7	48.60
Door casing, 1 x 6 pine	161	L.F.	5.2	154.56
Paint, stain shingles, 2 coats, oil base, brushwork	400	S.F.	6.7	48.00
Door and window trim, primer, oil base, brushwork	50	S.F.	0.6	1.20
Trim, primer, oil base, brushwork	116	L.F.	1.4	2.78
Door and window trim, 2 coats, oil base, brushwork	50	S.F.	1.0	3.00
Trim, 2 coats, oil base, brushwork	116	L.F.	1.4	2.78
Totals			78.9	$2,939.08

Contractor's Fee, Including Materials: **$8,539**

Trash Can Shed

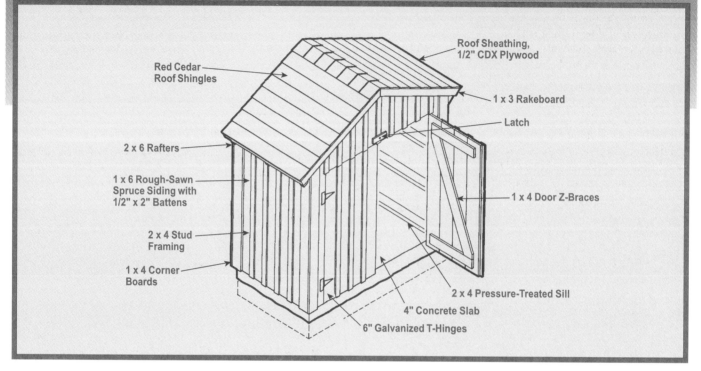

Red Cedar Roof Shingles

Roof Sheathing, 1/2" CDX Plywood

1 x 3 Rakeboard

Latch

2 x 6 Rafters

1 x 6 Rough-Sawn Spruce Siding with 1/2" x 2" Battens

1 x 4 Door Z-Braces

2 x 4 Stud Framing

1 x 4 Corner Boards

2 x 4 Pressure-Treated Sill

4" Concrete Slab

6" Galvanized T-Hinges

According to the Environmental Protection Agency, recycling programs have doubled in the last 15 years, as more and more homeowners take advantage of the opportunity to save the earth's resources and reduce industrial pollution. To do your part without the inconvenience of storing recyclables in your kitchen, mudroom, or garage, you might consider building a small shed like the one in this model project. It can hide trash cans and store recyclables in between pickups or trips to your community recycling station.

Materials & Tools

Pre-fabricated sheds in kit form can be purchased from many home centers and building supply houses. Most will deliver and build them for you. This is convenient, but limits your selection of designs and sizes, and will cost more than if you build it yourself.

The shed in this project is wide enough to hold two 32-gallon (standard-sized) trash cans, leaving about 24" for a stack of plastic recycling bins. The shed's rough-sawn board-and-batten design is

traditional, and can be stained to match your house and other outbuildings, or to blend into a background of trees and shrubs.

You'll also need concrete for the shed's slab foundation, and lumber to build the form that will hold the concrete. The last item for the shed itself is a good quality wood preservative, stain, or paint to protect and preserve your shed for years to come.

For sorting and storing recyclables, sturdy, plastic bins with interlocking feet can be stacked up to six high. Communities have different recycling requirements, but four categories— paper, glass, tin/aluminum, and plastic—are typical. The bins in this project have an 11-gallon capacity, which should suffice for an average family.

To build this shed, you'll need basic carpentry tools such as a hammer, a saw, a level, and a tape measure, and a bucket and trowel to mix and form the poured concrete slab. A shovel and a strong back are needed to dig and prepare your site for the slab.

What's Involved

The site chosen for the shed should be a level area of about 15 square feet, located close to the back or side entrance to your house. Prepare the area and build a form, properly squared and braced, for the poured concrete slab. Insert six anchor bolts in the wet concrete for the 2 x 4 sills. Once the concrete has cured, the three walls can be framed and nailed together, and the roof framed. The siding is of rough-sawn 1 x 6 spruce with 2" battens, attached to the framing with 6d hot-dipped galvanized nails.

Once the front gable is boarded up (horizontally, without battens), the 1/2" plywood roof sheathing can be cut and nailed, allowing a 3" overhang front and back, and a 1" overhang on the rafter ends. Install red cedar shingles, in courses about 6" to the weather, over a layer of roofing felt, using 3d galvanized box nails. Leaving the roof overhang unenclosed by a soffit saves materials and labor, and has the benefit of helping to ventilate the shed.

The doors are built by nailing boards to a Z-frame of 1 x 4's, and are hung with galvanized T-hinges. A combination of a wooden leaf latch and a metal barrel bolt will keep the doors securely shut. The siding should be stained with a wood preservative; the roof shingles can be left to weather naturally.

If more space is needed in the shed itself for more or larger plastic recycling bins, the overall dimensions can, of course, be increased, or storage-enhancing alterations made, such as installing an "attic" floor shelf in the gable peak that can be reached through a small door added to the front gable.

Level of Difficulty

Like the utility shed project, this little building is perfect for the beginner to learn a number of construction skills that can later be applied to larger, more ambitious home improvements. The walls, which use horizontal nailers, are not the usual stud framing. Nevertheless, the ability to measure carefully, cut accurately, and build walls of any type, plumb, square, and level are essential elements of the carpenter's craft, and are as necessary to good results in this tiny shed as they are in the largest house.

Roof framing is a challenge for a novice, but the use of 45° angles makes cutting the rafters somewhat easier than it otherwise might be, and, of course, there are only six of them. You could hardly expect to find an easier roof to shingle, and although cedar shingles are more expensive and take longer to install than asphalt strip shingles, they are much more attractive, especially when seen close up. If the side walls are perfectly plumb and in line with one another, hanging the door should also be quite easy.

This is a low-risk, very forgiving project for a beginner to undertake. With the possible exceptions of the digging and the pouring of the concrete slab, there is no heavy, messy, drudge work. As is the case with most new construction, the tasks involved are instantly rewarding.

A beginner working alone should add 200% to the time estimated for completion. An intermediate-level do-it-yourselfer should add 50%, and an expert, 20%.

What to Watch Out For

A good building starts with a good foundation. Be sure the form you build for the concrete slab is perfectly rectangular and level. Use braces and packed earth to ensure that it will not move, bulge, or rack out of square. If you plan and build the form carefully, its top edges can serve as your screeding guide.

Screeding concrete means to pack and level it by moving a straightedge board across the surface, with a back-and-forth sawing motion, while at the same time pushing the board forward. In this way, the concrete is lightly agitated to make it settle and fill any air pockets that may have formed, and the excess concrete is scraped off, leaving a smooth surface level with the top of the form.

Insert the sill anchors in the appropriate locations, and cover the slab with a sheet of plastic, leaving it for a few days to slowly cure. Once the concrete has set, the forms can be removed and cleaned for future use, and the area around the slab backfilled and tamped.

Before cutting the lumber for the rafters, measure, mark, and make cuts for the ridge, tail, and "bird's mouth" on a piece of 1 x 6 spruce. If you're going to make mistakes, make them on the less expensive lumber, so you can correct them before making the final cuts on the more expensive 2 x 6s.

See also:
Utility Shed.

Trash Can Shed with Recycling Bins, 2'-10" x 5'

Description	Quantity/ Unit	Labor-Hours	Material
Excavation, by hand, for slab area	0.25 C.Y.	0.5	
Fine grading, for slab on grade, confined area	1.50 S.Y.	0.1	
Forms in place, slab on grade, edge forms, to 6" high	32 L.F.	1.7	10.37
Concrete slab on grade, 4" thick	0.20 C.Y.	0.3	24.96
Sill anchor bolt, 1/2" diam., 6" long, including nut and washer	6 Ea.	0.5	6.77
Sill, 2 x 4 x 8', pressure-treated	16 L.F.	0.5	9.41
Framing, 2 x 4 x 8', KD studs	15 Ea.	1.5	48.96
Rafters, 2 x 6 x 8'	24 L.F.	0.5	15.55
Roof sheathing, 1/2" CDX plywood, 4' x 8' sheets	32 S.F.	0.5	23.81
Asphalt felt sheathing paper, 15 lb.	15 S.F.	0.1	0.54
Red cedar wood roofing shingles, no. 1 grade	1 Sq.	3.2	195.60
Siding, spruce, rough-sawn, 1 x 8, natural	330 L.F.	9.6	265.32
Bins, plastic, 11-gallon, 18-1/2" W x 20" D x 12" H	4 Ea.	0.1	50.16
Door hardware, 6" galvanized T-hinge	2 Pr.		61.20
4" galvanized barrel bolt	1 Ea.	0.1	3.70
Galvanized handle	1 Ea.	0.3	2.28
Nails, 6d H.D. galvanized	5 lb.		8.34
Nails, 3d galvanized box	2 lb.		3.34
Gypsum wallboard screws, 2" galvanized	1,000 Ea.		11.22
Gypsum wallboard screws, 1-1/4" galvanized	1,000 Ea.		8.58
Stain, exterior, 2 coats, oil base, brushwork	120 S.F.	2.0	14.40
Totals		21.5	$764.51

Contractor's Fee, Including Materials: **$2,292**

Play Structure

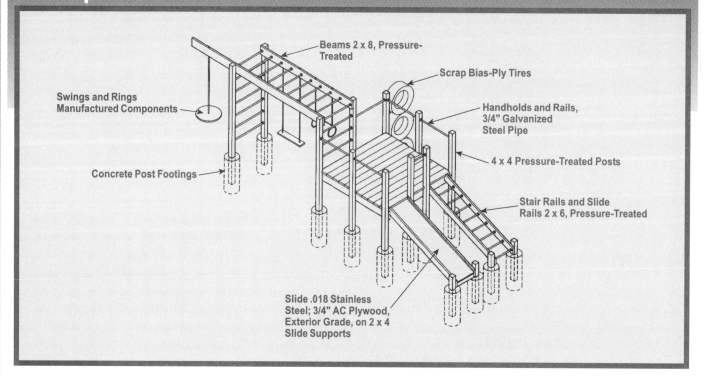

Swings and Rings Manufactured Components

Concrete Post Footings

Beams 2 x 8, Pressure-Treated

Scrap Bias-Ply Tires

Handholds and Rails, 3/4" Galvanized Steel Pipe

4 x 4 Pressure-Treated Posts

Stair Rails and Slide Rails 2 x 6, Pressure-Treated

Slide .018 Stainless Steel; 3/4" AC Plywood, Exterior Grade, on 2 x 4 Slide Supports

If you have active young children and a spacious, level yard, you already have two of the primary components of a backyard play structure. A carefully designed, well-built structure will provide years of safe, healthy fun. Made from wood, the structure in this project can also blend in nicely with your existing landscape.

Materials & Tools

These days the variety of play structures is enormous—made from all kinds of wood and plastic, and in virtually any imaginable configuration, from standard swings, slide, ladders, and monkey bars, to tree houses and forts. While this particular design offers the traditional components, you can adapt it to your child's preferences and to the dimensions of your yard.

You can build a play structure from scratch with basic lumber and other materials, or you can purchase a kit complete with instructions, fasteners, hangers, ladder rungs, swing seats and chains, and pre-cut, pre-drilled lumber. "Hardware only" kits provide

everything listed above except the lumber. Purchasing a high-quality kit can be a good investment for a homeowner who lacks the tools, skills, and/or time to design and build a play structure from scratch. The best kit designs have been field-tested and approved for strength and safety by independent rating organizations.

If you build your own play structure or work from a hardware-only kit, you must decide on the type of lumber to use. Standard construction-grade stock is not recommended because, even if coated with preservative, it's prone to rot from ground contact and constant weather exposure. This deterioration might not always be visible, and could go so far as to weaken the wood sufficiently to cause collapse or breaking, which might well result in serious injury to a child.

The best types of weather-resistant wood are cedar, redwood, and pressure-treated lumber, usually yellow pine. Treated lumber will last longest in damp ground or concrete, and thus is the best choice for posts. Because of health

concerns associated with CCA (chromated copper arsenate), previously used in treated lumber, this preservative is being replaced. Supplies of CCA lumber are expected to have run out by mid-2004.

Any well-stocked lumberyard will offer a wide selection of lengths and sizes. Select pieces that are relatively straight and, most importantly, free of large knots, which affect the wood's structural strength. Whichever lumber you choose, make sure to round over all corners and edges to a 1/4" radius, and regularly monitor the condition of the lumber, touching it up whenever necessary to prevent splintering.

To prepare the site where your play structure will go, you'll need a pick and shovel to excavate and then spread soft fill, such as sand or shredded bark. Any grass in this area will quickly be worn away, and the soil compacted to the consistency and hardness of asphalt. You'll also need a post-hole digger to install the supporting posts, a drill, ratchet and socket set, measuring tape, level, chalkline, and circular saw.

What's Involved

Designing and building your own custom play structure is a fairly complex and time-consuming undertaking. Excavating the site to a depth of 10" and filling it with soft landscaping material—is necessary for any play set, and the size of the area will determine the amount of time it will take. The next step is digging holes for and anchoring the posts and footings in concrete—for the structure's support.

The kit should include step-by-step installation instructions for building the structure, its platforms, and accessories. For safety's sake, all sharp hardware, such as bolt ends and nuts, should be countersunk so they don't protrude above the surface of the wood.

Once you have completed the basic structure, you can equip it with manufactured components: swings, gliders, rings, trapeze, cargo net, rope ladder, tire swing, and whatever else can be acquired through catalogs or at playground supply centers. Any chains used for swings should be encased in a length of rubber or plastic hose to prevent pinched fingers and strangulation. The 3/4" galvanized pipe used for the monkey bars should be pinned into the 2 x 8 rails by drilling a pilot hole through the top of the rail, into the pipe, and then driving a 3" galvanized screw. This will keep them from turning in a child's grasp and causing a fall.

Level of Difficulty

Building a play structure from a complete kit is the easiest route to take. All you need do is follow the directions and assemble the parts in sequence. Some kits are designed with bases and angled supports that require no digging and need no concrete. This saves a lot of labor, but the structure would lack the rock-solid stability of posts anchored in the ground. Working with a hardware-only kit means you have to purchase the lumber and then measure, cut, drill, and assemble it according to the plans supplied.

A beginner could do site preparation and assemble either type of kit, but should work under the guidance of an experienced hand in constructing a custom play structure, adding about 200% to the professional time estimates. An intermediate-level do-it-yourselfer should add about 100%, and an expert about 20%.

What to Watch Out For

Safety is a key factor in this project. Consult the Consumer Product Safety Council, the American Society of Testing and Materials, or the National Playground Safety Institute for recent information about the hazards you should consider when building your play structure. Although these organizations deal mostly with public playground safety, the information they provide is useful for anyone interested in constructing such a structure.

From time to time, and especially during the first several weeks of use, check all bolts and fasteners for tightness. The wood tends to shrink, and regular use can cause loosening of bolts.

Play Structure

Description	Quantity/ Unit	Labor- Hours	Material
Layout, clearing and excavating, by hand	5 C.Y.	10.0	
Compaction, vibratory, 8" lifts, common fill	5 C.Y.	0.2	
Fine sand	5 C.Y.		115.50
Backfill sand, by hand, light soil	5 C.Y.	2.9	
Forms in place, columns, round fiber tube, 8" diameter	60 L.F.	12.4	108.72
Concrete, redi-mix, 2000 psi	3.25 C.Y.		302.25
Footings, spread type, placing of concrete	3.25 C.Y.	2.8	
Posts, 4 x 4 x 10', treated	140 L.F.	5.0	226.80
Rim joists, 2 x 8 x 12', treated	24 L.F.	0.9	28.80
Deck joists and beam supports, 2 x 6 x 12', treated	36 L.F.	1.3	43.20
Decking, 2 x 6 x 8', treated	120 L.F.	4.3	144.00
Beams, 2 x 8 x 10', treated	20 L.F.	0.7	24.00
Slide rails, 2 x 6 x 10', treated	40 L.F.	1.4	48.00
Ladder rungs, slide supports, 2 x 4, treated	20 L.F.	0.6	11.76
Joist hangers, galvanized, 2 x 6 to 2 x 10 joists	8 Ea.	0.4	6.43
Bolts, square head, nut and washers incl., 3/8" x 4", galvanized	75 Ea.	5.0	20.70
Nails, 12d common, galvanized	5 Lb.		8.34
3/4" galvanized steel pipe	90 L.F.	11.8	176.04
3/4" galvanized elbows	4 Ea.	0.6	91.68
3/4" plywood, 4' x 8' sheets	32 S.F.	0.4	36.10
0.018 stainless steel, 2' x 10' sheets	20 S.F.	1.0	78.00
Tires, bias-ply, scrap	4 Ea.	4.0	11.38
Swings, manufactured	1 Ea.	1.5	88.20
Trapeze with rings	1 Ea.	1.8	55.20
Tire swing, manufactured	1 Ea.	1.0	2.84
Totals		70.0	$1,627.94

Contractor's Fee, Including Materials: $6,028

Redwood Hot Tub

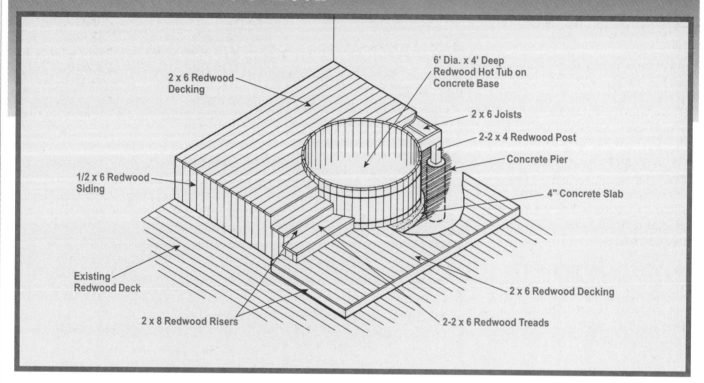

2 x 6 Redwood Decking

6' Dia. x 4' Deep Redwood Hot Tub on Concrete Base

2 x 6 Joists

2-2 x 4 Redwood Post

Concrete Pier

1/2 x 6 Redwood Siding

4" Concrete Slab

Existing Redwood Deck

2 x 6 Redwood Decking

2 x 8 Redwood Risers

2-2 x 6 Redwood Treads

Hot tubs have proven to be more than just a fad, as baby-boomers join athletes and others in appreciating the benefits for aging muscles and joints. Jetted hot water relieves physical and mental stress, stimulates blood flow and relaxes muscles, and soothes the pain of chronic arthritis, muscle pains, and neuralgia.

Hot tubs can be installed in interior or exterior settings and often complement another facility, such as a deluxe bath, exercise room, sauna, sun deck, or swimming pool. This plan involves installation of a 6' factory-prepared redwood hot tub on an existing redwood deck. The project estimate covers the basic materials, procedures, and considerations for a medium-sized exterior installation.

Materials

The materials list starts with redwood support members and decking. These are required to alter and prepare the existing deck to receive the hot tub. The remaining components are the pre-cut and pre-finished 6' hot tub kit and

incidental finish products. The framing and decking lumber are available from most lumberyards, as are some of the finish materials. (Redwood may have to be special-ordered in some areas of the country.)

The tub kit can be purchased from a specialty retailer, but some general building suppliers also carry hot tubs and spas on a special-order basis. Because the redwood deck materials and hot tub are expensive products, you should thoroughly research the options before finalizing the elements in your project.

The 6' hot tub suggested in this plan weighs well over three tons when filled and in use. Consequently, a considerable amount of support must be added to the existing deck if it is to bear the load. Each installation will present a different set of conditions, and the requirements for reinforcing your deck's support system may vary from what is recommended in this plan. Consult local building codes, follow the manufacturer's guidelines, and get a professional opinion before you

determine the arrangement and size of the deck joists and supports. In some cases, extra joists and bridging may be enough; in others, additional footings, posts, and supports may be required.

Most manufacturers of hot tubs recommend an independent ground support system for the unit, separate from the deck. When possible, this method of installation should be employed, even though it may involve considerable extra work and expense. When placing the kneewall and partial perimeter deck structure around the hot tub, be careful not to drive the fasteners into the side of the unit. Also, leave a small space for expansion and drainage between the tub and the deck frame. Use quality hot-dipped galvanized nails and lag screws for the decking and supports to prevent corrosion and rusting.

Many different sizes, shapes, and material options are available for hot tub kits. This plan features standard-grade bottom pieces and redwood staves collared with adjustable stainless steel hoops. The kit is precision cut and fitted at the factory, and delivered unassembled to the consumer. Also included in the price of the kit are the basic plumbing and electrical support packages for the operation of the facility.

In addition to redwood, hot tubs are available in teak, cedar, mahogany, and cypress. Coated fiberglass, stainless steel, and other materials are also used. Hot tubs come in different sizes and are designed as freestanding, exposed units or as inserts for redwood or other raised perimeter decks. Still others can be recessed into the raised section of an interior floor. Alternative sizes, designs, and settings will, of course, significantly affect the cost.

What's Involved

Many specialty retailers will provide, at extra charge, an on-site assembly service for the tub itself. Intermediate and novice do-it-yourselfers might consider this service, as the hot tub is fairly complicated and must be accurately fitted together so that it will be sturdy and watertight. The cost covers only the basic tub and its attached components, however, and does not include plumbing and electrical work.

After the support system, perimeter decking, and hot tub are in place, the installation can be left as is or finished with stain or sealer. If left untreated, the redwood will eventually weather to a tanned leather color, but it will not rot or deteriorate. However, because of the dampness, a sealer with a mildewcide additive is recommended for the exposed exterior surfaces of the tub and the surrounding deck. If the deck and tub are made from different materials, one or both may be stained to blend the new and old installations.

In most cases, a qualified plumber and electrician should be hired to tie in the support packages to the house utilities. Because the water pump, heater, filters, and other tub support facilities are often remotely located, extra expense may be incurred for additional piping and underground placement. Increased cost is to be expected for convenience options such as air switches, therapy

jets, chemical feeders, timers, lighting, and solar heat packages. Other extras include shelves, steps, and ladders.

Level of Difficulty

This project calls for some expensive materials and precise installation requirements, making it a challenge for all do-it-yourselfers. Beginners and most intermediates might want to stick to the more basic tasks, like the preliminary site preparation and the surface finishing. Experts who opt to assemble the kit themselves should follow the manufacturer's instructions precisely and consult a professional before tackling the operation. Experts should add 25% to the labor-hours for all carpentry work. Beginners and intermediates should add 100% and 50%, respectively, to the time estimate for the basic tasks that they attempt. All do-it-yourselfers should hire qualified tradespeople to accomplish the electrical and plumbing work.

What to Watch Out For

Make provisions in the planning stages of this project for proper drainage piping and protection against the effects of freezing temperatures. You should be able to completely drain the water during periods when the facility is not in use. In cold climates where the tub is to be used year-round, an optional freeze protection package can be purchased at extra cost. Floating or attached covers made of wood and plastic will reduce heat loss and protect the tub. If you plan to use the tub in cooler weather, you might consider installing a wind screen. All of these components will increase the cost of the project, but will also extend the time you can use it.

See also:
Elevated Deck, Deluxe Master Bath*.
*In *Interior Home Improvement Costs*

Redwood Hot Tub, 12' x 14'

Description	Quantity/ Unit	Labor- Hours	Material
Remove portion of existing deck, per 5 S.F.	6 Ea.	9.6	
Reinforce existing deck, 2 x 8, redwood stock	24 L.F.	0.9	28.80
Base support slab, hand grading	5.56 S.Y.	0.2	
Forms in place, edge forms for slab on grade, 6" high	24 L.F.	1.3	7.78
Concrete, ready mix, 3000 psi	1 C.Y.		97.20
Placing concrete, slab on grade, 4" thick	1 C.Y.	0.4	
Finishing concrete, broom finish	1 C.Y.	0.1	
Platform framing, redwood, posts, doubled, 2 x 4 stock	48 L.F.	1.4	28.22
Joists, 2 x 6 stock	130 L.F.	4.2	120.12
Decking, 2 x 6 x 12'	252 L.F.	8.1	232.85
Stair stringers, 2 x 8 stock	12 L.F.	0.4	14.40
Stair treads, 2 x 6 stock	18 L.F.	0.6	16.63
Stair risers, 2 x 8 stock	14 L.F.	0.5	16.80
Wood siding boards, vertical grain, clear beveled 1/2" x 6"	48 L.F.	1.5	122.11
Redwood hot tub kit, 6'dia., 4' deep w/circ. pump & drain system	1 Ea.	20.0	3,090.00
Plumbing, rough-in and supply	1 Ea.	7.7	171.60
Electrical rough-in, hook-up	1 Ea.	1.6	102.00
Totals		58.5	$4,048.51

Contractor's Fee, Including Materials:	**$8,934**

Porches and Entryways

Adding or enhancing a new entryway or porch, especially on the front of your house, has a dramatic effect on its appearance. In fact, porches and new entryways are the top elements used by architects to enhance a home's curb appeal.

Since these can be major renovations, a lot of thought should go into the planning and design. An architect can help ensure that the new construction complements and upgrades the look of your existing house. If you're also working with a landscape designer or contractor to improve your front yard, you'll want to coordinate his or her plan with the design of the new porch or entryway. Following are a few more tips for planning your project.

- Entry doors are made of various types of wood, steel, or fiberglass. Talk to your supplier about the kind of door that will be most appropriate for your situation. Steel doors are energy-efficient, durable, and offer the best security. Follow the manufacturer's painting instructions.
- A storm door prolongs the life of your entry door by protecting it from the elements. When choosing a storm door, look for models that feature a solid inner core, a low-maintenance finish, and a seamless outer shell.
- A storm door takes more abuse than the door it protects. Properly installing and securing it may require use of larger (e.g., 3") screws to ensure that the door is secured through the trim into the jamb.
- Check the recommendations and restrictions for your entry door manufacturer before buying a storm door.

- The size and angle of porch roof overhangs are based on the existing house roof and design, the angle of the sun, and the local climate. In very hot, sunny climates, a wide porch overhang can significantly reduce air conditioning use, and energy costs.
- For porch railings, trim, and lattice (to cover the gap between porch and ground), consider vinyl, molded polymer, and recycled plastic composite products for durability, low maintenance, and decay and insect resistance.
- Even if your porch is not enclosed, consider GFI outlets and wiring for lamps, ceiling fans, and overhead lights—for both outdoor living and security.
- Before beginning a project to enclose a breezeway or porch, inspect the condition of the existing roof and support system. Get an expert evaluation if you aren't knowledgeable in this area.
- In colder climates, insulate the floor of a converted room for comfort and energy savings.
- Screen panel options include conventional wood-framed panels, aluminum-framed panels, fiberglass screening, copper screening, and "invisible."
- A wheelchair ramp must meet specifications for the correct pitch, width, etc., and it must be well-constructed for proper and safe function. But you also want it to be attractive, as it will be a very visible feature of your home. Give it the kinds of considerations you would a new porch, especially if the ramp is planned for long-term use. Check the railing and trim options, and tie the ramp into the design of your home.

Finished Breezeway with Screens

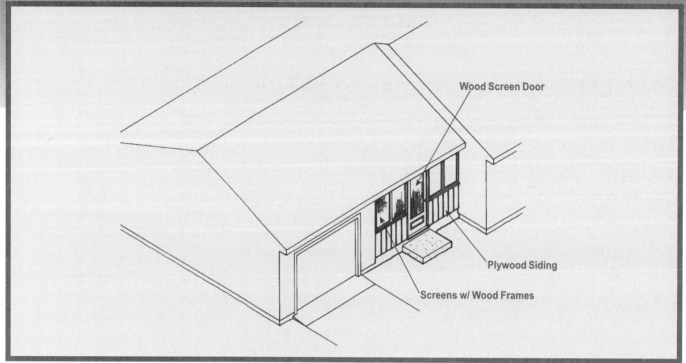

Wood Screen Door

Plywood Siding

Screens w/ Wood Frames

If you have an existing breezeway connecting your house and garage, you can gain new outdoor living area by enclosing it with screened walls. Since the roof, side walls, and floor are already in place, the cost is considerably less than building a new porch.

Because breezeways are generally at the first-floor level, the project site is easily accessible and can usually be framed and screened with few logistical problems. Like deck projects, breezeway conversions normally cause little or no disruption to the home's routine and, therefore, can be completed at a slower pace. For these reasons, this project provides an excellent opportunity to cut costs by completing most of the work on your own.

Materials & Tools

The materials needed to complete this project include standard 2 x 4s to frame the knee wall and door opening; 2 x 6s

for the ceiling joists (if needed); appropriate coverings for the knee wall, ceiling, and floor; and the screen panels and door. All of these materials are readily available from building supply outlets, so it pays to compare prices from several different retailers.

The screen panels are specialty items that may have to be custom-fitted to the opening created by the knee wall frame. Plan to include delivery as part of the purchase order for all of the materials, as the lumber needed to frame the enclosure, the knee wall, and ceiling coverings is cumbersome and will have to be trucked.

Pre-fabricated panels are included in this estimate, but you could also make your own if you have the skill and the right tools. If you intend to build the panels on your own, develop a master plan for all of the screening to determine the most cost-effective width for the individual panels, and then base the order for the screening and wood frame on that factor. Fiberglass and

aluminum screening come in different grades and costs.

This project requires standard hand and power tools, such as a hammer, drill, tape measure, and level, along with a circular saw, and a carpenter's square.

What's Involved

Precise framing is required for the knee wall, as it serves as the primary support for the screen panels. If it's not level and plumb, the panels won't fit correctly, and there will be spaces along the abutting edges. If the existing wood or concrete floor of the breezeway is uneven or slanted, shim or angle the bottom of the kneewall frame to attain a level surface on the top.

After the kneewall is in place and the doorway has been framed, rough-sawn plywood siding is applied to the exterior. The ceiling is prepared with 2 x 6 joists, spaced 24" on center, with 3/8" rough-sawn fir plywood for the ceiling surface.

Planning, building, and installing the screen panels requires special care. Tension of the screening is important for both the appearance and longevity of the panels. Get some instruction from a knowledgeable person on the ins and outs of screen panel manufacture and installation.

Consider installing the screen panels in such a way that they can be conveniently removed for periodic cleaning, painting, and storage during the off-season. If you hire a contractor to build the panels, it might be a good idea to have him or her complete the job by installing them.

Level of Difficulty

This project is a manageable undertaking for most do-it-yourselfers who have a basic knowledge of carpentry and experience in the use of tools. Do-it-yourselfers who are inexperienced in framing procedures should get some assistance for the knee wall construction, as its placement must be correct for the rest of the job to go smoothly.

As noted earlier, planning, building, and installing screen panels should be left to professionals or expert do-it-yourselfers. Hanging the screen door also requires advanced carpentry skills.

Generally, beginners should not attempt the frame, joist, door, carpeting, and screening tasks. They should double the professional time estimates for all other jobs and should seek professional help on unfamiliar tasks.

Intermediate-level do-it-yourselfers should stop short of the screen panel fabrication and carpeting, but should be able to complete the rest of the project, adding about 40% to the professional time. Experts should allow an extra 10% to the time for all tasks, and more for the screen panels.

What to Watch Out For

Include the additional expense for any electrical work, such as adding or replacing a ceiling light fixture, door lamp, and watertight duplex outlets. If you plan to use the space at night, be sure adequate lighting is planned. If you wait until the new materials are in place, the labor cost for electrical work will be higher than if it's done beforehand.

The details of the wood-framed screen panel installation will vary according to the desired finish appearance. In a basic installation, the panels can be secured at the top and bottom to the interior or exterior face of the new wall. For a more finished look, the panels can be secured with stops made from trim material, such as quarter-round molding or 1 x 1 pine strips.

Brass or galvanized wood screws can be used as fasteners to allow for easy removal of the panels for maintenance or storage during the off-season. This type of installation will require additional materials and more time to accomplish, but will provide a lot of convenience.

Options

Prefabricated aluminum screen panels can be used in place of the suggested wood-framed units, but may have to be custom-made at extra cost.

Rough-sawn 3/8" fir plywood is included for the ceiling surface, but other materials can be used. The price of indoor/outdoor carpeting is included in the model project, but other floor materials, such as weather-resistant tiling and exterior floor paint, can also be used. The same is true for the fir plywood used for the knee wall in this project.

See also:
Finished Breezeway with Windows, New Breezeway.

Finished Breezeway with Screens, 24' x 12'

Description	Quantity/Unit		Labor-Hours	Material
Framing for knee wall, plates 2 x 4 stock	96	L.F.	1.9	39.17
Framing for knee wall, studs, 2 x 4 x 3'	60	L.F.	1.3	18.72
Framing for doorways, 2 x 4 x 8'	72	L.F.	1.0	29.38
Ceiling joists, 2 x 6 x 24', 24" O.C.	240	L.F.	3.1	155.52
Ceiling, rough-sawn cedar, plywood, 3/8" thick, pre-stained	288	S.F.	6.8	414.72
Siding, knee wall, 2 sides 5/8" rough-sawn cedar, pre-stained	64	S.F.	1.5	92.16
Wood screens, 1-1/8" frames	200	S.F.	8.5	1,428.00
Screen door, wood, 2'-8" x 6'-9"	2	Ea.	2.7	278.40
Paint, inside wall and trim, primer	200	S.F.	1.4	12.00
Paint, inside wall and trim, 2 coats	200	S.F.	2.4	21.60
Indoor/outdoor carpet, 3/8" thick	32	S.Y.	25.6	1,251.07
Totals			56.2	$3,740.74

Contractor's Fee, Including Materials: **$8,412**

Finished Breezeway with Windows

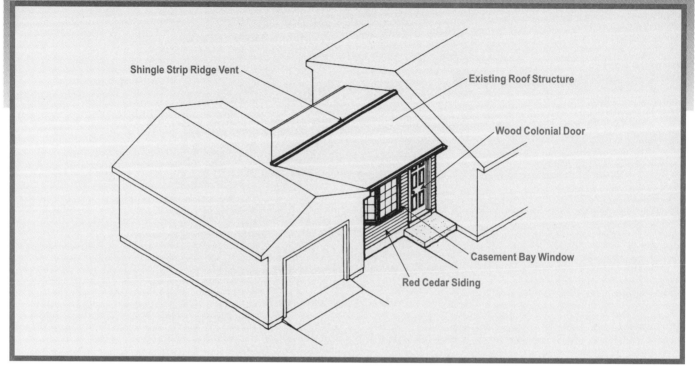

Shingle Strip Ridge Vent

Existing Roof Structure

Wood Colonial Door

Casement Bay Window

Red Cedar Siding

Converting a breezeway into interior living space takes the previous project one step further, giving you year-round use of the area. Linking your house and garage with a finished room results in more interior square footage (and therefore added value to your home). Having two walls, a roof, a floor, and a support structure already in place keeps the cost in line, especially if you're able to do some of the work yourself.

Materials & Tools

This project starts with framing and sheathing lumber for the two open walls, strapping, insulation, a door, windows, and house-wrap. The door and windows in this model plan are one option, which can be altered according to the style of your home. The three-lite bay window, for example, could be replaced with two double-hung windows. A quality window is generally worth the extra investment as it will pay for itself over years of energy efficiency, easy operation, and low maintenance.

The exterior door selected for this project is a traditional, two-lite Colonial style, but other options are available to give your project a custom look. *(See the Main Entry Door project.)*

The interior materials include 2 x 6 ceiling joists (if none are in place), gypsum wallboard for the ceiling and walls, and carpeting for the floor. You could use other wall and floor materials, based on personal preference and how you will use the room. Tile is a popular choice, especially if the new space is going to double as a mudroom. Trim materials to match your home, paint, and a minimal electrical installation complete the list of supplies for the interior work.

Standard hand and power tools, a dust mask, eye protection, gloves, and painting tools are also needed. For more specific information on the materials needed for the window, see the window projects earlier in this book.

What's Involved

This project requires framing the two open sides of the breezeway, installing a door and windows, and finish work, both exterior (for the two new walls) and interior (for all four walls). Because breezeways are on the first floor with easy access from the yard and garage, there shouldn't be any problems with materials storage and transport.

The first step is framing the walls of the breezeway with 2 x 4s set 16" on center with a single bottom and double top plate. The door and window rough openings should be built into the partition using jack studs and 2 x 8 headers. With the framing in place, 1/2" plywood sheathing can be nailed in place.

Before you attempt to install the door and windows, line up a helper, because these items are heavy, awkward to manipulate, and expensive to repair if they're dropped or mishandled. When the door and windows have been installed, the siding can be applied. Try to match closely the siding on the new walls to that of the rest of the house by using the same material, following existing course lines, and painting and finishing it in the same style and colors. Additional cost can be anticipated for matching special types of siding.

If portions of the exterior house and garage walls are included in the new room, they must be stripped of their exterior siding, and the walls covered with gypsum wallboard or other suitable interior wall material. The garage wall may also have to be insulated, and the garage doorway dressed up or relocated. All of these operations will add cost and time to the job.

When the breezeway has been closed in, it's time for the interior finish work. Before the wall and ceiling coverings are applied, the rough electrical wiring and boxes should be installed. This plan calls for basic electrical service to four wall outlets and a switch box. If you plan to install an overhead light fixture or a door lamp, the cost will increase.

Hire an electrician to do the work if you're not an expert in electrical installations. (Do-it-yourselfers should not attempt to tie in the new wiring to the panel, even if they're experienced enough to do all other parts of the electrical installation.)

Level of Difficulty

Experts, as well as intermediates with a command of basic carpentry and remodeling skills, should be able to complete most of this project. The site is convenient, the extent of remodeling fairly limited, and the tasks routine, with the exception of the electrical and carpeting installations. Specialty contractors should be hired to perform this work in most cases. Even experts should weigh the savings of doing these jobs on their own against the benefits of fast, high-quality, professional work.

For all tasks, except the electrical and flooring work, intermediates should add 40% to the professional time, and experts, 10%. Beginners should hire a contractor to do all of the exterior work, including the window and door installations. With some instruction and guidance, they can perform some of the insulation, wallboard, painting, and trim work. They should leave the placement of the ceiling joists, electrical installation, and carpeting to professionals. Beginners should add 80% to the professional time estimate for any interior jobs they can take on.

What to Watch Out For

Although this project is an economical way to gain interior living space, it does have the potential for problems that could add to its cost and increase the installation time. For example, the doorway into the house will require attention, because it should be converted into an interior opening and trimmed with the same materials used in the new room. If the doorway is one step above the new floor level, a suitable interior step should be built. The old door from this location can be used for the new exterior installation as a cost-cutting measure.

See also:
Window Projects, Finished Breezeway with Screens, New Breezeway, Main Entry Door, Skylights.

Finished Breezeway with Windows, 12' x 12'

Description	Quantity/Unit		Labor-Hours	Material
Wall framing: studs, 2 x 4, 16" O.C., 8' long	240	L.F.	7.8	339.84
Plates, single bottom, double top, 2 x 4	80	L.F.	1.6	32.64
Headers, over windows & door, 2 x 8	30	L.F.	1.4	29.88
Sheathing, plywood, 1/2" thick, 4' x 8' sheets	192	S.F.	2.7	142.85
Housewrap, spun bonded polypropylene	240	S.F.	0.5	46.08
Ceiling joists, 2 x 6, 16" O.C., 12' long	12	Ea.	0.2	7.78
Ceiling furring, 1 x 3 strapping, 16" O.C.	150	L.F.	3.4	50.40
Insulation, ceiling, 6" thick, R-19, paper-backed	150	S.F.	1.2	63.00
Insulation, walls, 3-1/2" thick, R-11, paper-backed	200	S.F.	1.4	64.80
Gypsum wallboard, walls & ceilings, 1/2" thk., taped & finished	576	S.F.	9.6	172.80
Door, exterior, 2-lite, colonial, 2'-8" x 6'-8"	1	Ea.	1.0	534.00
Window, 3-lite bay, casement type, vinyl-clad, premium	1	Ea.	2.6	1,920.00
Window, 2-lite casement window, plastic-clad, premium	2	Ea.	1.8	667.79
Siding, #1 red cedar shingles, 7-1/2" exposure	2	Sq.	7.8	259.20
Ridge vent, molded polyethylene	12	L.F.	0.6	36.72
Trim, colonial casing, windows, door, and base	120	L.F.	4.5	174.24
Paint, walls, ceiling, and trim, primer	575	S.F.	4.0	34.50
Paint, walls, ceiling, and trim, 2 coats	575	S.F.	6.8	62.10
Carpet, nylon anti-static	16	S.Y.	2.6	416.45
Pad, sponge rubber	16	S.Y.	0.9	60.48
Electrical work, 1 switch	1	Ea.	1.4	34.80
Electrical work, 4 plugs	4	Ea.	6.0	129.60
Totals			69.8	$5,279.95

Contractor's Fee, Including Materials:	**$11,482**

New Breezeway

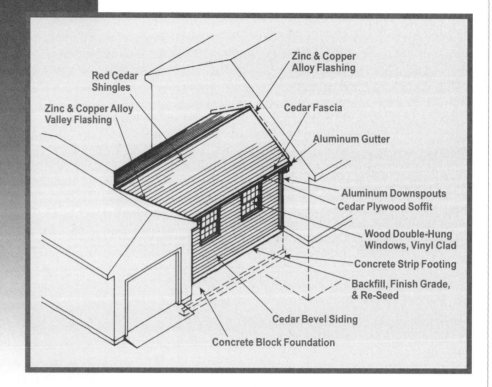

Red Cedar Shingles

Zinc & Copper Alloy Valley Flashing

Zinc & Copper Alloy Flashing

Cedar Fascia

Aluminum Gutter

Aluminum Downspouts
Cedar Plywood Soffit

Wood Double-Hung Windows, Vinyl Clad

Concrete Strip Footing

Backfill, Finish Grade, & Re-Seed

Cedar Bevel Siding

Concrete Block Foundation

This project provides new interior living space by means of an addition between a house and a detached garage. It's a challenging project for do-it-yourselfers and involves a variety of construction skills.

Materials & Tools

The exterior components of this breezeway are essentially the same as those used to construct the shell of a complete house. A foundation will support the load of the new structure. Framing, wall, and roof materials will complete the shell and make it weathertight.

Materials for finishing the interior have not been listed in the plan, but you can get estimates of their cost and installation time from the finished breezeway plans in this section.

The foundation materials consist of a concrete footing and concrete block wall. The structure above requires 2 x 8 joists, plywood decking for the floor, a vapor barrier, insulation, sheathing materials, 2 x 6 ceiling joists, and roofing materials.

This plan suggests a high-quality vinyl-clad sliding glass door unit and two

matching thermopane windows. Costs will vary considerably with the choice of alternative door and window products.

You'll need standard hand and power tools for this project, including a hammer, level, saws, and drill, as well as a shovel and a sidewall stripper.

What's Involved

Before the footing and wall are placed, the site needs to be excavated. Because of the close quarters between the house and garage, much of the digging may have to be done by hand to the necessary 4' depth, so plan accordingly.

After the concrete block foundation wall is in place, install 2 x 8 joists and plywood decking to frame the rough floor of the new room. The foundation and floor must be correctly installed to ensure a firm and level starting point for the rest of the project. Before placing the joists, cover the area with a vapor barrier, then insulate between the joists before laying the sub-floor.

Before framing the new walls, allow time to prepare the adjacent walls of the house and garage to receive the new materials. The siding must be stripped

and, in some cases, a door or window may have to be removed or relocated.

The new partitions should be carefully laid out, since they must be erected to a predetermined height and length as established by the existing garage and house. This plan suggests 25/32" wood fiberboard for sheathing, but 5/8" CDX plywood can also be used at a slight increase in cost. Make sure the wall framing is even with the face of the concrete to allow the siding to extend over the lip of the foundation for proper weather protection.

When the rough walls have been erected, the 2 x 6 ceiling joists should be installed to keep the new walls in place before the roof is framed. Placing the rafters is the most challenging part of the project. Advanced framing skills and know-how are required, so get help if you haven't done roof framing before.

Check the condition of the garage shingles, support, and sheathing before you begin framing the new roof. Additional cost and time will be required if you're careless in stripping the old roofing.

After 1/2" plywood has been placed on the rafters, the roofing material is installed to make the new structure weathertight. The cost for the roofing material and the time to install it vary with different products. Because roofing is a specialty trade, even intermediates and experts might consider hiring a professional who can do the work faster and better.

After the roof and soffit are in place, the door and windows can be installed, and the exterior walls sided. Like the roofing, the siding material (and paint or stain) for the new addition should match or complement the material on the house and garage. Gutters and downspouts are an integral part of the weatherproofing so don't cut corners by leaving them out.

Level of Difficulty

All of the basic operations in exterior house construction are required from foundation work and framing to exterior finish work like siding and gutter installations. Inexperienced do-it-yourselfers should consider hiring a professional to install the foundation.

Beginners should add 100% to the professional time for the basic tasks, such as excavation, sheathing, painting, and landscaping. Intermediates should add 40% to the time for the basic tasks and wall framing, and 60% for roof framing. Experts should add 10% to the professional time for all tasks, except the

foundation work and roofing. Adjust the time increase based on the amount of experience you have in these areas.

What to Watch Out For

Hiring specialty contractors often makes sense even if you have experience and skill enough to complete the job on your own. The roofing for this project is a good example. A capable roofer can close in this structure quickly, even though there is a considerable amount of flashing involved. Correct installation of the valley is important to prevent leaking.

See also:
Bow/Bay Window, Finished Breezeway with Screens, Finished Breezeway with Windows, Skylights, New Window.

New Breezeway, 10' x 16'

Description	Quantity/ Unit	Labor-Hours	Material
Excavate by hand for footing, 4' wide, 4' deep	19 C.Y.	19.0	
Footings, 8" thick x 16" wide x 32' long	2 C.Y.	5.6	242.40
Foundation, 8" concrete block, 3'-4" high	160 S.F.	14.1	320.64
Vapor barrier, 6 mil polyethylene	1.60 Sq.	0.3	5.28
Floor framing, sill plate, 2 x 6	32 L.F.	0.9	20.74
Joists, 2 x 8 x 10', 16" O.C.	160 L.F.	2.3	159.36
Ceiling and floor insulation, 6" thick, R-19	320 S.F.	1.9	172.80
Subfloor, 5/8" x 4' x 8' plywood	160 S.F.	1.9	149.76
Wall insulation, fiberglass, kraft faced, 3-1/2" thick, R-11	256 S.F.	1.3	82.94
Wall framing, 2 x 4 x 8', 16" O.C.	32 L.F.	5.1	117.12
Headers over openings, 2 x 8	48 L.F.	2.3	47.81
Sheathing for walls, wood fiberboard 25/32" thick, 4' x 8' sheets	256 S.F.	3.4	86.02
Housewrap, spun bonded polypropylene	320 S.F.	0.7	61.44
Roof framing, joists, 2 x 6 x 10', 16" O.C.	130 L.F.	1.7	84.24
Ridge board, 1 x 8 x 12'	24 L.F.	0.7	35.14
Rafters, 2 x 6 x 12', 16" O.C.	312 L.F.	5.0	202.18
Valley rafters, 2 x 6	32 L.F.	0.7	20.74
Sub-fascia, 2 x 8 x 16'	32 L.F.	2.3	31.87
Sheathing for roof, 1/2" thick, 4' x 8' sheets	240 S.F.	2.7	178.56
Felt paper, 15 lb.	240 S.F.	0.3	9.82
Shingles, #1 red cedar, 5-1/2" exposure, on roof	3 Sq.	9.6	586.80
Ridge vent, polyethylene	16 L.F.	0.8	48.96
Flashing, zinc & copper alloy, 0.020" thick	140 S.F.	7.2	547.68
Fascia, 1 x 8 rough-sawn cedar	32 L.F.	1.1	39.55
Soffit, 3/8" rough-sawn cedar plywood, stained	48 S.F.	1.1	69.12
Vented drip edge, aluminum	32 L.F.	0.6	59.14
Gutters & downspouts, aluminum	52 L.F.	3.5	77.38
Windows, wood, vinyl-clad, insulated 3' x 4'	2 Ea.	1.8	780.00
Sliding glass door, wood, vinyl-clad, 6' x 6'-10"	1 Ea.	4.0	1,380.00
Siding, cedar-beveled, rough-sawn, stained	220 S.F.	7.3	704.88
Backfill, finish grade and re-seed	1 C.Y.	0.6	
Totals		109.8	$6,322.37

Contractor's Fee, Including Materials: $14,925

Enclosed Porch with Screens

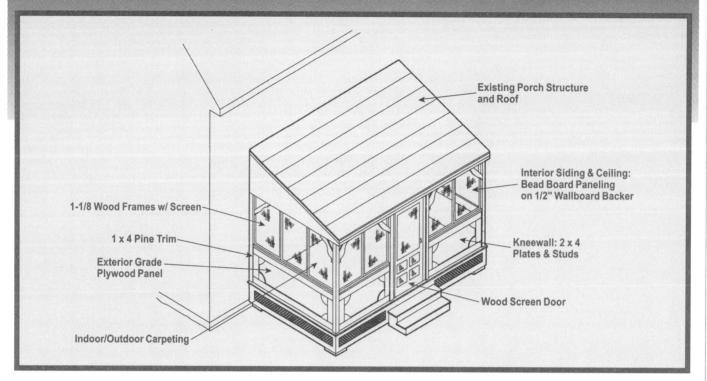

Existing Porch Structure and Roof

Interior Siding & Ceiling: Bead Board Paneling on 1/2" Wallboard Backer

1-1/8 Wood Frames w/ Screen

1 x 4 Pine Trim

Exterior Grade Plywood Panel

Kneewall: 2 x 4 Plates & Studs

Wood Screen Door

Indoor/Outdoor Carpeting

A screened-in porch can provide years of enjoyment for you and your family, and greatly increase your home's living and entertaining space. If you have a porch with a roof and support structure already in place, screening it in can be done at a reasonable cost by most do-it-yourselfers.

Materials & Tools

As in other porch and breezeway conversions, the amount and type of materials are determined by the fixed dimensions of the structure and the style of your house. If your dimensions differ from this model, the cost will have to be adjusted. All of the products are conventional construction materials, such as 2 x 4s, sheathing materials, 2 x 6s, and interior finish products that can be purchased at competitive prices from most building supply outlets.

The screen panels for the new porch are the most important components. They deserve special consideration before you decide on their design and the best installation method. Beginners might

consider hiring a carpenter to manufacture the panels, but they are also available in standard factory-prepared models with both wood and aluminum frames. Special ordering may be necessary. If you intend to make the panels on your own, be sure to read up on the techniques involved and consult a knowledgeable person for tips.

In most cases, removable wood-framed panels are best. They may cost more and take longer to fabricate than permanently installed ones, but they have many advantages. They can be removed and stored when the porch is not in use and can be cleaned and painted easily on the ground and then put in place.

Indoor/outdoor carpeting has been chosen for the flooring in this project, but you could opt for tile or even wood. Along with trim and paint, flooring can create a custom look. Basic framing and carpentry tools are also required, most of which can be found in a typical tool box.

What's Involved

Before the new materials are installed, the condition of the existing porch should be thoroughly evaluated. If the roof or floor is in need of repair, it should be fixed, which may add to the costs incurred. Any extra work, like installing a gutter and downspout, finishing the fascia and soffit, or building a new set of stairs, should also be completed at this stage.

The frame for the knee wall of the porch is constructed between the existing roof support posts. Unless there is good reason to move them, all support posts should be kept in position to maintain the structural integrity of the roof support system. In situations where a post has to be relocated or replaced, temporary bracing must be provided, and extra cost and time added. Inexperienced do-it-yourselfers should consult a contractor if extensive post relocation or replacement is required.

Once the framing has been completed, the exterior walls are covered with a single layer of sheet siding, or sheathed and then covered with appropriate cedar, vinyl, or aluminum exterior finish.

Inside the enclosure, the 2 x 6 ceiling joists are put into place, followed by the finish coverings for the ceiling, walls, and floor. The moderate finish work suggested in this plan provides covering for the ceiling, knee wall, and floor, but not the interior wall.

Level of Difficulty

The carpentry skills required are within the capabilities of most do-it-yourselfers. The heavy exterior work is already done, and the filling in of the shell is all that is required. Professional help may be needed if extensive reconditioning is necessary for the roof, floor, or support system, but minor problems can be corrected by most do-it-yourselfers.

Beginners should consider hiring a contractor or supplier who specializes in screen fabrication to manufacture and install the screen panels, as these tasks require precision carpentry skills. For other tasks they should add 100% to the professional time estimates. Intermediates and experts who tackle the screen panel manufacture should add extra time to their normal work rate to allow for the precise nature of the operation. For all other procedures, they should add 50% and 20%, respectively.

What to Watch Out For

Before you decide on the screen panels, investigate the options. Conventional wood-framed panels are durable, but require maintenance and are prone to bowing and warping if poorly made or installed. Aluminum-framed panels may not be as attractive as wood-framed units, but will last longer, and periodic cleaning is the only required maintenance.

Aluminum screen material is also long-lasting and capable of withstanding abuse, but may discolor with corrosion

after a few years. Fiberglass screening tends to stretch and bow more easily than aluminum, but it retains its original appearance longer. Copper screening performs and looks better than both aluminum and fiberglass, but is very expensive.

Options

Many options are available for the type and quality of materials used in the porch's interior, depending on the degree of finish work desired. If you want a rustic look, the ceiling can be left open to the rafters, the inside of the exterior walls left uncovered, and the existing floor repainted or left as is. The cost of the project will be reduced significantly in this case.

A decorative cap with trim for the kneewall is an extra-cost option. With some ingenuity, the trim for the cap can be designed to serve as the bottom stop for the screen panels.

Sheet goods, like rough-sawn cedar plywood, install quickly and provide an attractive, low-maintenance finish. Barnboard, tongue-and-groove, beadboard, and plank coverings are other options—usually more expensive and requiring a bit more time to install.

Electrical work is not included in this plan, but you might consider installing outdoor-approved outlets and an overhead or wall lighting fixture.

See also:
Enclosed Porch with Glass Doors, Enclosed Porch with Windows, Entryway Steps, Porch.

Enclosed Porch with Screens, 8' x 14'

Description	Quantity/ Unit		Labor-Hours	Material
Framing for knee wall, 2 x 4 plates	96	L.F.	2.4	39.17
Studs, 2 x 4 x 3', 16" O.C.	84	L.F.	1.2	34.27
Framing, for doorway, 2 x 4 x 8'	32	L.F.	1.0	13.06
Ceiling joists, 2 x 6 x 8', 16" O.C.	88	L.F.	1.1	57.02
Ceiling & kneewall stiffener, gypsum board, 3/8"	224	S.F.	1.8	53.76
Ceiling and knee walls, bead board paneling, 1/4", prefinished	224	S.F.	8.5	325.25
Siding, exterior, for knee wall, plywood panels, 7/16", smooth	72	S.F.	1.6	63.07
Wood screens, 1-1/8" frames, pre-fabricated	140	S.F.	6.0	999.60
Screen door, wood, 2'-8" x 6'-9"	1	Ea.	1.8	313.20
Trim, 1 x 4 pine	72	L.F.	2.9	58.75
Indoor/outdoor carpet, 3/8" thick	112	S.F.	2.0	323.90
Paint, screens, primer and 2 coats, oil base, brushwork	10	Ea.	13.3	29.52
Panels, 2 coats, oil base, brushwork	72	S.F.	0.9	9.50
Door, primer coat, oil base, brushwork	1	Ea.	0.8	2.54
Door, 2 coats, oil base, brushwork	4	Ea.	5.3	16.46
Totals			50.6	$2,339.07

Contractor's Fee, Including Materials: **$6,106**

Enclosed Porch with Glass Doors

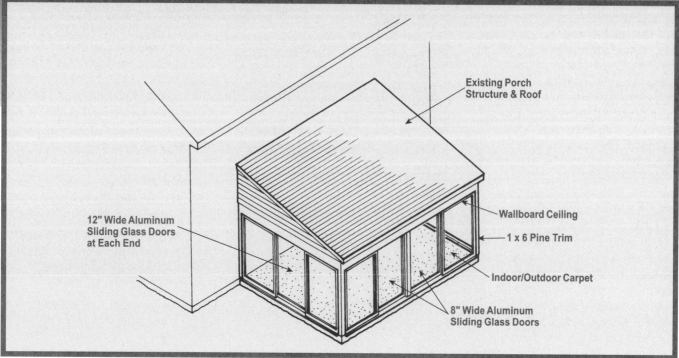

Existing Porch Structure & Roof

12" Wide Aluminum Sliding Glass Doors at Each End

Wallboard Ceiling

1 x 6 Pine Trim

Indoor/Outdoor Carpet

8" Wide Aluminum Sliding Glass Doors

Open, attached porches provide an economical opportunity to increase your living space, with a glassed-in, finished sunroom. Generally, this is a moderate-to-difficult project because of its size and the specialized skills required for several of the tasks. Even if you hire a professional for most of the work though, the cost is reasonable, considering the year-round use you will get from the space.

Materials & Tools

As in other porch plans in this section, it's assumed that a weathertight, sturdy roof and a solid foundation are already in place. The materials, therefore, include items that are required to fill in the needed walls, finish the interior surfaces, and otherwise enhance the porch's function and appearance.

Because of variables in size, materials, and roof style, as well as the condition of your particular porch, the estimate will have to be modified for individual projects. The type and extent of finish work will also affect the price.

The sliding glass doors are primary components of this conversion, so give careful consideration to their selection and installation. The units are available in various standard sizes, including the 6' and 8' models used in this plan. Generally, the extra expense for top-quality units is worthwhile, as they will provide years of reliable, weathertight service. Plan in extra time if the model you want requires special ordering.

The framing lumber, wood trim, gypsum wallboard, insulation, and paint are readily available from building supply retailers. Standard hand and power tools are also required.

What's Involved

The most important aspect of the sliding door installation is the correct sizing and precise leveling, plumbing, and squaring of the rough openings.

Because of their expense, they should be handled carefully, with plenty of time allowed for installation. *(See the Patio Doors project earlier in this book for more information on installation.)*

Filling in and finishing the walls require basic carpentry skills. These tasks can be accomplished by most do-it-yourselfers, with patience guidance from a pro. Work carefully, as even minor mistakes will be very noticeable once the wallboard is put on and the interior is finished. When the walls are complete and the door is in place, the interior and exterior will need to be finished to match your home.

The doorway from the porch area to the house may require some attention after the sliders and interior finish are in place. If the porch area is to become a new room, you may want to enlarge the existing interior doorway, build a rough frame with appropriate jack studs and headers, and then trim the opening. A slider unit or French door can also be installed in this location. If the porch is to be winterized, but will still be considered more as a warm-weather-use area, the old doorway can be left as is with no additional work required. The exterior siding and any existing windows can be left in place or removed, depending on the level of finish treatment desired in the new enclosed porch. Your choice of treatment for this area will have a substantial impact on the cost of the project.

Level of Difficulty

Because of the specialty work involved in the sliding glass door installation, beginners and intermediates with limited remodeling experience should consider hiring a professional to complete this task. Even experienced intermediates and experts should seek some assistance. The components are heavy and expensive, and precise workmanship is required.

The rest of the project should be within the reach of most do-it-yourselfers. Beginners should double the professional time for all work they can do and should seek advice when needed. Intermediates and experts should add 50% and 20%, respectively, to the estimated time for basic procedures, and 60% and 30% for the sliding door installations. All do-it-yourselfers will have to add both time and expense for extra interior finish work if the porch is to be integrated into the first-floor plan of the house.

What to Watch Out For

Before any new materials are installed, the existing structure should be thoroughly inspected and evaluated. The roof should be dry, and its sheathing solid and free from rot or deterioration. Carefully check the flashing along the seam between the roof and the wall of the house, and examine the underside of the roof's subsurface and rafters. If the roof is held up by a post-and-beam support system, check the condition and alignment of the supporting members.

This plan assumes a concrete slab as the floor and foundation for the porch, but some porches have a concrete block or poured-in-place foundation with a wood floor. Regardless of the type, evaluate the foundation's condition at the start. If deficiencies or weaknesses are found, make sure they're corrected before proceeding with the project. Extensive reconditioning may require the services of a roofing or foundation specialist.

If the new glassed-in porch is to be used as an interior room, your local building code may require a minimum number of electrical outlets. Extra lighting fixtures and wall switches may also be desirable. These should be provided for before the ceiling and walls are framed and covered. Space may be required between the rough framing members of the slider units if wiring and switch or outlet boxes are to be installed in this location. Also, supply wires for the wall switches, outlets, and overhead lighting fixtures must be run above the finished ceiling and then fed to the electrical service panel. Unless you're an expert in electrical work, hire a professional.

Options

If you want to further increase the light and ventilation in the room, you might invest a bit more by adding skylights.

See also:

Electrical System Upgrade*, Enclosed Porch projects, Porch, Interior Doors*, Light Fixture Installation/Upgrade*, Patio Doors, Skylights.

*In *Interior Home Improvement Costs*

Enclosed Porch with Glass Doors

Description	Quantity/ Unit	Labor- Hours	Material
Demolition, existing door 3' x 7', and enlarge opening	1 Ea.	0.5	
Door frame demolition, including trim, wood	1 Ea.	0.5	
Frame opening into house, 2 x 4 studs	32 L.F.	1.0	13.06
Header over opening, 2 x 10 stock	20 L.F.	1.0	28.32
Trim at opening, 1 x 6 jamb, stock pine	24 L.F.	0.8	18.72
Casing, stock pine, 11/16" x 2-1/2"	24 L.F.	0.8	36.00
Sliding glass doors, aluminum, 5/8" insul. glass, 8' wide	2 Ea.	10.7	3,120.00
Sliding glass doors, aluminum, 5/8" insul. glass, 12' wide	2 Ea.	12.8	5,160.00
Miscellaneous wood blocking for ceiling, 2 x 4 stock	48 L.F.	1.5	19.58
Insulation, fiberglass, kraft-faced batts, 6" thick, R-19	192 S.F.	1.1	103.68
Gypsum wallboard, ceiling, 1/2" thick, taped and finished	192 S.F.	3.2	57.60
Trim, cove molding, 9/16" x 1-3/4"	60 L.F.	1.8	43.20
Indoor/outdoor carpet, 3/8" thick	189 S.F.	3.4	546.59
Paint, trim, wood, incl. puttying, primer, oil base, brushwork	50 L.F.	0.6	1.20
Trim, wood, incl. puttying, 1 coat, oil base, brushwork	50 L.F.	0.6	1.80
Ceiling, primer, oil base, roller	192 S.F.	0.8	11.52
Ceiling, 1 coat, oil base, roller	192 S.F.	1.2	11.52
Totals		42.3	$9,172.79

Contractor's Fee, Including Materials:	**$15,430**

Enclosed Porch/Mudroom

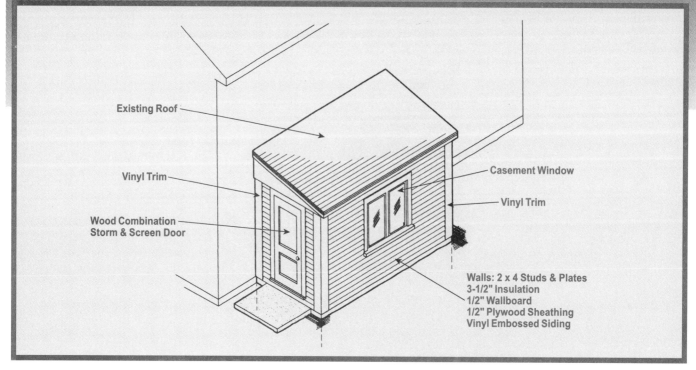

Existing Roof

Vinyl Trim

Wood Combination Storm & Screen Door

Casement Window

Vinyl Trim

**Walls: 2 x 4 Studs & Plates
3-1/2" Insulation
1/2" Wallboard
1/2" Plywood Sheathing
Vinyl Embossed Siding**

Muddy shoes are inevitable on rainy days, and can wreak havoc on clean floors and carpets. Many homeowners have chosen to avoid this problem by building mudrooms, complete with shelving and storage, durable and easily cleaned flooring, and even laundry facilities. If an entirely new addition is not in your budget, you can convert a porch or roofed-over entryway.

This model project converts a 6' x 12' porch into a mud- and laundry room. As the roof, floor, and support system are already in place, the project can be accomplished quickly and economically by many do-it-yourselfers.

Materials & Tools

The amount of required materials will depend on the dimensions of the space, the style and design of the house, and the extent to which the porch must be prepared and reconditioned. This plan includes a medium-sized thermopane casement window, but double-hung or jalousie models may be more appropriate to your house. Embossed vinyl siding is used in the model

project, but you'll need to choose a material that matches or coordinates with your house. The total cost will vary accordingly.

Other project materials include lumber and plywood sheathing for the basic framing; siding; finish materials, such as gypsum wallboard, paint, and insulation; and finish flooring. Refurbishing or changing the exterior trim or adding a gutter and downspout, if required, will also add time and expense to the project, as will adding fixtures for plumbing and additional electrical work for laundry facilities.

Standard hand and power tools will be needed for this project, along with a few specialized items, such as a sidewall stripper.

What's Involved

Be sure that the existing porch roof, floor, and foundation are in good condition before installing any new material. After the porch has been reconditioned, the 2 x 4 walls are framed, and the exterior of the structure

is sheathed. The window and door are installed before the siding.

In most cases, the framing will fill areas between the existing roof support posts, so allow extra time, and possibly cost, for cutting and measuring. If any of the posts have to be replaced or relocated, be sure to provide temporary roof support while framing.

Within the enclosure, the old exterior wall must be prepared for new materials. Ceiling joists and insulation must be installed, and finish coverings applied to the ceiling, walls, and floor. The wall preparation involves stripping the old siding and reconditioning the existing entryway for the new interior door. The exterior ends of the new ceiling joists must be tapered to fit the existing roof line, and the interior ends plated and lagged into the side of the house. The ceiling and new exterior walls should be insulated to the recommended R-value ratings for your climate.

Before gypsum wallboard or paneling is applied, all rough electrical and plumbing work should be completed. You can place the underlayment for the carpeting or vinyl flooring directly on the old porch floorboards, but be sure to use threaded flooring nails and then level the surface before the finish is applied. A layer of building paper between the old floor and underlayment will reduce the chance of moisture penetration, while helping to eliminate squeaks. Install an air infiltration barrier before the new siding.

Level of Difficulty

All levels of do-it-yourselfers should be able to perform at least some of the tasks in this project. Because the site is conveniently located, and much of the structure is already in place, the project is considerably less difficult and expensive than a complete room addition of the same size.

Assuming that only minimal reconditioning is needed, the most challenging operations are the wall framing and ceiling joist placement. If you're unfamiliar with framing, get some instruction from an experienced carpenter.

Beginners should double the professional time estimates throughout the project and get help before they begin tasks that are new to them. Intermediates should add 50% to the estimated labor hours, and experts 20%. Hire an electrician and a plumber if the project calls for such work, unless you're an expert.

What to Watch Out For

Two critical areas of the existing porch should be examined before construction: the roof and the support system. Check the flashing and general condition of the roof covering and rafters for signs of leakage and deterioration.

Make a thorough inspection of the floor joists, decking, and foundation. Rot, insect infestation, moisture, and

weakness of the frame and flooring are problems that must be corrected before the project continues.

These preliminary measures take time and may involve extra expense if reconditioning work is called for. If a problem looks serious, hire a specialist and correct the situation.

Installing rough plumbing and insulating the floor can be challenging, even if the floor is accessible from underneath the structure. Space limitations may require the use of batts, rather than rolled insulation.

If the floor is inaccessible from below, styrofoam sheets can be placed between sleepers made from strapping or 2 x 4s on the old floor surface. With this method, however, new decking as well as underlayment must be installed before the finished floor is laid. In this case, you may also have to make adjustments for the new floor height at both entryways.

See also:
Enclosed Porch with Screens, Enclosed Porch with Glass Doors, Porch, Flooring*, New Window.

*In *Interior Home Improvement Costs*

Enclosed Porch/Mudroom, 6' x 12'

Description	Quantity/Unit		Labor-Hours	Material
Door demolition, exterior, 3' x 7' high, single	1	Ea.	2.0	
Door frame demolition, including trim, wood	1	Ea.	1.0	
Siding demolition, shingles	96	S.F.	2.2	
Wall framing, studs and plates, 2 x 4 x 8', 16" O.C.	24	L.F.	3.8	96.48
Headers over openings, 2 x 6 stock	24	L.F.	1.1	15.55
Sheathing, plywood, 1/2" thick, 4' x 8' sheets	192	S.F.	2.7	142.85
Housewrap, spun bonded polypropylene	240	S.F.	0.5	46.08
Ceiling joists, 2 x 6 x 12', 16" O.C.	60	L.F.	0.8	38.88
Insulation, ceiling & floor, 6" thick, R-19, foil faced	144	S.F.	0.7	70.85
Walls, 3-1/2" thick, R-11, foil faced	160	S.F.	0.8	76.80
Gypsum wallboard, walls & ceiling, 1/2" thk., taped & finished	324	S.F.	5.4	97.20
Door, wood entrance, birch, solid-core, 2'-8" x 6'-8"	1	Ea.	1.0	109.20
Lockset, residential, exterior	1	Ea.	1.0	158.40
Door, prehung exterior, comb. storm and screen, 6'-9" x 2'-8"	1	Ea.	1.1	112.20
Windows, wood casement, insulating glass, 4' x 4'	1	Ea.	0.7	732.00
Trim, colonial casing, windows, door and base	70	L.F.	2.8	137.76
Siding, vinyl embossed, white, 8" wide, w/trim	40	S.F.	1.3	26.40
Flooring, vinyl sheet, 0.065" thick	72	S.F.	2.9	228.10
Flooring, adhesive cement, 1 gallon covers 200 to 300 S.F.	1	Gal.		19.62
Recessed box with washer/dryer hook-up	1	Ea.	0.4	48.00
Rough-in, supply, waste and vent for washer box	1	Ea.	2.3	122.40
Dryer vent kit	1	Ea.	0.4	8.04
Electric baseboard heat	4	L.F.	4.8	196.80
Paint, trim, wood, incl. puttying, primer coat, oil base, brushwork	88	S.F.	1.1	2.11
Trim, wood, incl. puttying, 2 coats, oil base, brushwork	88	S.F.	1.8	5.28
Walls and ceilings, primer, oil base, brushwork, smooth finish	384	S.F.	2.7	23.04
Walls and ceilings, 2 coats, oil base, brushwork, smooth finish	384	S.F.	4.5	41.47
Totals			49.8	$2,555.51

Contractor's Fee, Including Materials:	**$6,438**

Porch

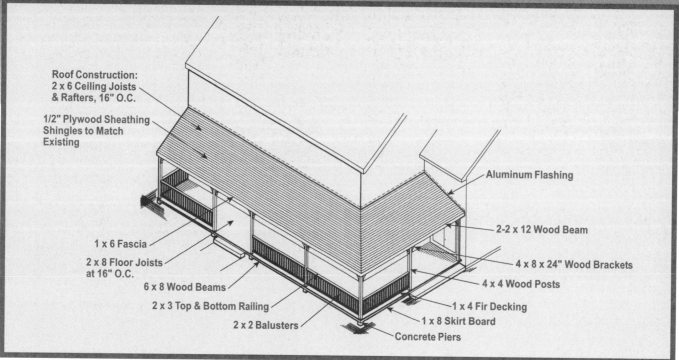

Roof Construction:
2 x 6 Ceiling Joists & Rafters, 16" O.C.

1/2" Plywood Sheathing
Shingles to Match Existing

Aluminum Flashing

2-2 x 12 Wood Beam

4 x 8 x 24" Wood Brackets

4 x 4 Wood Posts

1 x 4 Fir Decking

1 x 8 Skirt Board

Concrete Piers

2 x 2 Balusters

2 x 3 Top & Bottom Railing

6 x 8 Wood Beams

2 x 8 Floor Joists at 16" O.C.

1 x 6 Fascia

Before air conditioning and television, people spent a great deal of time sitting outside on their porches to keep cool in summer and stay connected with neighbors and passers-by. As new technology emerged, people spent more time indoors, and porches were built less often—until a strong revival began in the 1980s. Today, half of all new single-family homes are being built with porches. In addition to providing charm, character, and curb appeal, porches give you outdoor living space—a very strong trend in new construction and home remodeling.

Invest in proper planning for your porch design to make sure it not only achieves your goals for looks and function, but is in harmony with the style of your house. A front porch has such a big impact on your home's appearance and value that it may be wise to consult an architect to design it and a professional contractor to build it.

Materials & Tools

The materials start with basic decking and framing lumber. These and any ornamental trim pieces come in a variety of woods, and prices. You'll also need roofing materials to match your existing roof, and concrete for the footings. Select the lumber yourself if possible, looking for straight, unblemished deck boards and sound support members. Most suppliers will deliver free of charge for a complete order.

The support system for the porch consists of footings, 4 x 4 posts, and the frame, which is comprised of 2 x 6 joists and headers. The decking in this plan is 1 x 4 fir, but decking of another thickness, width, or edge type can also be used.

This porch plan includes a basic handrail and baluster-type railing. Fancy Victorian-style details, such as gingerbread trim, can be purchased at most home centers or renovation supply stores and easily added to the porch. For additional expense, you could substitute the posts with round or square columns and add decorative trim.

Placing the concrete footings requires tube forms to house the concrete while

it sets. A polyethylene barrier (if you choose to add one—see "What to Watch Out For") will be needed beneath the deck, along with gravel fill.

If you have a professional do the concrete work, you'll need a shovel for the excavation and carpentry tools for the deck framing and roof. If you paint or stain the railing and trim, you'll need those supplies as well.

What's Involved

Before the footing excavations begin, their locations must be accurately laid out to ensure a square and level support system for the rest of the structure. Several methods can be used to establish the location of the support posts and footings, but one of the most accurate is to approximate the layout with batter boards at the four corners, then use string-lines to mark equal diagonals. Failure to place the footings and posts precisely at square with each other and with the side of your house will adversely affect the project.

Double-check the post locations before you set them in concrete, and use

Courtesy of David & Chrissy Hoadley

cylindrical tube forms for the foundations. Also check the layout of the 2 x 6 frame to be sure the double joists are correctly aligned with the 4 x 4 supports. Lay the decking across the frame using spacers to guarantee the boards are separated equally, and nail it in place.

Once the frame, deck, and posts are installed, the roof is built. This is among the most difficult aspects of this project. If you're considering building the porch yourself, it's recommended that you bring in a roofer for this part of the project. The porch roof must be tied into the existing house roof, and the roofing material on your house will need to be extended to cover the porch. A consistent and neat appearance requires professional workmanship.

The amount of work that needs to be done to the house wall will depend on where and how the new porch roof structure will attach to it. Securing the porch roof to a flat house wall requires removal of siding, and cutting and patching to match existing surfaces. Plan the pitch of the porch roof so that there's room for proper flashing below second-floor windows.

Level of Difficulty

This project should be undertaken only by expert do-it-yourselfers. Beginners and intermediates should hire a professional contractor. If you're a beginner working with someone more skilled on any aspects of the work, add 300% to the labor-hours. Intermediates should add 100%-200%.

What to Watch Out For

Because the porch in this model project is built at ground level, it's a good idea to add a polyethylene barrier to prevent weeds and hardy grasses growing there.

See also:

Deck Projects, Enclosed Porches, Entryway Steps, Main Entry Door, Re-Roofing with Asphalt Shingles.

Alternate Materials

Roofing Options
Cost per Square, Installed

Asphalt roll roofing	$46.50
Asphalt shingles	$91.50
Cedar shingles	$310.00
Clay tile	$805.00

Porch, 24' x 16'

Description	Quantity/Unit	Labor-Hours	Material
Layout, excavate post holes	1 C.Y.	2.5	
Concrete, field mix. 1 C.F. per bag, for posts	10 Bags		69.60
Forms, round fiber tube, 1 use, 8" diameter	30 L.F.	6.2	54.36
Porch material, fir posts, 4 x 4 x 8'	48 L.F.	1.7	133.06
Wood beam, 6 x 8	40 L.F.	1.1	119.04
Joists, 2 x 8	192 L.F.	8.5	178.18
Fir decking, 1 x 4	256 S.F.	7.4	580.61
Double 2 x 12 wood beam	32 L.F.	0.8	61.44
Rafters, 2 x 6, 16" O.C.	216 L.F.	3.5	139.97
Ceiling joists, 2 x 6, 16" O.C.	216 L.F.	2.4	139.97
Plywood sheathing, 1/2" CDX	360 S.F.	3.1	280.80
Fascia, 1 x 6, pine	32 L.F.	1.1	22.27
Skirt board, 1 x 8, pine	32 L.F.	1.1	59.90
Wood brackets	10 Ea.	1.9	378.00
Railing material, 2 x 3, fir, top and bottom rails	48 L.F.	1.3	112.32
Roofing material, asphalt felt paper, 15#	360 S.F.	0.8	12.96
Shingles, asphalt strip, organic, class C, 235 lb. per square	3.60 Sq.	5.8	174.96
Flashing, aluminum, 0.019" thick	40 S.F.	2.2	31.20
Brick steps	16 L.F.	5.3	42.60
Joist and beam hangers, 18 ga. galvanized	48 Ea.	2.3	38.59
Nails, galvanized	50 lb.		83.40
Bolts, 1/2" x 7-1/2" lag bolts, square head, w/nut and washer	24 Ea.	1.5	18.14
Totals		60.5	$2,731.37

Contractor's Fee, Including Materials: **$7,231**

Wood Entryway Steps

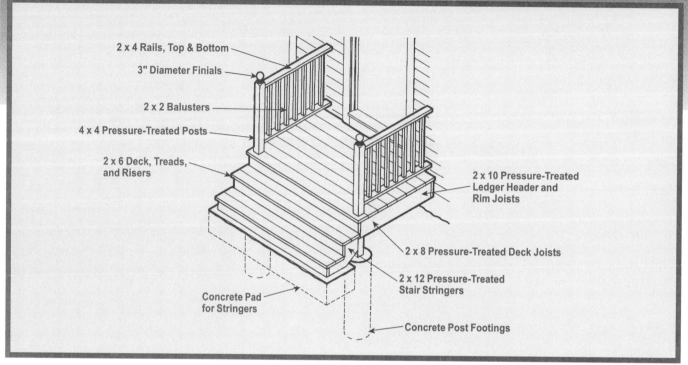

- 2 x 4 Rails, Top & Bottom
- 3" Diameter Finials
- 2 x 2 Balusters
- 4 x 4 Pressure-Treated Posts
- 2 x 6 Deck, Treads, and Risers
- 2 x 10 Pressure-Treated Ledger Header and Rim Joists
- 2 x 8 Pressure-Treated Deck Joists
- 2 x 12 Pressure-Treated Stair Stringers
- Concrete Pad for Stringers
- Concrete Post Footings

Replacing or adding a set of entryway steps improves your home's curb appeal and makes the entrance safer for your family and guests. If your house is over 25 years old, its wood porch, deck, and steps could very well be in need of repair or replacement. Prior to the availability of pressure-treated wood, outdoor wood steps and porches were built from standard construction-grade lumber, which is eventually destroyed by years of weather and wood-boring insects.

The stairs in this project are a sort of mini-deck, which can be adapted in numerous ways to blend with a variety of architectural layouts and styles. This structure can be built by many do-it-yourselfers.

Materials & Tools

You'll need two types of wood for this project—pressure-treated lumber for the steps' framing and support, and higher-grade redwood or cedar for the decking, railings, and trim. If you're concerned about the safety of pressure-treated wood (which previously used chromated copper arsenate, or CCA,

as a preservative), this problem was eliminated by the end of 2004, when old material stocks were used up, and safer preservative materials have completely replaced them.

A post-hole digger and concrete are needed to anchor the posts in place, as with any deck. Concrete is also necessary to form the footing for the stair stringers. You'll need a shovel, drill, and a carbide-tipped saw for cutting the pressure-treated lumber, as well a hammer, nails, and screws. Purchase a good quality wood preservative to protect your entry steps over time.

What's Involved

If you're replacing an existing set of steps, the best approach may be duplicating their construction. Take the lines and measurements from the structure before you demolish it. Changes or improvements can be made easily on paper beforehand. If you're building an entirely new set of steps, lay out the dimensions directly on the house wall below the door opening.

Draw this design on paper, indicating all measurements and dimensions, visualizing as you do so the exact building sequence and noting potential trouble spots.

Dig the post holes and set the points in place, plumbing and bracing them diagonally with lengths of 1 x 3 strapping and stakes. Mix and pour the concrete around the posts to anchor them securely. Taking careful measurements, lay out the lines for the concrete pad that will serve as the footing for the stair stringers. After excavating, build a level form around the opening, and mix and pour the concrete.

Attach the ledgerboard to the house wall below the door opening, securing it with 6" lag screws. Construct the perimeter of the frame (end joists and header), and install the inner joists using steel hangers, 24" on center. Mark, cut, and attach the four stair stringers to the header, using the metal framing anchors.

The cedar risers, treads, and deck boards can now be cut and installed. Blunting the point of a nail before driving it will help prevent splitting when nailing the ends of the deck boards. Also, driving the nails at a slight angle will increase their holding power. To provide for surface drainage, leave a space of about 1/8" between the deck boards using 16d nails as spacers.

The two balustrades each consist of a top and bottom rail, with a series of 2 x 2 balusters laid out with a spacing of about 1-1/2" between each of them. Once assembled, the balustrades are attached to the house wall and to the posts, which can then be cut off at the appropriate height (about 38" above the decking) and topped with the finials.

After sanding the wood to ease the sharp edges and to smooth the surface, a coat of clear wood preservative should be applied and allowed to soak in for about a week. The structure can then be finished with a top coat of opaque stain to match or blend with the color of the house or its trim.

Level of Difficulty

Deck building is a very popular do-it-yourself project, and this wooden step structure is really nothing more than a miniature deck. Careful planning and layout, as always, are essential to successful results. A beginner should have experienced help during this phase, and at least some general guidance during the construction, especially when marking and cutting the stair stringers.

Framing a deck is basic rough carpentry. Pressure-treated lumber can be difficult to work with; it tends to be splintery, is somewhat hard to cut, and nailing into it can result in a lot of bent nails and frustration (Use a 20- or 25-ounce hammer.) As is the case with many outdoor, heavy lumber structures, the finish work required is not quite as fussy as interior finish carpentry can be. This is not to excuse sloppy workmanship, but decks are somewhat forgiving, and you can get away with less than perfect cuts and joints that would be

Built by J.P. Gallagher, Hanover, MA

unacceptable inside the house. Adding a roof to this design would make it quite a bit more challenging, but still a potential do-it-yourself project.

Beginners should have some help with this project, and should add about 150% to the estimated times. Intermediate-level do-it-yourselfers should add about 50%, and experts, about 15%.

What to Watch Out For

For all exterior construction, you should use corrosion-resistant nails—stainless steel, aluminum alloy, or high-quality, hot-dipped galvanized. Avoid electroplated galvanized nails. Poor quality nails can react to the chemicals in pressure-treated lumber and to the natural resins of such woods as cedar or redwood. The resulting corrosion can produce unsightly stains and ultimately compromise the nail's holding power.

Options

There are many variations that can be worked out on this basic design, depending on the style of your house. One of the more functional ones would be the addition of a small pitched roof over the deck, to provide some weather protection overhead.

See also:

Brick Entryway Steps, Elevated Deck, Porch, Ground-Level Deck, Main Entry Door, Wheelchair Ramp.

Wood Entryway Steps

Description	Quantity/Unit		Labor-Hours	Material
Pre-mixed concrete, for post anchors & footing pad, 70# bags	12	bags		83.52
Place concrete, spread footing, under 1 C.Y.	12	bags	0.9	
Ledger, end joists, and header, 2 x 10	22	L.F.	0.4	31.15
Inner joists, 2 x 8	8	L.F.	0.1	7.97
Posts, 4 x 4	16	L.F.	0.7	24.00
Stair stringer, 2 x 12	10	L.F.	1.2	19.20
Joist and beam hangers, galvanized, 2 x 6 to 2 x 10 joists	4	Ea.	0.2	3.22
Bolts, square head, nuts and washers incl., 1/2" x 6", galvanized	4	Ea.	0.2	3.02
Nails, 16d common, H.D. galvanized	10	Lb.		16.68
Decking, stair treads, risers, 2 x 6	90	L.F.	4.4	228.96
Rails and balusters	16	L.F.	5.8	271.68
Trim for top rail, cove molding, 3/8" x 5/8"	16	L.F.	0.5	8.64
Finials, 3" diameter	2	Ea.	0.5	13.44
Paint, railings, newels & spindles, stain, oil base, brushwork	32	L.F.	2.8	5.76
Stair stringers, rough sawn wood, stain, oil base, brushwork	116	L.F.	10.3	20.88
Totals			28.0	$738.12

Contractor's Fee, Including Materials: **$2,570**

Brick Entryway Steps

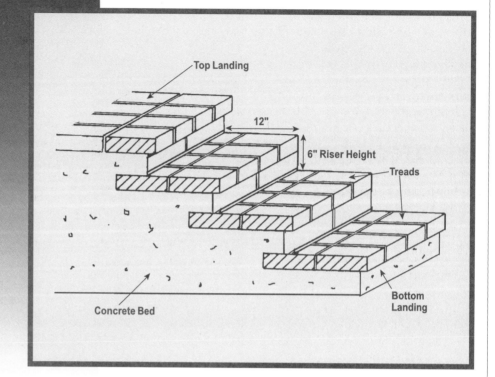

Top Landing

12"

6" Riser Height

Treads

Concrete Bed

Bottom Landing

Brick or stone steps add elegance to a home's main entrance can also accommodate elevation changes in the yard—from a patio or walkway, or a steep lot to the street. For a steep slope, consider sets of a few steps with short, flat landings between them, rather than one long, unbroken set of stairs (which also requires a handrail). The more gradual elevation change allows for changes of direction and opportunities to landscape with shrubs, trees, and flowers along the way. This project includes estimates for stairs with both brick- and stone-faced treads.

Materials & Tools

A variety of brick or stone materials, with widely varying prices, can be used with a concrete bedding material. Because loose stones or bricks create a tripping hazard, all finish materials should be embedded in concrete, not dry-set. The brick or stone color should complement your home and surrounding features. Purchase some extra pieces for future repairs, since it can be difficult to find an exact match at a later date. Even brick made with the same material will vary in color from batch to batch, and stone color varies by

when and where it is cut in the rock formation, as well as the quarry region.

The stair location and size must be determined before you purchase materials. The vertical distance between steps is the "riser" height. Building codes require that risers be the same height for each step in a set of stairs, typically between 6" and 7-3/4". This allows a smooth, natural movement from one level to another and helps prevent falls. If your project is subject to city review, riser height difference is not permitted to vary by more than 1/8" from what is specified. The top surface is called the "tread." A depth (or front-to-back) dimension for stone or brick treads is often 12"—a comfortable space for most people. The width of the treads, and the set of stairs as a whole, is typically 3'-4'.

Once you've decided on the stair location and measured the elevation changes, you can determine the number of risers needed. Two-riser stairs are the most common for a home entrance or change of elevation. Five-riser stairs may be needed for steeper elevation changes. Handrails are not required for most exterior stairs, but

may be desirable for elderly residents or those with limited mobility.

Generally, 10.5 bricks are allowed per square foot of tread area. This leaves about 3" of exposed concrete on the bottom of the riser for each step. Extra bricks (5% of the total planned for the steps) will be required for breakage and cutting to complete the treads. If you're using stone or slate, the pieces may be irregular. Seek professional guidance to calculate the quantity. Determine the square footage of tread area by multiplying the length times the width. (For example, 4' x 1' equals 4 square feet of area for a single tread, times the number of treads equals the square footage.) Added to this amount is the square footage for any landings. If you want the exposed riser face to be stone, you'll have to cut the stone to fit this dimension or find smaller stone pieces. Unless you're working with pre-cut stone pieces, there will be some exposed concrete in the finished surface areas.

Solid stone steps are available from some suppliers in granite and limestone. Since they're so heavy, you'll need an outside contractor with personnel and lifting equipment to lift and place them. In some cases, you might have to rent a piece of lifting equipment. Otherwise, simple tools are required for this job: a hand compactor to tamp the gravel in place at the base, a shovel and pick for excavation, a hammer and saw, a carpenter's level, and a tape measure for building the formwork. A rubber mallet is needed to tap the brick or stone into place, and a chisel to cut bricks to exactly fit. Safety goggles are a must when working with, and especially cutting, stone or brick, along with gloves, and protective clothing and footwear.

What's Involved

You may need to remove existing wood or concrete stairs. If the old stairs are wood, you'll probably only need to use a crowbar to remove them. If they're concrete, you'll have to bust them up with a sledgehammer and remove them in pieces you can handle.

The first task in building the new steps is providing a foundation. This involves excavating the area, compacting the soil, and adding sand or gravel for the stair base. If the brick stair is for a front entrance, only minimal excavation may be needed, because there is probably already a height between the entrance door and the ground.

You'll need to excavate where the gravel base will be installed, but not for the stairs themselves. Dig at least 6" below ground, removing the soil and replacing it with gravel, hand-compacted in place. Excavate at least 6" wider on each side of the desired stair width to allow for the formwork. This means that if you have a 4' wide stair, excavate a 5' width. The depth of the excavation will be the height of the risers plus 6" for the gravel base. This means that a stair with two 7" risers will need to be excavated to 14" plus 6", for a total depth of 20". The length of the excavation will be the depth (front to back) of the treads times the number of steps. For example, if the tread depth is 12", excavate 24" in length for the two treads in a two-riser stair. Once excavation is complete, fill the bottom 6" with gravel and tamp it in place to provide a well-drained foundation.

The next task is installing forms (solid sides and stepped riser heights) and placing the concrete bedding and finish material. For a patio area or path, a small landing is desirable on the top and bottom of the stairs. This will mean additional excavation for the gravel base and concrete bedding for the brick or stone. The top landing can be formed along with the stairs, but the bottom landing will have to be installed after you remove the stair formwork.

Next, pour the concrete and place the brick or stone. After 24 hours, you should be able to remove the forms and backfill the soil around the area where the forms were removed. (Don't pour concrete if the temperature is expected to drop below freezing.) A bottom landing can now be excavated, concrete poured for it, and stone or brick installed.

Level of Difficulty

The physical demands are moderate. The task demanding the most construction skill is building the stair forms. Beginners should seek expert advice from a local building supplier. They should add 150%-200% to the professional installer's time in the estimate. Intermediates should add 100%-150%, and skilled do-it-yourselfers 75% more time.

What to Watch Out For

If you're working with brick, you'll have to cut some pieces to fit with a chisel. This is a learned skill and should be done on a level surface like a wood plank. The amount of cutting will depend on the stair dimensions. For best results, set the bricks on edge. For stone steps, get some help lifting stone pieces. Large stones (including pre-cut pieces in 4' lengths), can weigh up to several hundred pounds and will require lifting equipment. Smaller pieces can be moved by two people sharing the load, using good lifting practices (straight back, bent knees).

See also:
Wood Entryway Steps, Brick Walkway, Concrete Patio, Brick and Flagstone Patio.

Alternate Materials

Material Options
Cost per Each, Installed

Stone top pieces	$102.00
Cut stone treads	$102.00

Brick Entryway Steps, 2 Steps, 4' Wide

Description	Quantity/Unit	Labor-Hours	Material
Excavate for stair base & top thread	0.37 C.Y.	0.7	
Crushed gravel delivered	0.19 C.Y.	0.1	6.27
Backfill gravel material into stair base	0.19 C.Y.	0.1	
Hand tamp gravel material into stair base	0.19 C.Y.	0.1	
Concrete forms for stairs	24 S.F.	4.8	12.67
Concrete forms for brick shelf	20 SFCA	3.2	20.88
Reinforcing for concrete base	0.04 Ton	0.6	36.48
Concrete for stairs	1.19 C.Y.	1.0	119.95
Brick laid on edge	20.66 S.F.	4.7	58.51
Brick concrete bed and joints	20.66 S.F.	0.6	22.06
Cleanup, remove forms and backfill around stairs	1 Ea.	4.0	
Totals		19.9	$276.82

Contractor's Fee, Including Materials: $1,374

Wheelchair Ramp

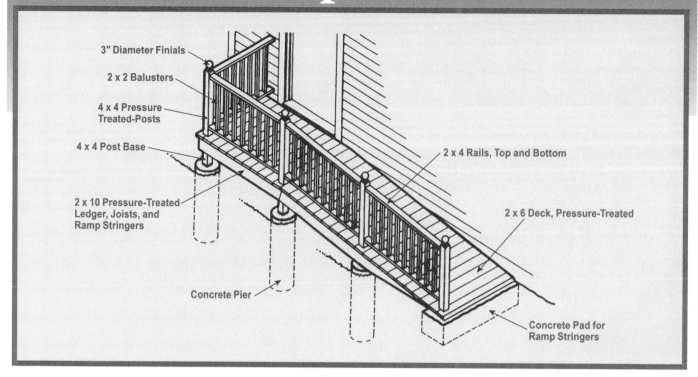

3" Diameter Finials

2 x 2 Balusters

4 x 4 Pressure Treated-Posts

4 x 4 Post Base

2 x 10 Pressure-Treated Ledger, Joists, and Ramp Stringers

2 x 4 Rails, Top and Bottom

2 x 6 Deck, Pressure-Treated

Concrete Pier

Concrete Pad for Ramp Stringers

High-tech innovations have improved the design and capabilities of modern wheelchairs, and these in turn have given their owners a high degree of mobility and independence. Ramps and lifts are, however, still necessary for wheelchair operators to get from one level to another. A thoughtfully designed and well-constructed ramp can blend very well with the house, and it will greatly enhance the life of the person who uses it.

Materials & Tools

Most houses typically have two or three entry doors—front, back, and/or side. Where to locate the wheelchair ramp depends on a number of factors, mainly accessibility, appearance, and ease of installation. The latter is largely determined by the height of the door threshold above grade and whether the wheelchair is manually propelled, motorized, or pushed by an assistant.

To determine the required length of the ramp, consider the height of the door. Unassisted wheelchair use requires a ramp slope no steeper than 1" in 12" (1" of rise for every 12" of "run"). Thus, a door 18" above-grade calls for a ramp length (not including a landing) of 18' for unassisted use. The principle is simple: the lower the door, the shorter the ramp.

Pressure-treated lumber is used for the posts and framing, while the more attractive cedar is used for the decking and balustrade. Other types of wood for the decking are possible, as long as their final finish is slip-resistant and allows for water to run off properly. You'll need basic carpentry tools for this project, along with the lumber, and—to finish the ramp—a wood preservative and sand paint for traction.

What's Involved

First determine where you'll build the ramp. Pragmatic considerations should rule, but appearances should not be overlooked. Even if it means some extra work, the best choice for a ramp

location, as far as appearance is concerned, is a side or back entry. In terms of accessibility, a paved and fairly level approach to the door is an obvious necessity, so building the ramp off a driveway or walkway is ideal.

This design is similar to that of the entryway steps project that appears earlier in this book. The basic construction process is also similar to that used for the steps. First, lay out the landing. To provide space for the wheelchair and its turning radius, the landing should be at least 36" x 60" in front of an in-swing door, or 60" x 60" if the door swings out. Most exterior entry doors swing in, allowing for the installation of a storm door, which always swings out.

To avoid building a more massive deck, it might be worthwhile to eliminate a storm door and, if necessary, replace an entry door with a good quality, weathertight, insulated model. *(Refer to the Main Entry Door project earlier in this book.)*

Next, lay out the posts, and build a form for the concrete pad that will support the bottoms of the ramp stringers. When you've finished excavating, align the posts square and parallel to the house. Plumb and brace them, and then pour the concrete. Attach the 2 x 10 ledgerboard to the house at the proper distance below the threshold, using nails and lag screws. Working out from the ledger, install the rim and deck joists.

Determining the precise angles, top and bottom, for the ramp stringers can be a challenge. One way to do it is to stretch a length of mason's cord along the house wall to indicate the top edge of the stringer. (Snapping a chalk line is likely to be difficult, if not impossible, because of the irregularity of the side wall and foundation surfaces). Make trial cuts on scraps of wood to get the proper angles.

Once you've figured out the angles, measure and cut one stringer. Test it for fit at each of the layout locations, and if it works, use it as a template for marking the others. Once you've cut the angles on the ramp stringers, attach them to the header joist with framing anchors. Use a straightedge to make sure the tops of the stringers line up. Nail solid blocking between the stringers, making sure to keep them parallel to the house.

After laying the deck boards, trim the tops of the posts about 38" above the deck. Assemble and install the balustrade sections, and cap the posts with wooden finials. In general, bolts and mechanical fasteners are recommended, as a nailed joint may loosen up, and there's no way to tighten joints that shrink. After applying clear wood preservative, the balustrade can be stained, and the decking coated with sand paint to provide good traction.

Level of Difficulty

As mentioned above, the height of the door above grade determines the length of the ramp. A rise of 6" calls for a ramp of only 6'. Any number of variables could increase the difficulty of your specific project (for example, the need to remove shrubs or install a paved

walkway), but building this mini-deck and ramp is not, in itself, that difficult.

A beginner would, no doubt, need some help with the design and layout, and with figuring the angles on the stringers, but once the relationship of the components is understood, the physical work of assembling them calls for no more than fairly basic carpentry, with the usual reminder to measure twice, so you can cut only once. Beginners should add about 150% to the time estimates. Intermediate-level do-it-yourselfers should add 75%, and experts, 20%.

What to Watch Out For

Prior to starting this project, it's recommended that you consult with your local building inspector. Because of the permanent nature of most ramps and the large amount of area they take

up, permits may be rquired. Most state and local building codes include sections on handicapped requirements, or may refer you to the 2004 *Americans with Disabilities Act Accessibility Guidelines*. This is also an excellent source of information for developing handrail heights and platform sizes.

The deck lumber of the landing can be no lower than 1/2" below the threshold. If you lay the deck flush to the top of the threshold, you almost guarantee that rainwater from the deck will run in under the door. Be sure to check the finish level against the threshold before laying out the landing. Also, a small roof or overhang above the landing will help protect it, the door, and persons using the entrance from rain and snow.

See also:
Main Entry Door, Driveway, Brick Walkway.

Wheelchair Ramp, 4' x 15'

Description	Quantity/ Unit	Labor- Hours	Material
Layout, excavate post holes	0.50 C.Y.	0.5	
Forms, round fiber tube, 8″ diameter, for posts	12 L.F.	2.5	21.74
Concrete, field mix, 1 C.F. per bag, for posts	4 Ea.		27.84
Placing concrete, footings, under 1 C.Y.	0.50 C.Y.	0.4	
Post base, 4 x 4	4 Ea.	0.2	26.16
Framing, for ramp, 4 x 4 posts, treated	32 L.F.	1.3	48.00
Joists, 2 x 10, treated	80 L.F.	1.4	113.28
Joist hangers, galvanized, 2 x 10	4 Ea.	0.2	3.22
Post base 16 ga. galvanized, 2-piece, for 4 x 4	4 Ea.	0.4	39.24
Lag screws, 1/2″ x 7-1/2″, galvanized	5 Ea.	0.3	2.16
Decking, 2 x 6, treated	112 L.F.	5.4	284.93
Rails and balusters	36 L.F.	13.1	611.28
Finials	4 Ea.	1.0	26.88
Nails, 16d common, h.d. galvanized	10 Lb.		16.68
Nails, 10d box, h.d. galvanized	10 Lb.		16.68
Waterproofing, silicone, on decking, sprayed	60 S.F.	0.2	45.36
Paint, railings, cap, baluster, newels & spindles, stain, brushwork	40 L.F.	3.6	7.20
Deck, oil base, brushwork, 1 coat	112 L.F.	0.8	9.41
Sand paint, oil base, brushwork, 1 coat, for no-slip decking	112 L.F.	6.0	12.10
Totals		37.3	$1,312.16

Contractor's Fee, Including Materials: **$3,916**

Roofing, Siding, and Gutters

If your roof or siding is old and has problems, such as leaks, worn sections, or missing or cupped and bowed shingles, it may be time for a total replacement. Here are a few tips:

- Many roofing and siding materials are bought and sold by the *square*. One square is equal to an area that measures 100 square feet, or 10' x 10'.

- When replacing a section of siding, repeat the detailing and scale of the original siding as much as possible.

- Different siding textures and profiles can make a big difference in appearance, but not necessarily in cost. Check out all of your options. Most natural wood siding will discolor (age) differently depending on the exposure. Applying stain or preservatives can ensure consistent color. Rough-textured siding tends to hold stain better.

- Surfaces exposed to direct sun may weather more quickly than faces exposed to moderate shade. Quality workmanship is even more crucial in these circumstances.

- When re-siding, consider adding rigid insulation if your house is under-insulated and drafty.

- If you're replacing siding on an older house, it may be difficult to find matching material. Your supplier may be able to special order replacement siding or suggest a source in your area.

- Where different materials meet—at flashing penetrations on a roof, for example—sealants and caulk provide extra protection against water penetration.

- Asphalt shingles are the most common material. They're easy to install, economical, and come in a variety of colors and textures. They are generally guaranteed for up to 20 years. Some brands have a 30-year guarantee.

- Asphalt shingles should be applied on a minimum slope of 2" in 12". Wood shingles should be applied on a minimum slope of 3" in 12", and wood shakes, 4" in 12".

- Before removing existing roofing material, be sure to check on the costs and appropriate methods of disposal, as well as any permits or other requirements of the town building department.

- If your existing roof has asbestos shingles, hire a professional.

- Check the condition of flashings. Replace them if they're damaged.

- Re-nail sheathing if necessary before applying new roofing or siding. If sheathing was not properly nailed down when originally installed, it may have warped over time.

- Avoid working on roofs in extreme heat or cold, as scuffing may occur.

- When working above the ground level where staging is required, be sure to use secure planks and brackets. When working from a ladder, place the ladder so the feet are away from the house at a distance equal to 1/4 the height of the ladder.

Aluminum Gutters

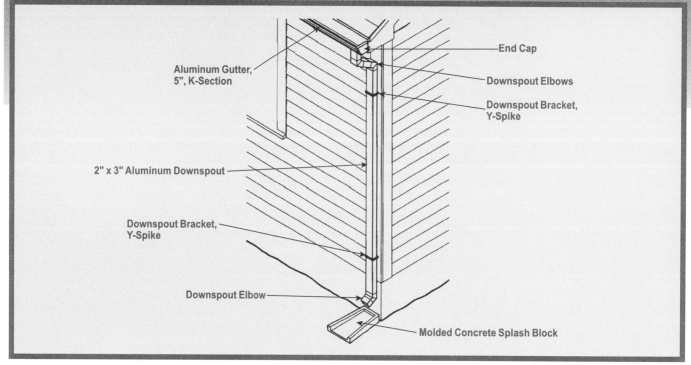

Aluminum Gutter, 5", K-Section

End Cap

Downspout Elbows

Downspout Bracket, Y-Spike

2" x 3" Aluminum Downspout

Downspout Bracket, Y-Spike

Downspout Elbow

Molded Concrete Splash Block

Few things do more damage to your home's exterior than water. Rainwater that collects in old or clogged gutters can slowly rot the surrounding wood and allow access for carpenter ants, termites, and other unwanted guests. Properly maintained, gutters help divert rainwater away from your home's roof and foundation. When chosen and installed correctly, gutters can also blend into your home's façade and be almost invisible.

Materials & Tools

Aluminum gutters are by far the most popular for residential construction. They're relatively inexpensive and not overly difficult to install. They come in several shapes: rectangular, beveled, K-section, and half round, and in a variety of sizes from 2-1/2" to 6" in height and from 3" to 8" in width. Stock lengths also vary, with 10' being the most common. Many building suppliers have machines that can roll-form gutter

sections of virtually any length, and gutter contractors will usually bring one of these machines to the job site.

Aluminum gutter components include inside and outside miter corners, joint connectors, outlets, end caps, and various elbows. Gutters can be attached with fascia brackets, strap hangers, or spikes and ferrules. If you're doing the installation yourself, be prepared with a ladder, tape measure, drill, hammer, chalk line, and some aluminum sealant, in addition to the gutter itself.

What's Involved

Gutters are often run level for appearance, but a slight slope—about 1" in 15'—toward the downspout is desirable for drainage. Downspouts should be spaced at no less than 20' intervals, and no more than 50' apart. This is a factor of roof area and pitch; a large roof with a steep pitch will require more downspouts to handle the volume and rate of flow than will a smaller, shallow-pitched roof. On a short run, a gutter can be installed to slope in one

direction to a single downspout. On longer runs, the gutter should be crowned, or positioned slightly higher, in the center to drain the water to downspouts at both ends.

All connections should be made using aluminum sealant to prevent seepage through the joints. For spike-and-ferrule attachment, locate the rafter ends and determine their spacing. Mark these on the front lip of the gutter and drill holes for the spikes. The top outside edge of the gutter should be in line with the plane of the roof. Snap a line on the fascia board to mark the desired pitch for the bottom of the gutter. Hold the gutter to the line and drive the spikes through the ferrules into the rafter ends. Drive the Y-spikes into the corner board, using two for every 10' of downspout. Connect the downspout to the gutter outlet and wire the spout to the brackets. Attach the elbow to the bottom of the downspout and place the splash block under it. You are now ready for rain.

Level of Difficulty

Installing gutters of any kind is a two-person job, and very long lengths (over 20') are more safely handled by three people. The gutters themselves are lightweight and require little preparation before they're ready to go up. Spike-and-ferrule installation is the quickest; fascia brackets and strap hangers take a little more time to attach.

Installing gutters means working from a ladder high above the ground. If you're inexperienced or uneasy about this aspect of the job, hire someone to do it.

If you decide to install more complicated gutters, such as wood, it is advisable you hire a professional unless you are an expert. Beginners should add at least 100% to the professional time estimates for aluminum gutters; intermediates, at least 50%. Experts should figure on about 20% more time.

What to Watch Out For

Sometimes the backfill or material used to fill the excavation around a house's foundation settles, creating a low spot. Introducing shrub beds next to a foundation can also create water collection problems against the foundation. By installing gutters, you can prevent large amounts or run-off water from landing next to the foundation, as the downspouts discharge the water away. One gutter section may require two or more downspouts, depending on the area of the roof involved. Care should be taken not to discharge downspouts into shrub or flower beds located next to the foundation.

An important part of a gutter replacement project is assessing the condition of the cornice—fascia, soffit, look-outs, and rafter ends—once the old gutter has been removed. Rotted wood must be replaced to prevent further damage and to provide solid backing to which to secure the new gutter. If you hire a professional, make it clear to the contractor that inspecting the cornice and replacing rotted wood is part of the job, and that you're willing to pay for the necessary work and materials over and above the cost of the gutter. Money spent on a thorough job is never wasted.

To make your new gutter last longer and provide trouble-free service, check with your local lumberyard for gutter accessories such as wire strainers and gutter screens that will eliminate clogging of downspouts and of the gutter itself. Keep in mind that even if you have these devices, you should still check your gutters periodically and remove any debris.

If you're adding gutters to a section of roof that currently does not have a gutter, keep in mind that you'll be directing the total volume of water run-off to one location. Anticipate the amount of water you're redirecting and be sure it won't create a safety hazard or drainage problem at the end point of the gutter system.

Clean your gutters out twice a year. Remove accumulated leaves, twigs, and other debris by hand, and flush the entire system out with water afterward.

Options

Aluminum is the most popular choice of material for do-it-yourself gutter systems, but there are other options at different price ranges. Wood gutters are expensive and are the most difficult to install, but for some houses of traditional design and construction, they may be the only architecturally correct choice.

Copper gutters, also costly, can be custom-made, and, when treated with an acid solution to darken them, can virtually disappear against the trim of a dark-stained house. Vinyl or plastic C-shaped PVC (polyvinyl chloride) gutters with round downspouts offer durability. Talk with a contractor about the look you desire within your budget to determine which material best suits your needs.

Aluminum Gutter, 20 L.F.

Description	Quantity/Unit	Labor-Hours	Material
Remove old gutter and downspout	20 L.F.	0.7	
Gutter, alum., stock, 5" box, 0.026", plain, incl. ends & outlets	20 L.F.	1.3	29.76
Downspout, aluminum, 2" x 3" including brackets	20 L.F.	0.8	16.32
Wire strainer, rectangular, 2" x 3"	1 Ea.	0.1	1.94
Elbows, aluminum, 2" x 3"	3 Ea.	0.2	3.28
Totals		3.1	$51.30

Contractor's Fee, Including Materials: $257

Attic Ventilation

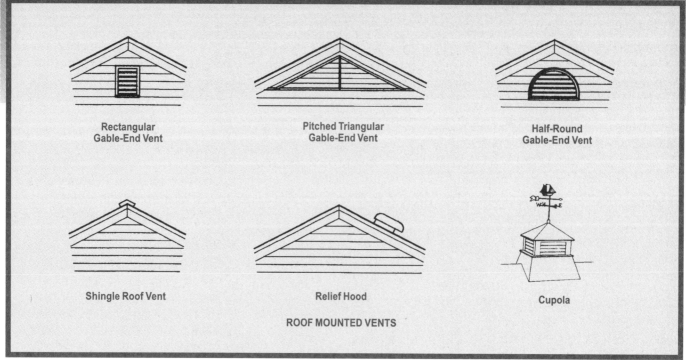

Rectangular Gable-End Vent

Pitched Triangular Gable-End Vent

Half-Round Gable-End Vent

Shingle Roof Vent

Relief Hood

Cupola

ROOF MOUNTED VENTS

In an era of escalating energy costs, builders and homeowners are acutely aware of the need for sufficient insulation, draft-free windows, and other energy-saving measures. Often overlooked, however, is the equally important need for adequate ventilation to prevent moisture condensation in a tightly built, heavily insulated house.

Moisture buildup can cause a number of problems, including wet (and therefore compacted and ineffective) insulation, rotted sheathing and framing members, blistered and peeling exterior paint, ice dams, and leaks. Mold, another by-product of inadequate ventilation, can cause odors and health problems, and can prevent homeowners from obtaining insurance in some regions.

Materials, Tools, & Equipment

There are many types of ventilators. Wood louvered units are generally installed in the end walls of gable roofs, as close to the ridge as possible. Hip roofs should have air inlets at the soffit, and outlets at or near the peak of the roof. Roof-mounted ventilators are designed to remove air from the attic without the use of motor-driven fans, either by convection or external wind action. Relief hoods may also be used for air exhaust. Hoods and ventilators are usually made of galvanized steel or aluminum.

Soffit material may be 3/4" solid lumber, matched boards, plywood, corrugated and perforated aluminum, or special material with pre-cut screened openings. Choose the system that best matches your house. Soffits are not highly visible from a distance, but can

be noticed up-close and should be properly installed to look right. You may also need some roofing and siding material to repair damage caused around holes that have been cut.

To complete any of these installations, you'll need access to your attic from both the inside and outside (which will require an extension ladder). You'll also need a tape measure, level, carpenter's square, a jigsaw or reciprocating saw, plywood, and basic carpentry tools like a hammer and a screwdriver. Wear goggles and a dust mask to protect yourself from dust and debris.

What's Involved

You'll need to determine the minimum free-air area (ventilator openings) for attic vents, based on the ceiling area of the rooms below. The free-air area should be 1/300 of the ceiling area. For example, if the ceiling area is 1,200 square feet, the minimum total free-air area of the ventilators should be 4 square feet. The actual area must be increased to allow for any restrictions such as louvers or wire screen mesh.

Coarse screen should be used because dirt and airborne debris tend to clog fine mesh screens. Fiberglass screening may be easier to install than metal types. It also does not corrode or oxidize. When painting, be careful not to contact the screen and clog the mesh with paint. Ventilators should never be closed. In cold climates, it's especially important to leave them open in winter as well as summer.

The soffit, sometimes called the planchia, is the finished underside of the box cornice—that part of the roof that overhangs the exterior wall. If hole-type soffit vents are, for whatever reason, inadequate, and the slot-type system is required, it will often be easier to remove and replace the soffit rather than attempt cutting out the proper-sized strip. The simplest installation is to screen the soffit opening and nail up two narrow boards so that you leave a 2"-wide open strip the full length of each soffit. Make sure there's no insulation in the soffit; the insulation on the attic floor should end at the outer wall, not extend into the overhang.

To guarantee the unobstructed flow of air from soffit vent to ridge vent, cardboard or styrofoam forms (pans) may be introduced to slightly depress the insulation at the wall line, providing a natural duct for air flow. Framing bays (the spaces between rafters) are sometimes isolated by skylights or other roof penetrations. If possible, air flow should also be provided to these bays by boring a series of small holes through the rafter.

Level of Difficulty

The main difficulty involved in this project lies in the fact that most of the work has to be done from an extension ladder or on the roof. Safety considerations alone would put this work beyond recommendation for the beginner or even the moderately competent intermediate do-it-yourselfer. Using hand and power tools at ground level and with solid footing can be hazardous enough, but working with them at heights and in awkward positions is best left to experts or professional contractors.

This project also requires cutting holes in the soffits, walls, and/or roof. Any improper installation of the vent unit or finished surface materials could result in leaks and potentially serious damage that would be costly to repair, further reasons why this work should not be undertaken by the novice.

The level of difficulty, least to most, for the various installation tasks is as follows: hole-type soffit vents, gable vents, slot-type soffit vents, roof-mounted vents, and ridge vents. Even if you're experienced enough to undertake all or part of this project, seek the advice of a professional to determine what type and size ventilating devices are best suited to your situation. It would be most discouraging to complete the job only to discover that the venting system you labored mightily to install is wrong or undersized. Experts should add about 30% to the professional time for all tasks.

What to Watch Out For

Any time a project involves disturbing existing roofing or siding material, the cutting and matching may affect a much greater area than you might have expected. Examine the condition of existing material before starting to ensure that you have enough materials to replace any area that might be disturbed during the renovation.

Options

To maximize the benefits of efficient attic ventilation, you could install a whole-house fan in the attic floor to help cool the house in warm weather. The fan draws cool air in through open windows in the rooms below and exhausts hot indoor air through the attic vents to the outside. Whole-house fans can minimize the need for air conditioning in many climates.

Solar-powered fans are often used for attic ventilation in warm, sunny parts of the country. With the addition of a thermostat, the fan can be set to turn on and off based on temperature fluctuations.

Attic Ventilation

Description	Quantity/ Unit	Labor-Hours	Material
Remove ridge shingles and cut vent opening at ridge and soffit	1 Job	4.0	
Ridge vent strip, polyethylene	40 L.F.	2.0	122.40
Under eaves vent, aluminum, mill finish, 16" x 4"	10 Ea.	1.7	21.36
Totals		7.7	$143.76

Contractor's Fee, Including Materials:	$575

Exterior Painting or Staining

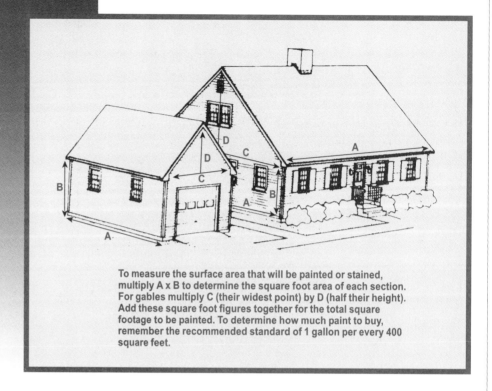

To measure the surface area that will be painted or stained, multiply A x B to determine the square foot area of each section. For gables multiply C (their widest point) by D (half their height). Add these square foot figures together for the total square footage to be painted. To determine how much paint to buy, remember the recommended standard of 1 gallon per every 400 square feet.

A peeling, blistering, faded exterior can make even an architecturally beautiful home look shabby. While there's no denying the hefty expense and labor that can be involved in painting a house's exterior, it's one of the best ways to spruce up your home—and an investment in both curb appeal and overall value. If you tackle this project yourself, and have the patience to follow professional painting practices, the savings can be substantial.

Materials & Tools

There are a variety of choices for covering and protecting the exterior wood siding on your home, all of which offer their own benefits and drawbacks. Discuss the options with a professional before purchasing your materials to decide which will best suit your needs and the look you want to accomplish.

The two main paint choices are acrylic latex-based and alkyd oil-based, both of which should last about five years before needing reapplication. There are advantages and disadvantages to each. Latex paint, which has become more popular in recent years, has better color retention and won't fade as quickly as oil-based paint. It's also more flexible after it dries, as opposed to oil paint, which dries harder, tends to be brittle, and thus is more likely to crack and peel. Latex dries faster, cleans up with water, and is somewhat easier and more pleasant to work with.

Oil-based paint tends to flow more easily off the brush, and levels better, leaving a smoother surface than latex. Over the years, oil paint chalks, or oxidizes, so that a new coat doesn't add much to the total thickness of the paint film. Latex, on the other hand, builds up with each repainting, resulting in a thicker film that becomes increasingly less flexible and resilient, and thus more apt to flake and chip. On bare wood, the linseed oil in oil paint permeates the fibers and helps prevent the siding's drying out. Latex, however, tends to bond to just the surface of the wood,

without adding moisture-restoring elements to it.

Stain is becoming a popular choice, especially for new, rough-sawn siding. It has a more "watery" consistency and goes on easier than paint. It also does not peel and should not need scraping. There are two types of stain: semi-transparent, usually oil-based, and opaque, usually latex. The former allows more of the grain and texture of the wood to show through; the latter is denser, having the consistency of very thin paint. Stain leaves little film on the wood's surface, and thus protects it less from sun and moisture than does paint; also, stain lasts only about half as long. Stained wood on the sunny side of a house should be given a new coating as often as every three years.

To prepare for painting, you'll need a paint scraper, sander, mild cleaning detergent, caulk and a caulk gun, and primer. For painting or staining, you'll need a variety of paint brushes and possible rollers, paint trays, a ladder or scaffolding, drop cloths, tape, and a dust mask and safety glasses.

What's Involved

The key to a good paint job is good preparation. Even the best paint, if applied to an unprepared surface, will begin to flake off in a couple of years, or sooner. Preparation involves scraping all loose, cracked, peeling, and blistering paint, and feathering the edges with sandpaper to blend the paint surface to the bare wood. If you use a power sander, be careful not to gouge or otherwise damage the siding or the trim. After scraping and sanding, the siding should be washed down with household detergent (or bleach and water where there is mold) and thoroughly rinsed off.

When the surface is dry, caulk any openings or cracks that might allow water, air, or insects to penetrate the walls. Also, thoroughly clean out any small areas of rot and cover them with exterior-grade spackling or epoxy filler. Badly rotted boards should be removed and replaced with new lumber. Remove any loose or cracked putty around the window panes, prime the bare wood, and reputty with glazing compound.

Spot-prime all bare wood on the siding and trim, and wherever there is evidence of chalking. Use the type of primer recommended on the label of the topcoat paint. Place drop cloths over plants, shrubs, walkways, and other exterior features that you don't want splattered with paint, and mask or remove such items as door hardware.

Use a 2-1/2" angled sash brush for the windows and small trim. Paint windows, working from the glass outwards to the frame. Use a 3" or 4" brush for the wider trim and siding. Work from the top down, and try to paint entire areas between natural breaks, such as between windows. In this way you'll avoid uneven application and lap marks, which occur when wet paint is brushed over paint that has already started to dry. Paint the bottom edge of the siding first, then the face. On wood shingles, get plenty of paint on their bottom edges to seal the porous end grain.

If there is little or no color change between the existing paint and the new coat, you may get by with one coat. Otherwise you'll need a second coat for complete coverage and to give the paint job a professional finish.

Level of Difficulty

Preparing and painting the exterior of a house, even a smaller one, are physically demanding and time-consuming tasks. Before you decide to do it yourself, calculate the time required to prepare and paint your whole house, taking into account the number of doors, windows, and shutters; the number of different colors involved; the height and accessibility of the walls; the tools and equipment needed; the speed at which you can efficiently work; and the amount of time (assuming good weather) you have available to do it.

Painting a house is a challenging job that requires time spent at heights and climbing up and down a ladder. Beginners should add at least 200% to the times estimated. Intermediates should add 100%, and experts, 50%.

What to Watch Out For

If you're thinking of using a power-washer to help remove loose paint, consult a paint professional first for guidance. Proceed with caution, as power-washing can damage wood. Any wood that is safe to power-wash must be given ample time to thoroughly dry before caulking and re-painting.

Spray-painting machines have become very popular over the last few years, but using one requires a good deal of preparation and careful application to prevent the paint from hitting unintended targets, like the trees or your car. You can save yourself time and effort by spray-painting the shutters, but generally, large-area spray painting should be left to professionals.

When spot-priming where a dark color will be used for the finish coat, you may want to tint the primer by adding a bit of the finish paint to it. This will make the finish coats cover more evenly.

Although professionals will tell you latex can be put over oil, and vice versa, you are safer using the same type of paint as is already on the house.

Exterior Painting, 2,000 S.F.

Description	Quantity/ Unit	Labor-Hours	Material
Exterior siding, clapboard, oil base, primer, brushwork	2,000 S.F.	24.6	216.00
Exterior siding, clapboard, oil base, 2 coats, brushwork	2,000 S.F.	39.5	384.00
Panel door and frame, oil base, primer, brushwork	2 Ea.	2.7	4.37
Panel door and frame, oil base, 2 coats, brushwork	2 Ea.	5.3	14.16
Windows, oil base, primer, brushwork	16 Ea.	10.7	13.82
Windows, oil base, 2 coats, brushwork	16 Ea.	18.3	29.95
Gutters and downspouts, oil base, primer, brushwork	80 L.F.	1.0	10.56
Gutters and downspouts, oil base, 2 coats, brushwork	80 L.F.	1.6	21.12
Totals		103.7	$693.98

Contractor's Fee, Including Materials:	**$6,274**

Re-Roofing with Asphalt Shingles

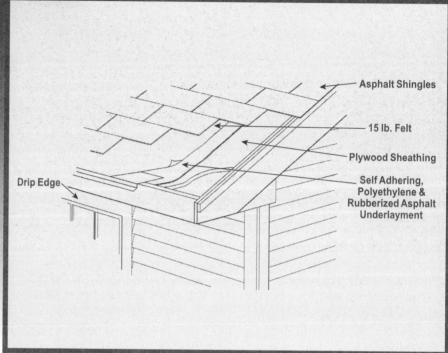

Asphalt Shingles

15 lb. Felt

Plywood Sheathing

Self Adhering,
Polyethylene &
Rubberized Asphalt
Underlayment

Drip Edge

A well-designed roof, clad with high-quality, durable materials, can last for many years with minimal maintenance. When that time runs out, however, homeowners are wise to replace the roofing before leaks damage inside ceilings and walls. In many areas, insurance companies will not cover homes that have a history of water damage. Evidence of rainwater penetrating the roof—such as stained ceilings and walls, peeling paint, or crumbling plaster—should never be ignored.

Materials & Tools

The choice of materials and method of installation are determined by the style of your home, cost, roof slope, local climate, and the material's longevity, energy-efficiency, and wind- and fire-resistance. If you're re-roofing, you might want to stick with the same material you're replacing, unless it proved unsatisfactory. If you're roofing a newly built addition, you'll want to match the rest of the house. Get advice from experienced local roofers, note the roofing materials prevalent in your area, and beware of investing in unproven products.

To estimate roofing materials, find the area of the roof in square feet and divide the total by 100 to determine the number of squares needed. For a simple, unobstructed roof, add another 10% for waste and cuts. A complicated roof—with many dormers, valleys, and obstructions—requires more materials.

Because they combine cost efficiency with acceptable looks, asphalt/fiberglass shingles are the choice for 80% of homes in North America. They're produced from a fiberglass mat, coated with asphalt, and surfaced with variously colored mineral granules. They'll last from 15-30 years, depending on their design, weight, and local climatic conditions.

Shingles come in several grades or weights, based on the thickness of the mineral surface and the weight of the base mat. Standard roof shingles weigh around 220 pounds per square (100 square feet of roof area—3 bundles); heavyweight shingles, from 290-330 pounds per square. Shingles—either wood or asphalt—can be laid on any roof with a minimum slope of 4:12; that is, a roof with a vertical rise of 4" for every 12" of horizontal run. Roofs with

less slope, or those that are flat, require some form of roll or sheet roofing.

Roofing projects are difficult for several reasons and best left to professionals. If you're confident in your capabilities for this job, you'll need some equipment including a ladder, sturdy staging, and a pump and roof jacks, and you might need to rent a dumpster for the old roofing debris. Small tools include a utility knife, tape measure, hammer, caulk gun, and tin snips.

A pneumatic roofing nail gun, powered by an air compressor, will cut one third of the time versus hand-driving nails with a hammer, but inexperienced users may sacrifice consistency and accuracy if they move too fast. Learn how to properly use a nail gun to operate it safely.

What's Involved

The first step in re-roofing is to remove the old materials. Stripping old shingles and cleaning up the resulting mess is hard, dirty work, and can take a day or two alone. Re-roofing usually begins at the eaves with a metal drip edge, followed by overlapping layers of 15-lb. roofing felt. In snowy climates, a wide strip of neoprene roofing membrane

wear and weathering. For this reason, tab shingles have a shorter life than the tabless variety. If your roof is shallow-pitched, screened by overhanging trees, or otherwise not highly visible, you might consider the durability of tabless shingles as an acceptable trade-off for their somewhat less attractive appearance.

Options

If slate is considered as a substitute for other materials or is simply replacing old, worn-out slate, the house's structural support system should be thoroughly evaluated beforehand, as the weight of this material is so much greater than most other roofing materials.

Terra cotta like slate, requires appropriate structural support, and has particular installation requirements. Both slate and tile roofs are best installed by professional contractors. Certain custom designs, sheet materials such as roll roofing, galvanized steel, aluminum, copper, and tin are other specialties.

should be added along the eaves, extending far enough up the roof to seal it against leaks caused by ice dams. On a roof of average pitch, a 3'-wide strip usually suffices.

Beginning with a double starter course, the shingles are laid up the roof, staggering the vertical joints and maintaining straight lines in both directions, horizontally along the butts and vertically along the cut-outs. At all intersections or obstructions—hips, valleys, walls, chimneys, skylights, soil stacks, etc.—an appropriate membrane or flashing should be installed, whether of roofing felt, aluminum, copper, or lead, and, in some cases, should also be sealed with caulk or roofing tar. The ridge is capped with either cut shingles or a ridge vent.

Level of Difficulty

Roofing should be done only by expert do-it-yourselfers or professionals. Beginners and intermediates should hire a contractor, or work with expert guidance and assistance.

Roofing involves working in high places and awkward positions, and is physically demanding. If you tackle any part of this yourself, you'll be climbing ladders carrying bundles of shingles weighing upwards of 70 pounds, and moving them around a sloping roof. You'll have to work for long periods of time bent over in the elements. You

must be able to measure, cut, and nail shingles quickly and accurately, while keeping your balance.

Beginners and intermediates should add 150% and 75%, respectively, for any tasks they can do. Expert do-it-yourselfers should have at least one helper and add 25% to the time estimated to complete the project.

What to Watch Out For

Narrow slots called "cut-outs" divide asphalt strip shingles into two or three tabs that, when laid, give them the appearance of separate pieces. The vertical lines of the cutouts add visual interest to the roof, but expose the shingles directly below them to more

Alternate Materials

Material Options
Cost per Square, Installed

Material	Cost
Asphalt roll roofing	$46.50
Asphalt shingles	$91.50
Cedar shingles	$310.00
Clay tile	$805.00

Re-Roofing with Asphalt Shinges, 1,000 S.F.

Description	Quantity/ Unit	Labor-Hours	Material
Roofing demolition, remove asphalt strip shingles	1,000 S.F.	11.4	
Self adhering polyethylene & rubberized asphalt underlayment	2.10 Sq.	0.8	105.84
Shingles, standard strip, class A, 210-235 lb. per square	10 Sq.	14.5	372.00
Asphalt felt sheathing paper, 15 lb.	1,100 S.F.	2.4	39.60
Drip edge, galvanized	70 L.F.	1.4	18.48
Nails, roofing, threaded, galvanized	50 Lb.		81.00
Totals		30.5	$616.92

Contractor's Fee, Including Materials:	**$2,398**

Re-Siding: Wood Clapboard

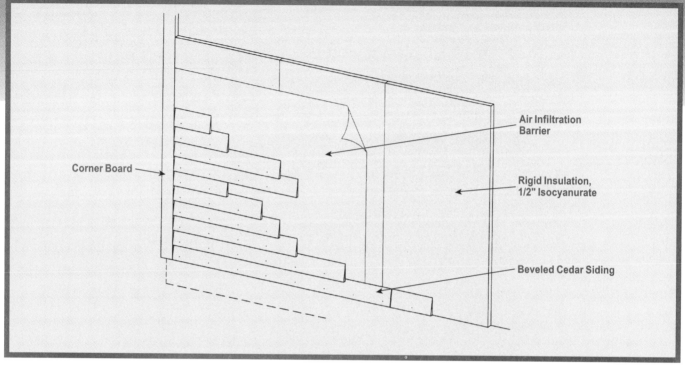

Corner Board

Air Infiltration Barrier

Rigid Insulation, 1/2" Isocyanurate

Beveled Cedar Siding

A house's siding is usually its most visible component. To do its part in the curb appeal equation, it must be attractive, fit the architecture, hold its finish, withstand the weather, and protect everything behind it from the elements. Over time, the ravages of sun, wind, rain, snow, and airborne pollution must be countered with continued maintenance.

There comes a time when siding, especially wood clapboards or shingles, must be replaced—because it's badly deteriorated, or because the style of the old siding is not attractive or appropriate to the house. Re-siding, or removing the existing siding and replacing it with new material, can dramatically transform your house, giving it a totally new appearance.

Materials & Tools

Choosing siding styles and materials is a matter of looks, performance, and cost. Consider the siding's history of durability, the sort of maintenance it requires, and its appropriateness to local climate. Wood, vinyl, and metal siding all have advantages and disadvantages, and the question of which is best is hotly debated by professionals. Real wood is versatile and easily matches most architectural styles. Cedar was chosen for this model project, for its resistance to decay , though redwood, Douglas fir, Ponderosa pine, and local species can also be good choices. Wood imitations are available in vinyl, aluminum, and other synthetic materials.

Remember that the results of your decision may outlast your ownership of the house, and will certainly affect its value for better or for worse. Read up on pros and cons of each type of siding

material and consult with knowledgeable local building professionals, not just materials salespeople, to get their views on the best siding for your house.

The most common traditional form of wood siding is clapboard. When properly installed, clapboards are inherently waterproof and, if properly maintained with paint or stain, can last for well over a hundred years. A high quality 1/2" x 6" vertical grain (VG) red cedar clapboard is about 1/2" thick at the bottom and 1/4" thick where the next board overlaps its upper edge. A thin board like this, cut to expose its vertical grain, is less likely to warp or split than a thicker board or one cut to produce a flat-grained surface. Cedar clapboards are very light and easy to work with. They can be installed smooth side out for painting, or rough side out for staining.

To estimate how much siding is needed, calculate the entire exterior surface area, deducting the area of all openings (window and doors) from the total. To calculate the area of a gable, multiply the length by the width, and divide the result by two. Be sure to factor in some additional material for waste.

In order to re-side your home, you will first need to remove the old siding, which requires the use of a special tool called a sidewall stripper. An air infiltration barrier of some sort is needed, and staples to hold it in place on the sheathing. For tools, you'll need a saw, a level, a chalk line, a hammer and nails, and scaffolding or ladders.

What's Involved

Before you can install new clapboards, you need to take down the original siding. Use the sidewall stripper and work from the top down. After making any necessary repairs to the exposed sheathing, cover it with an air infiltration barrier.

The new clapboards can be installed from the ground up, as the top clapboard overlaps the one beneath it to create an overhang and to hide the nails. The first row is the most important, as it will affect all successive rows. Be sure to check measurements every few courses against the first row to ensure a level installation.

It's good practice to back-prime (paint the back side of) clapboards and other exterior wood trim pieces before putting them up. This prevents moisture from penetrating the boards from back to front, and helps to extend the life of both lumber and finish.

Despite the instructions found in many manuals, it's not good practice to place the nails high enough on the clapboard

so they miss the top of the piece below. The idea that inspired this practice is sound enough: to let the boards expand and contract independently of one another. The problem is that the section of the clapboard is unsupported from behind, and you'll split almost as many boards as you nail, no matter how careful you are. It is better to use 5d hot-dipped galvanized box nails and place them about 1/2" from the bottom edge of the clapboard, nailing through the board below into the sheathing.

Level of Difficulty

Re-siding with wood clapboards is not a job for a beginner working alone, but is certainly within the abilities of a competent, intermediate-level do-it-yourselfer. Putting up the air infiltration barrier, taking measurements, snapping chalk lines, and handling long pieces of clapboard are all much easier with a helper and using pump jack scaffolding instead of ladders, which have to be constantly moved and adjusted. Having a third helper to make the cuts on a fine-

toothed chop saw will save you even more time and effort.

Beginners should have proper guidance from preparation through completion, and should add 150% to the estimated times. Intermediates and experts will find the job much easier to accomplish with at least one helper. Intermediates should add about 75% to the times; experts, about 25%.

What to Watch Out For

A significant advantage to replacing the siding on an older house is that you can reduce drafts by installing a good air infiltration barrier (house-wrap). Heat loss can be further reduced by adding 1/2" of rigid insulation.

Once your siding is installed, it's crucial to maintain it with a coat of stain or paint every few years—every 2-5 years for stain, and every 7 or so years for paint.

See also:
Exterior Painting or Staining.

Re-Siding with Wood Clapboard, 16'-4" x 135'

Description	Quantity/ Unit		Labor-Hours	Material
Siding demolition, clapboards, horizontal	2,200	S.F.	46.3	
Siding, 1/2" x 6" beveled cedar, A grade	2,200	S.F.	70.4	8,659.20
Housewrap, spun bonded polypropylene	2,200	S.F.	4.6	422.40
Insulation board, isocyanurate, 1/2", R-3.9	2,200	S.F.	22.0	765.60
Paint, back prime clapboards, oil base, primer, brushwork	2,200	S.F.	27.1	237.60
Clapboards, oil base, primer, brushwork	2,200	S.F.	27.1	237.60
Clapboards, oil base, 1 coat, brushwork	2,200	S.F.	43.5	422.40
Totals			241.0	$10,744.80

Contractor's Fee, Including Materials: **$27,818**

Room Additions

A room addition is clearly one of the biggest projects one can take on, not only in physical size, complexity, and cost, but in the planning needed to make sure it blends properly with the existing house. This is a case when it pays to protect your investment by consulting an architect. He or she can help you define the dimensions and features of your addition so that it will meet your needs (within your budget, and within building code and zoning requirements) and match your home's exterior and interior trim details, roof slopes, and window and dormer styles and spacing.

Following are some additional items to consider.

- Check building codes and any zoning restrictions for permit requirements and other issues that may affect your building plans.

- Make a rough sketch of what you have in mind for the addition before consulting with a builder or architect.

- During the design stage, go out in the yard and mark off the addition with stakes and string—get a feel for how it will affect your yard, your view, and the neighboring properties.

- In an addition that includes a public area like a family room, allow for plenty of natural light.

- If you need to remove trees when clearing the site for an addition, check with your local building department to make sure there are no restrictions. Certain trees may have a landscape resale or firewood value. Stump disposal can be expensive if the stumps cannot be buried on-site.

- If excavating for the new addition involves large quantities of cut (removal of soil) or fill (bringing in additional soil), be aware of the effects this may have on site drainage and utilities.

- Contact your local utility companies to verify the location of—and avoid damage to—underground electrical, gas, water, and septic lines.

- If you plan and estimate your project well in advance of construction, get a cost update before work actually begins. The price of lumber can change dramatically in a relatively short time, and this will have a big effect on a materials list for a sizeable addition.

- Depending on the wall spans and heights, and the grade of lumber used, some waste of materials is inevitable. A rule of thumb for lumber waste is 5% to 10%, depending on material quality and the complexity of the project.

8' x 8' Entry Addition

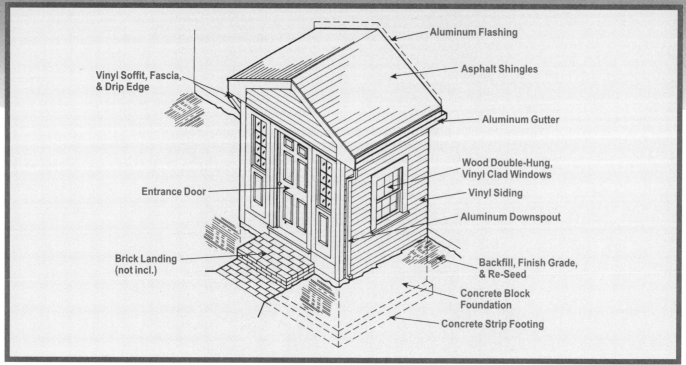

- Vinyl Soffit, Fascia, & Drip Edge
- Aluminum Flashing
- Asphalt Shingles
- Aluminum Gutter
- Entrance Door
- Wood Double-Hung, Vinyl Clad Windows
- Vinyl Siding
- Aluminum Downspout
- Brick Landing (not incl.)
- Backfill, Finish Grade, & Re-Seed
- Concrete Block Foundation
- Concrete Strip Footing

One of the first things visitors see when they come to your home is the entryway. If your home has a flat, uninspired front entrance, you might consider building a foyer addition—for curb appeal and a convenient space to enter the house, take off your coat, and transition to the living area. Because this project is small and at first-floor level, do-it-yourselfers can take a hand in its construction and enjoy the benefits of lower costs.

Materials, Tools, & Equipment

Room additions, like other projects involving attached structures, should complement the style and design of the house, while meeting the need for new living area. The structure requires framing, sheathing, and finish materials common to wood-framed houses, which tie in with the existing house. The support for the floor of the room consists of joists and plywood sheathing.

For the roof, this plan includes economical asphalt strip shingles, vinyl siding, and aluminum soffits and fascias. Wood siding or shingles and many other finish products can be substituted to blend with the style of your home. These changes will alter the cost of materials and the estimated installation time.

Inside, many different surface coverings and trim materials can be used, according to your décor in the adjoining rooms, and practical requirements, such as tile flooring for a waterproof entrance. Two double-hung windows and a new steel door with sidelights are included in this model plan. If you're happy with your old exterior door, it can be relocated to the new entryway.

To build this addition, you'll need common carpentry tools for the framing, as well as more specialized items like a sidewall stripper to remove the siding from what will become interior wall, and shims to accurately hang the front door and any windows.

What's Involved

The excavation should be wide enough to accommodate the footing and foundation walls, and deep enough to rest on firm, frost-free subsoil. Once the footing has been carefully laid out and put in place, the concrete block wall can be built. A formed and poured concrete wall is a slightly more expensive option. Generally, do-it-yourselfers should leave the excavation and masonry tasks to a professional contractor, as specialized skills, equipment, and considerable physical exertion are involved.

Before the floor frame is placed, a 2 x 8 pressure-treated sill should be lagged or bolted into the top of the foundation wall. A vent should be included in the foundation as protection against dampness and deterioration.

When planning the roof frame, be sure to match or complement the pitch of the house roof and the design of its soffit and fascia. This part of the project requires substantial carpentry skills and a basic knowledge of framing procedures. Most do-it-yourselfers can accomplish the roof work for this addition if they're instructed in procedures for laying it out and flashing it correctly at the house wall.

The sheathing and exterior finish materials can be applied once the framing is complete. Be sure to take care of all the preliminary work, such as installing the electrical rough-in, insulation, and underlayment, before you apply interior finish surfaces.

Level of Difficulty

Because of its small size, much of this entry addition can be completed by do-it-yourselfers. As noted earlier, the foundation work should be left to a contractor, but the finish work and possibly the framing can be performed by those with moderate remodeling skills.

Siding, roofing, electrical work, and flooring can be done much faster and with better results by a specialty contractor.

Generally, beginners should add 100% to the time estimate for the basic tasks and hire a professional for the technically demanding specialty operations. Intermediates should add 50% to the time for basic tasks and carpentry procedures, and 70% for any specialty work. Experts should add about 20% to the professional time for all tasks. All do-it-yourselfers should hire an electrician for the electrical rough-in and circuit tie-in.

What to Watch Out For

Although windows and doors are some of the more expensive items in any home improvement project, it does not pay to compromise on their quality. Whenever possible, purchase well-made thermopane window units. Not only do they provide ongoing energy savings, but they usually include screens and do not require the extra expense of storm windows.

See also:
Main Entry Door, New Window.

Entry Addition, 8' x 8'

Description	Quantity/Unit	Labor-Hours	Material
Excavate trench, with machine, 4' deep	1 C.Y.	0.1	
Footing, 8" thick x 16" wide x 20' long	1 C.Y.	2.9	139.20
Foundation, 8" concrete block, not reinf., 1/2" parged, 4' high	96 S.F.	9.0	207.36
Frame, floor, 2 x 8 joists, 16" O.C., 8' long	80 L.F.	1.2	79.68
Floor sheathing, 5/8" plywood, 4' x 8' sheets	64 S.F.	0.8	59.90
Frame, walls, 2 x 4 studs & plates, 8' long	20 L.F.	4.8	118.08
Header over window, 2 x 8 stock	8 L.F.	0.4	7.97
Wall sheathing, 1/2" x 4' x 8' plywood	192 L.F.	2.7	142.85
Frame, roof, 2 x 6 rafters, 16" O.C.	80 L.F.	1.3	51.84
Ridge board & sub-fascia, 2 x 8 x 8'	24 L.F.	0.9	23.90
Ceiling, 2 x 6 joists, 16" O.C.	48 L.F.	0.6	31.10
Roof sheathing, 1/2" x 4' x 8' plywood	96 L.F.	1.1	71.42
Siding, vinyl embossed, 8" wide, with trim	180 S.F.	5.8	142.56
Soffit, 12" wide, vented, fascia & drip edge	35 L.F.	4.7	62.16
Roofing, standard strip shingles, class C, 4 bundles per square	1 Sq.	2.0	58.80
Housewrap, spun bonded polypropylene	320 S.F.	0.7	61.44
Flashing, aluminum stop flashing, 0.019" thick	20 S.F.	1.1	15.60
Gutters, 5" aluminum, 0.027" thick	16 L.F.	1.1	23.81
Downspouts, 2" x 3", embossed aluminum, 0.020" thick	20 L.F.	0.8	16.32
Wall insulation, fiberglass, 3-1/2" thick, R-11	180 S.F.	0.9	86.40
Floor insulation, fiberglass, kraft faced, 6" thick, R-19	64 S.F.	0.4	26.88
Ceiling, insulation, fiberglass, 9" thick, R-30	50 S.F.	0.3	24.60
Window, double hung, vinyl clad, 2'-4" x 5'-4"	2 Ea.	1.6	664.80
Exterior door, steel, 3'-0" x 6'-8", prehung	1 Ea.	1.0	564.00
Sidelights	2 Ea.	1.1	744.00
Lockset, standard duty, cylindrical, keyed	1 Ea.	0.8	84.60
Gypsum wallboard, walls, 1/2" thick, taped and finished	320 S.F.	5.3	96.00
Underlayment, 5/8" plywood, underlayment grade	64 S.F.	0.7	93.70
Flooring, vinyl sheet, 0.125" thick	50 S.F.	2.0	238.80
Trim, ranch base & casing	50 L.F.	1.7	91.80
Paint, walls & ceilings, primer, oil base, roller	320 S.F.	1.3	19.20
Door and frame, primer, oil base, brushwork	1 Ea.	1.3	2.18
Window, incl. frame and trim, primer, oil base, brushwork	2 Ea.	1.1	0.70
Trim, primer, oil base, brushwork	50 L.F.	0.6	1.20
Walls & ceilings, 2 coats, oil base, roller	320 S.F.	1.3	19.20
Door and frame, 2 coats, oil base, brushwork	1 Ea.	1.3	2.18
Window, incl. frame and trim, 2 coats, oil base, brushwork	2 Ea.	1.1	0.70
Trim, 2 coats, oil base, brushwork	50 L.F.	0.6	1.20
Wiring, 4 outlets	4 Ea.	6.0	129.60
Wiring, 1 switch	1 Ea.	1.4	34.80
Backfill, by hand, no compaction, light soil	1 C.Y.	0.6	
Fine grading and seeding, incl. lime, fertilizer & seed	7 S.Y.	0.3	1.34
Totals		74.7	$4,241.87

Contractor's Fee, Including Materials: $10,253

8' x 12' Room Addition

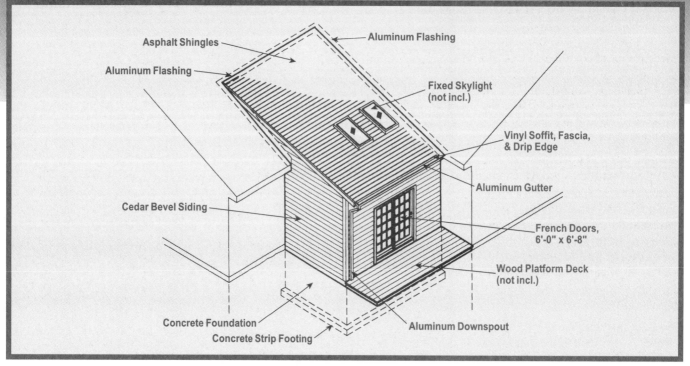

Asphalt Shingles
Aluminum Flashing
Aluminum Flashing
Fixed Skylight
(not incl.)
Vinyl Soffit, Fascia,
& Drip Edge
Aluminum Gutter
Cedar Bevel Siding
French Doors,
6'-0" x 6'-8"
Wood Platform Deck
(not incl.)
Concrete Foundation
Concrete Strip Footing
Aluminum Downspout

For homeowners who have tight property lines, but need more space for an office, kitchen or bedroom expansion, or playroom, an 8 x 12 addition may be the perfect solution. If you can plan it between two existing wings of your home, cost savings can be significant. While the structure is moderate in size, this is a challenging project for most do-it-yourselfers, as it calls for quite a few construction skills.

Materials & Tools

One benefit of placing an addition in a corner location is that two walls are already in place. Thus, you can save on the materials and labor normally required to frame and cover them. For the new construction, a concrete foundation is required on the remaining two sides.

The other materials are framing lumber, roofing, and interior and exterior finish products that match the style of the house and function of the room. Cost will vary accordingly. Because of the size of the order and weight of materials, arrange for delivery before you close the

deal. Standard hand and power tools are also needed if you're planning to do some of the work yourself.

What's Involved

First, the site must be cleared and excavated to a 4' depth, in preparation for the foundation. The foundation in this plan is constructed from poured concrete, but concrete blocks can also be used. Foundation vents are a good investment, as they help increase air circulation and prevent moisture from accumulating underneath the structure. In most cases, excavations and foundation installations of this size are best left to a contractor. Accurate leveling of the forms and skillful concrete placement are vital to the success of the project. Next, the sill, floor joists, and floor sheathing are placed.

The two exterior walls are then framed on the deck and raised into position. The sheathing for the walls in this plan is 3/4" rigid styrofoam, but standard 1/2" plywood can also be used.

Framing and flashing a roof for a corner structure can be tricky, so most do-it-yourselfers should consider hiring a professional for at least this part of the project. The roof requires plywood sheathing before it is covered with a layer of felt paper and asphalt shingles. When the roof structure is complete, the gable end is framed with individually placed 2 x 4s.

Before installing the interior finish materials, insulate the exterior walls and the ceiling at the recommended R-values for your location. In colder climates, the floor should also be insulated before the sheathing is placed. Additional materials costs will have to be added to the project, but your floors will be warmer in winter, and your heating bill lower.

Hire a contractor for the electrical work unless you're an expert. After it's been roughed in, the interior finish materials can be installed. Hanging and finishing the gypsum wallboard requires some know-how, but most intermediate-level do-it-yourselfers can do the job with assistance.

Level of Difficulty

This project requires a moderate to advanced level of skill in many construction procedures. Once the structure is closed in, interior tasks like painting and laying the flooring can be completed over an extended period. Beginners can approach this work at a relaxed pace, having left most of the exterior work to contractors.

For more basic tasks, like painting and sheathing, beginners should add 100% to the professional time. Intermediates should add 50% for all but the roof framing and finishing, and should leave the electrical and carpet installations to professionals. Experts should add 20% to all tasks. All do-it yourselfers should hire a contractor for the excavation and foundation work.

What to Watch Out For

Visit your local zoning board with the dimensions of your home and its proximity to your property lines before you design your addition. You'll need permits to build, and the options for size and location may still be limited by local zoning requirements.

See also:

Bow/Bay Window, Main Entry Door, Skylights, New Window, Exterior Painting or Staining, Flooring Projects*.

*In *Interior Home Improvement Costs*

Room Addition, 8' x 12'

Description	Quantity/ Unit	Labor- Hours	Material
Excavate, with machine, 4' deep	28 C.Y.	3.0	
Footing, 8" thick x 16" wide x 20' long	2 C.Y.	5.9	278.40
Foundation wall, 8" thick concrete block, 4' high	176 S.F.	16.6	380.16
Frame, floor, 2 x 10 joists, 16" O.C., 12' long	192 L.F.	3.4	271.87
Sill plates, 2 x 8, pressure treated	48 L.F.	1.7	57.60
Floor, sheathing, 5/8" plywood, 4' x 8' sheets	192 S.F.	2.3	179.71
Frame, walls, 2 x 4 plates, 12' long	144 L.F.	2.9	58.75
2 x 4 studs, 8' long	320 L.F.	4.7	130.56
Header over window & door, 2 x 8	36 L.F.	1.7	35.86
Wall sheathing, 1/2" plywood, 4' x 8' sheets	384 S.F.	5.5	285.70
Frame, roof, 2 x 6 rafters, 16" O.C., 8' long	208 L.F.	3.3	134.78
Ceiling, 2 x 6 joists, 16" O.C.	156 L.F.	2.0	101.09
Ridge board & sub-fascia, 2 x 8	50 L.F.	1.8	49.80
Roof sheathing, 5/8" plywood, 4' x 8' sheets	288 S.F.	3.5	269.57
Building paper, asphalt felt, 15 lb., on roof & walls	700 S.F.	1.5	25.20
Roofing, asphalt shingles, class A, 285 lb. premium	3 Sq.	6.9	196.20
Flashing at wall, 0.01 thick	16 S.F.	0.9	9.79
Ridge vent strip, polyethylene	16 L.F.	0.8	48.96
Siding, no. 1 red cedar shingles, 7-1/2" exposure, on walls	2.50 Sq.	9.8	324.00
Soffit, 12" wide vented, fascia & drip edge	48 L.F.	6.4	85.25
Gutters, 5" aluminum, 0.027" thick	34 L.F.	2.3	50.59
Downspouts, 2" x 3", embossed aluminum, 0.020" thick	20 L.F.	0.8	16.32
Wall insulation, fiberglass, 3-1/2" thick, R-11	400 S.F.	2.0	192.00
Ceiling insulation, fiberglass, 6" thick, R-19	200 S.F.	1.0	98.40
Sliding door, vinyl clad, 6'-0" x 6'-8"	1 Ea.	4.0	1,380.00
Dbl.-hung window, 36" x 48", insul., incl. frame, screen, & trim	3 Ea.	2.7	1,170.00
Gypsum wallboard, walls, 1/2" thick, taped and finished	384 S.F.	6.4	115.20
Paneling, 1/4" thick for wainscot, 4' x 8' sheets	160 S.F.	6.1	232.32
Plywood, 3/8" thick, backing for paneling, 4' x 8' sheets	160 S.F.	2.1	124.80
Carpet, 24 oz. nylon, light to medium traffic	207 S.F.	3.7	317.95
Prime quality, urethane pad for carpet	207 S.F.	1.2	62.10
Trim, ranch, base, casing & wainscoting	155 L.F.	5.2	284.58
Paint, walls & ceilings, primer, oil base, roller	420 S.F.	1.6	25.20
Door and frame, primer, oil base, brushwork	1 Ea.	1.3	2.18
Window, incl. frame and trim, primer, oil base, brushwork	3 Ea.	1.7	1.04
Trim, primer, oil base, brushwork	155 L.F.	1.9	3.72
Walls & ceilings, 2 coats, oil base, roller	420 S.F.	1.6	25.20
Door and frame, 2 coats, oil base, brushwork	1 Ea.	1.3	2.18
Window, incl. frame and trim, 2 coats, oil base, brushwork	3 Ea.	1.7	1.04
Trim, 2 coats, oil base, brushwork	155 L.F.	1.9	3.72
Wiring, 4 outlets	4 Ea.	6.0	129.60
Wiring, 1 switch	1 Ea.	1.4	34.80
Backfill, by hand, no compaction, light soil	2 C.Y.	1.1	
Fine grading and seeding, incl. lime, fertilizer & seed	21 S.Y.	1.0	4.03
Totals		144.6	$7,200.22

Contractor's Fee, Including Materials: $10,842

12' x 16' Room Addition

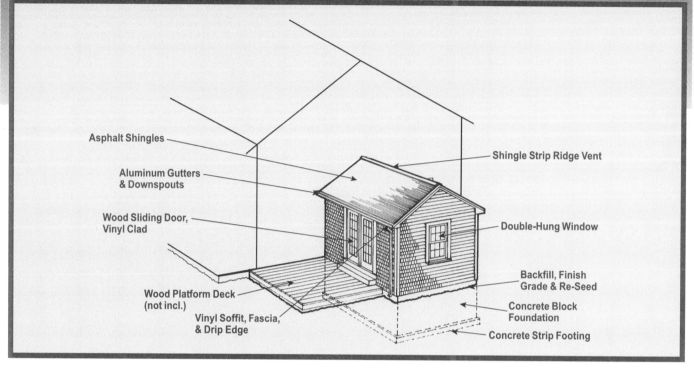

- Asphalt Shingles
- Aluminum Gutters & Downspouts
- Wood Sliding Door, Vinyl Clad
- Wood Platform Deck (not incl.)
- Vinyl Soffit, Fascia, & Drip Edge
- Shingle Strip Ridge Vent
- Double-Hung Window
- Backfill, Finish Grade & Re-Seed
- Concrete Block Foundation
- Concrete Strip Footing

An addition like the one in this model project is like a miniature house with its own foundation, frame, and interior and exterior finishes. Because it's a big project, beginners and intermediate-level do-it-yourselfers with limited remodeling experience should hire a contractor to do much of the work.

Materials & Tools

Your choice of inside and outside finish materials will depend on the room's function and the style of your house. On the exterior, it's particularly important to complement the features of the existing house, such as siding, roofing, and style of windows.

The support and subsurface materials are pretty similar regardless of the home's style. The materials list will include framing, sheathing, and finish products. If you plan to do some of the work yourself, you'll need standard hand and power tools. Experts who think they're up to the foundation work will need to rent an excavator.

What's Involved

Care should be taken in determining the location of the addition, with consideration of roof lines and exterior trim to ensure that it will blend with the existing house. Just as important are local zoning regulations, which should be checked before you design the addition or apply for a permit.

Most do-it-yourselfers should have the excavation and foundation work done by a contractor. Even if you have the knowledge and experience to install the foundation yourself, check with a professional before you start, to make sure your methods are correct. A poorly constructed or misaligned foundation can cause major problems.

After the foundation has been placed, the framing, sheathing, and exterior finishing tasks can be accomplished. A 2 x 8 sill plate is secured to the top of the foundation wall, and 2 x 10 floor joists installed. Use at least 5/8" plywood for the decking and stagger the termination seams for additional floor strengthening.

Standard 2 x 4 framing is covered with 1/2" sheathing, and appropriate headers and jack studs are used for the door and window openings. If you're not experienced in framing, get some help for this part of the wall and roof work.

The gable roof is basic in design, but requires skill to construct. The most critical aspect of the roof framing is the layout of the rafter template, which must duplicate or complement the roof pitch and overhang features of the existing house. This plan also includes a ridge vent strip, which should be installed as part of the finish roofing.

The final steps in the exterior work include finishing the soffit and trim, installing the door and windows, and applying the siding.

Within the new room, preliminary work on the floor, ceiling, and walls has to be completed before the finish products are placed. The sub-floor should be covered with appropriate underlayment if carpeting or resilient flooring is to be installed. The ceiling and walls should be insulated and roughed in for the desired electrical fixtures.

Courtesy of Robert and Susan Alexander

Level of Difficulty

Beginners and intermediate-level do-it-yourselfers can tackle the painting, landscaping, and other less demanding tasks, but advanced procedures like the foundation, roof framing, and finish work should be left to experts or a professional contractor.

Beginners should add at least 100% to the professional labor-hours for the tasks they undertake. Intermediates should be able to handle the interior work, except the carpeting and electrical, adding 50% to the professional time. Experts should add 20% for the work they do.

What to Watch Out For

The existing house wall that is included in the addition requires considerable preparation before it can be finished to blend in with the new room. The exterior siding has to be removed, and any structural alterations made before the wallboard or other finish material is applied. If an acceptable entryway from the house to the new room already exists, no structural work is required, and the rough opening can simply be trimmed. If there is no access, or an existing entryway has to be widened, the wall will have to be opened, and new structural members added.

See also:

Bow/Bay Window, Main Entry Door, Skylights, New Window, 8' x 12' Room Addition.

Room Addition, 12' x 16'

Description	Quantity/Unit	Labor-Hours	Material
Excavate, with machine, 4' deep	28 C.Y.	3.0	
Footing, 8" thick x 16" wide x 20' long	2 C.Y.	5.9	278.40
Foundation wall, 8" thick concrete block, 4' high	176 S.F.	16.6	380.16
Frame, floor, 2 x 10 joists, 16" O.C., 12' long	192 L.F.	3.4	271.87
Sill plates, 2 x 8, pressure treated	48 L.F.	1.7	57.60
Floor, sheathing, 5/8" plywood, 4' x 8' sheets	192 S.F.	2.3	179.71
Frame, walls, 2 x 4 plates, 12' long	144 L.F.	2.9	58.75
2 x 4 studs, 8' long	320 L.F.	4.7	130.56
Header over window & door, 2 x 8	36 L.F.	1.7	35.86
Wall sheathing, 1/2" plywood, 4' x 8' sheets	384 S.F.	5.5	285.70
Frame, roof, 2 x 6 rafters, 16" O.C., 8' long	208 L.F.	3.3	134.78
Ceiling, 2 x 6 joists, 16" O.C.	156 L.F.	2.0	101.09
Ridge board & sub-fascia, 2 x 8	50 L.F.	1.8	49.80
Roof sheathing, 5/8" plywood, 4' x 8' sheets	288 S.F.	3.5	269.57
Building paper, asphalt felt, 15 lb., on roof & walls	700 S.F.	1.5	25.20
Roofing, asphalt shingles, class A, 285 lb. premium	3 Sq.	6.9	196.20
Flashing at wall, 0.01 thick	16 S.F.	0.9	9.79
Ridge vent strip, polyethylene	16 L.F.	0.8	48.96
Siding, no. 1 red cedar shingles, 7-1/2" exposure, on walls	2.50 Sq.	9.8	324.00
Soffit, 12" wide vented, fascia & drip edge	48 L.F.	6.4	85.25
Gutters, 5" aluminum, 0.027" thick	34 L.F.	2.3	50.59
Downspouts, 2" x 3", embossed aluminum, 0.020" thick	20 L.F.	0.8	16.32
Wall insulation, fiberglass, 3-1/2" thick, R-11	400 S.F.	2.0	192.00
Ceiling insulation, fiberglass, 6" thick, R-19	200 S.F.	1.0	98.40
Sliding door, vinyl clad, 6'-0" x 6'-8"	1 Ea.	4.0	1,380.00
Dbl.-hung window, 36" x 48", insul., incl. frame, screen, & trim	3 Ea.	2.7	1,170.00
Gypsum wallboard, walls, 1/2" thick, taped and finished	384 S.F.	6.4	115.20
Paneling, 1/4" thick for wainscot, 4' x 8' sheets	160 S.F.	6.1	232.32
Plywood, 3/8" thick, backing for paneling, 4' x 8' sheets	160 S.F.	2.1	124.80
Carpet, 24 oz. nylon, light to medium traffic	207 S.F.	3.7	317.95
Prime quality, urethane pad for carpet	207 S.F.	1.2	62.10
Trim, ranch, base, casing & wainscoting	155 L.F.	5.2	284.58
Paint, walls & ceilings, primer, oil base, roller	420 S.F.	1.6	25.20
Door and frame, primer, oil base, brushwork	1 Ea.	1.3	2.18
Window, incl. frame and trim, primer, oil base, brushwork	3 Ea.	1.7	1.04
Trim, primer, oil base, brushwork	155 L.F.	1.9	3.72
Walls & ceilings, 2 coats, oil base, roller	420 S.F.	1.6	25.20
Door and frame, 2 coats, oil base, brushwork	1 Ea.	1.3	2.18
Window, incl. frame and trim, 2 coats, oil base, brushwork	3 Ea.	1.7	1.04
Trim, 2 coats, oil base, brushwork	155 L.F.	1.9	3.72
Wiring, 4 outlets	4 Ea.	6.0	129.60
Wiring, 1 switch	1 Ea.	1.4	34.80
Backfill, by hand, no compaction, light soil	2 C.Y.	1.1	
Fine grading and seeding, incl. lime, fertilizer & seed	21 S.Y.	1.0	4.03
Totals		144.6	$7,200.22

Contractor's Fee, Including Materials:	**$18,411**

20' x 24' Room Addition

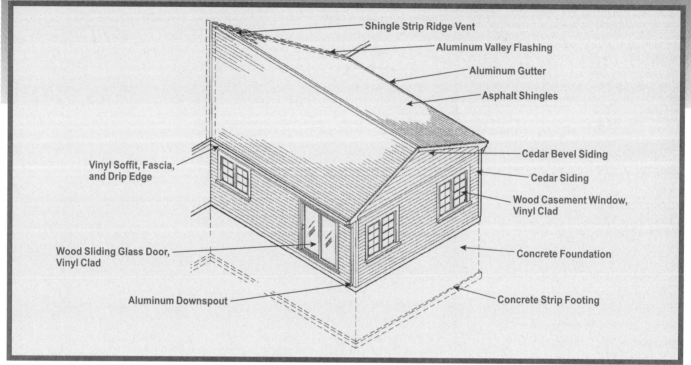

Shingle Strip Ridge Vent

Aluminum Valley Flashing

Aluminum Gutter

Asphalt Shingles

Cedar Bevel Siding

Cedar Siding

Wood Casement Window, Vinyl Clad

Concrete Foundation

Concrete Strip Footing

Vinyl Soffit, Fascia, and Drip Edge

Wood Sliding Glass Door, Vinyl Clad

Aluminum Downspout

If it's in proper scale with the other homes in your neighborhood, building a room addition can enhance the value of your home, while providing the extra space you need.

In this project, a large addition is attached to the eave-end of a house at first-floor level. It has over 400 square feet of floor area, a full 20' x 24' basement, and a roof structure tied into the existing house. This area is big enough to accommodate a new master bedroom suite or a family room/kitchen.

Materials, Tools, & Equipment

All exterior and interior finish materials should be selected to complement the style of your house. Included are concrete for the foundation, framing lumber, sheathing, an air-infiltration barrier, insulation, roofing and flooring, electrical materials, windows and doors, siding, trim, and finish materials, like gypsum wallboard and paint.

Unless you have a good-sized garage or sheltered enclosure, wait to have interior materials delivered until after the shell has been erected and made weathertight. Even the exterior items require a large outside area adjacent to the site. Ask the supplier if delivery can be staggered through the phases of construction.

For the foundation work, an excavator will be needed to dig 9' down for the foundation. The tools for this project are many and varied because of the numerous construction specialties.

What's Involved

This is a major project, best constructed by a professional contractor. The amount of work involved in the first step, site excavation, will vary depending on the topographical features of your yard. If the grade falls off to or below basement walk-out level, there will be less excavation. The basement floor slab can be placed at any time after the foundation has been completed.

A tripled 2 x 10 center support beam must be installed to bear the load of the floor and its frame. This primary structural component must be properly supported with lally columns or other suitable posts placed at regular intervals and set into a preformed notch in the foundation wall. After the sill has been placed and the 2 x 10 floor joists installed, 5/8" plywood sheathing will cover the frame.

The 2 x 4 walls are then laid out, built on the deck, positioned, fastened, and covered with 1/2" sheathing. The next step is placing and sheathing the ceiling joists and roof rafters. Then the gable end can be studded and sheathed to complete the framing. After the layout and placement of the new roof frame have been completed, the finished roofing, soffit and trim, window and doors, and siding are installed to close in the new structure.

The shell should be thoroughly insulated on the ceiling, walls, and, in colder areas, the floor. The electrical rough-in is installed, followed by the finish coverings. The gypsum wallboard for the walls and ceiling can be taped and finished or, at extra cost, professionally skim-coated with plaster. All of the interior finish products demand careful and patient installation; specialty contractors may have to be hired for this job.

Level of Difficulty

Most do-it-yourselfers should hire a general contractor or subcontractors to do the heavy exterior work and specialty installations. Beginners can probably handle the insulation, wallboard, trim work, and painting. They should add 100% to the professional time for these jobs and seek advice as they go.

Beginners and most intermediate-level do-it-yourselfers should not tackle the framing, and should hire a contractor for the wall framing and the roof. Intermediates should tack on 50% to the times for non-specialty tasks. Experts should add 20% to all procedures except the roof framing and specialty finish work, which take more time. All do-it-yourselfers should leave the foundation work and the basement floor to a contractor.

What to Watch Out For

Building a stairway to the new basement will take away valuable floor space and may require a partition and a door to enclose the stairwell. Exterior cellar access is an option.

See also:
Bow/Bay Window, Patio Doors, New Window, Room Additions, Skylights.

Alternate Materials

Door Options
Cost per Each, Installed

Steel entrance, 3' w	$265.00
Wood entrance, 3' w	$380.00
Sliding door, wood, 6' w	$1,300.00
Sliding door, alum, 6' w	$1,600.00
French door, wood, 5' w	$1,350.00

Room Addition, 20' x 24'

Description	Quantity/Unit	Labor-Hours	Material
Excavate, trench, with machine, 9' deep	160 C.Y.	11.4	
Haul-off excess material	135 C.Y.	15.4	
Footing, 8" thick x 16" wide x 20' long	3 C.Y.	8.8	417.60
Foundation walls, 8" thick C.I.P. concrete	12 C.Y.	23.8	2,073.60
Basement floor slab, 4" thick conc., text. finish, no reinf., no form	480 S.F.	9.4	604.80
Frame floor, 2 x 10 joists, 16" O.C., 12' long	540 L.F.	9.6	764.64
Center support beam, triple 2 x 10	24 L.F.	0.7	67.68
Floor, sheathing, 5/8" plywood, 4' x 8' sheets	480 S.F.	5.7	449.28
Frame & trim, basement stairs, complete	1 Ea.	5.3	1,044.00
Frame, walls, 2 x 4 plates, 12' long, 2 x 4 studs, 8' long	96 L.F.	19.2	472.32
Header over window & door, 2 x 8	60 L.F.	2.8	59.76
Wall sheathing, 3/4" x 4' x 8' plywood	544 S.F.	8.9	613.63
Housewrap, spun bonded polypropylene	680 S.F.	1.4	130.56
Corner bracing, 16 ga. steel straps, 10' long	80 L.F.	1.1	64.32
Frame, roof, 2 x 8 rafters, 16" O.C., 14' long	1,008 L.F.	17.0	1,003.97
Ridge board & sub-fascia, 2 x 10	38 L.F.	3.5	124.61
Roof sheathing, 1/2" x 4' x 8' plywood	800 S.F.	9.1	595.20
Roofing, asphalt shingles, 285 lb. premium	8 Sq.	18.3	523.20
Building paper, asphalt felt sheathing paper, 15 lb., on roof	800 S.F.	1.7	28.80
Ridge vent strip, polyethylene	20 L.F.	1.0	61.20
Siding, cedar board & batten, pre-stained	390 S.F.	17.3	1,445.04
Cedar bevel, 1/2" x 8", grade A, pre-stained	60 S.F.	1.5	175.20
Soffit, 12" wide vented, with fascia	48 L.F.	6.4	85.25
Drip edge, aluminum, 0.016" thick, mill finish	48 L.F.	1.0	14.40
Flashing, at valley, 0.019" thick	80 S.F.	4.4	62.40
Gutters, 5" aluminum, 0.016" thick	68 L.F.	4.5	101.18
Downspouts, 5" aluminum, 0.025" thick	68 L.F.	3.9	121.58
Wall insulation, fiberglass, 3-1/2" thick, R-11	550 S.F.	2.8	264.00
Ceiling, insulation, fiberglass, 9" thick, R-30	480 S.F.	2.8	408.96
Door, sliding glass, 6' x 6'-8", vinyl clad, insul. glass	1 Ea.	4.0	1,380.00
Casement window, 24" x 52", vinyl clad, thermopane	4 Ea.	3.6	1,584.00
Casement window, 24" x 36", vinyl clad, thermopane	2 Ea.	1.6	422.40
Gypsum wallboard, walls, 1/2" thick, taped and finished	1,184 S.F.	19.6	355.20
Carpet, 22-oz. nylon, light to medium traffic	486 S.F.	8.7	746.50
Prime urethane pad for carpet	486 S.F.	2.9	145.80
Trim, ranch, base & casing	168 L.F.	5.6	308.45
Paint, walls & ceilings, primer, oil base, roller	1,200 S.F.	4.7	72.00
Window, incl. frame and trim, primer, oil base, brushwork	6 Ea.	1.5	2.09
Trim, primer, oil base, brushwork	168 L.F.	2.1	4.03
Walls & ceilings, 2 coats, oil base, roller	1,200 S.F.	4.7	72.00
Window, incl. frame and trim, 2 coats, oil base, brushwork	6 Ea.	2.4	4.54
Trim, 2 coats, oil base, brushwork	168 L.F.	2.1	4.03
Wiring, 8 outlets	8 Ea.	12.0	259.20
Wiring, 2 switches	2 Ea.	2.8	69.60
Backfill, by hand, no compaction, light soil	10 C.Y.	5.7	
Fine grading and seeding, incl. lime, fertilizer & seed	53 S.Y.	2.5	10.18
Totals		305.2	$17,217.20

Contractor's Fee, Including Materials: $42,948

10' x 10' Second-Story Addition

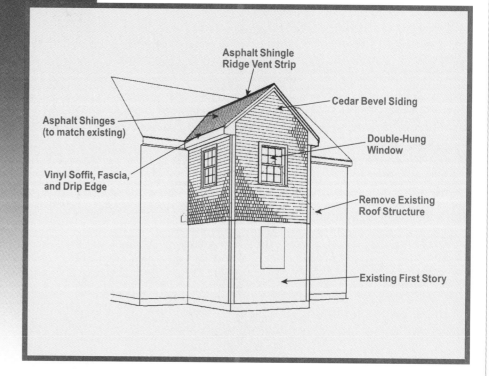

Asphalt Shingle
Ridge Vent Strip

Asphalt Shinges
(to match existing)

Cedar Bevel Siding

Double-Hung
Window

Vinyl Soffit, Fascia,
and Drip Edge

Remove Existing
Roof Structure

Existing First Story

Expanding your second-story living space is a good investment in your home's value, but this addition is an involved and challenging project. Most do-it-yourselfers should hire a contractor to perform the exterior work at the very least.

Materials & Tools

The materials used in this project are the same as those employed in first-floor room additions in this book, except that in this case, there is no requirement for a new foundation. Framing, sheathing, and finish products are installed on top of the supporting first-floor structure. The tools required for this project are numerous because of the variety of tasks involved.

What's Involved

This is a time-consuming project. Extensive preliminary planning and preparation of the old structure must be done beforehand. The first step is to

evaluate the condition of the existing wood structure and its foundation.

The original structure was not designed to support the extra load of this new addition, and even expert do-it-yourselfers should hire a qualified professional to determine if it's strong enough to support the new load. Foundation work and added lumber may have to be included in the estimate to reinforce the original structure before the new work begins

A part of your home's interior will be unprotected for at least several days while the old roof covering, sheathing, and rafters are removed. Have a large tarp on hand for protection in case of inclement weather. If you plan to do this part of the project yourself, be sure to have plenty of help on hand to move things along quickly.

After the roof has been removed, the existing ceiling joists become the floor joists for the new room. If they're 2 x 8s or larger, they're probably strong enough as they are, but if they're 2 x 6s, they'll need reinforcing.

The deck is then sheathed with 5/8" plywood before the wall partitions are framed. Be sure to prepare the area of the existing house wall to receive the new framing.

The roof is then framed with 2 x 6 rafters and a 2 x 8 ridge board tapered into the existing roof. The sheathing and finished roofing, selected to match the existing finish roof material, are applied next. After the window and siding are installed, the structure is ready for the interior work.

This plan assumes that access to the new room will come through a finished second-story room or hallway. If possible, an existing window opening should be used as part of the future rough opening for a door. If you can't use an existing window opening, then the wall will have to be opened at a convenient place, and headers and jack studs placed to rough-in the entryway. Additional cost and time may have to be included, according to its type, size, and conditions of placement.

Before finishing the walls and ceiling, the electrical wiring must be run, and the walls insulated. Before tapping into your wiring system, be certain that the circuit you use can handle the additional load. An electrician should always be hired to assess your current system and to tie in the new wiring. When the rough electrical work and insulating have been completed, the walls, ceiling, and floors can be finished.

Level of Difficulty

This project is one of the most challenging home improvements in this book. The exterior operations require not only advanced carpentry and tool handling skills, but also fast and efficient work at roof level. All do-it-yourselfers are encouraged to hire a specialty contractor for the exterior work. Even experts should consider professional help for the roof removal, and plan to have several experienced helpers on hand to assist with framing and roofing.

Beginners and intermediates should add 100% and 50%, respectively, to the professional time estimates for the interior work they can do. Experts should add 30% to the times for all exterior tasks they feel they can handle, and 20% for the interior operations.

What to Watch Out For

The new attached structure should blend with the house and complement its style, and not look "tacked on." Roofing and siding lines are particularly important in this blending process. The roof line of the new room should align with that of the existing house, and courses of asphalt roofing must meet precisely at the valley.

The new siding should follow the same spacing used in the structure below, and its course lines should align at the seam between the house and the addition. You may want to consider repainting your entire house after construction is complete.

See also:
Exterior Painting or Staining, Replacement Window, Room Additions, Skylights.

Second-Story Addition, 10' x 10'

Description	Quantity/ Unit	Labor-Hours	Material
Demolition, roofing built-up, shingles, asphalt strip	100 S.F.	1.1	
Demolition, gutters, aluminum or wood, edge hung	40 L.F.	1.3	
Demolition, roofing, sheathing, per 5 S.F.	20 Ea.	26.7	
Demolition, roofing, rafters, joists, & supports	100 S.F.	1.5	
Frame, boxing at existing joists, 2 x 6 stock	30 L.F.	0.4	19.44
Floor, sheathing, 5/8" plywood, 4' x 8' sheets	100 S.F.	1.2	93.60
Frame, walls, 2 x 4 studs & plates, 8' long	30 L.F.	6.0	147.60
Headers over windows, 2 x 8	24 L.F.	1.1	23.90
Gable end studs, 2 x 4	10 L.F.	0.2	4.08
Wall sheathing, 1/2" x 4' x 8' plywood	270 S.F.	3.8	200.88
Housewrap, spun bonded polypropylene	300 S.F.	0.6	57.60
Frame, roof, 2 x 6 rafters, 16" O.C.	192 L.F.	3.1	124.42
Sub-fascia, 2 x 8	20 L.F.	1.4	19.92
Ceiling, 2 x 6 joists, 16" O.C.	90 L.F.	1.2	58.32
Roof sheathing, 1/2" x 4' x 8' plywood	224 S.F.	2.6	166.66
Siding, cedar bevel, 1/2" x 8", A grade	270 S.F.	7.9	946.08
Soffit, 12" wide vented, fascia & drip edge	20 L.F.	2.7	35.52
Roofing, standard asphalt strip shingles	3 Sq.	5.3	154.80
Building paper, asphalt felt sheathing paper, 15 lb.	270 S.F.	0.6	9.72
Ridge vent strip, polyethylene	10 L.F.	0.5	30.60
Flashing, at valley	40 S.F.	2.2	31.20
Gutters, 5" aluminum, 0.027" thick	20 L.F.	1.3	29.76
Downspouts, 2" x 3", embossed aluminum	36 L.F.	1.5	29.38
Wall insulation, fiberglass, 3-1/2" thick, R-11	270 S.F.	1.4	129.60
Ceiling insulation, fiberglass, 9" thick, R-30	100 S.F.	0.7	85.20
Window, double-hung, 3' x 4'-6", thermopane, clad	2 Ea.	1.8	638.40
Gypsum wallboard, walls, 1/2" thick, taped and finished	416 S.F.	6.9	124.80
Underlayment, 3/8" particleboard	100 S.F.	1.1	74.40
Carpet, Olefin, 22 oz., light traffic	11 S.Y.	0.2	10.16
Carpet padding, sponge rubber pad	11 S.Y.	0.1	4.62
Trim, ranch, base & casing	70 L.F.	2.3	128.52
Paint, walls and ceiling, primer, oil base, roller	512 S.F.	2.0	30.72
Trim, primer, oil base, brushwork	110 L.F.	1.4	2.64
Walls and ceiling, 2 coats, oil base, roller	512 S.F.	5.1	61.44
Trim, 2 coats, oil base, brushwork	110 L.F.	2.2	6.60
Wiring, 4 outlets	4 Ea.	6.0	129.60
Wiring, 1 switch	1 Ea.	1.4	34.80
Totals		106.8	$3,644.98

Contractor's Fee, Including Materials: $10,821

20' x 24' Second-Story Addition

Sold

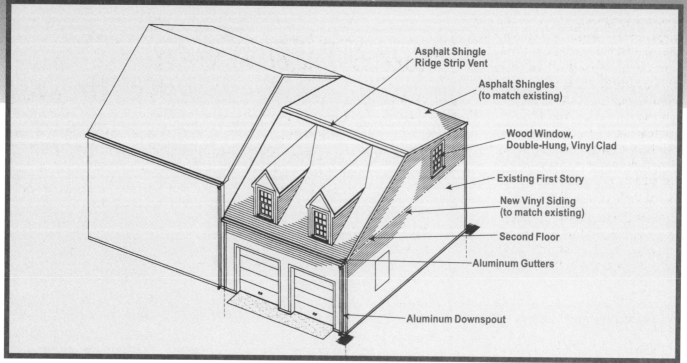

- Asphalt Shingle Ridge Strip Vent
- Asphalt Shingles (to match existing)
- Wood Window, Double-Hung, Vinyl Clad
- Existing First Story
- New Vinyl Siding (to match existing)
- Second Floor
- Aluminum Gutters
- Aluminum Downspout

This large addition provides nearly 500 square feet of new interior space above an attached garage. This space is big enough for a new master suite; a generous-sized area for office, play, or workshop space; or even an in-law apartment. With quality workmanship and a harmonious tie-in to your existing home, this addition could prove a good investment, if and when you sell your home. The size of the homes in your neighborhood will also affect the project's return on investment.

A project on this scale is a major commitment in terms of the time and expense, but it does have the advantage of being out of the way of normal household activities. The costs for removing the existing roof and strengthening the support system are offset by the savings from not having to excavate for and construct a foundation. Professional expertise is required for much of this project, particularly the exterior work.

Materials & Tools

The same materials and tools used to build other wood frame structures in this book are used for this addition. You'll need framing lumber, sheathing, an air-infiltration barrier, insulation, roofing and flooring materials, windows and a door, and siding, trim, and finish materials, like gypsum wallboard, trim, and paint. Professionals can advise you.

Because it will be a very large order, shop around for the best prices. In working out the delivery conditions, consider the size of the order and where you will store the materials. Unless your garage space is completely available, you might want to wait to have interior materials delivered until after the shell of the addition has been erected and made weathertight. Allow space outside to store the exterior items. Ask the supplier if delivery can be staggered throughout the construction.

Standard hand and power tools for carpentry, roofing, and painting are also needed if you're planning to do some of the work yourself.

What's Involved

It's a good idea to consult an architect to help you determine the design and structural plan. Care should be taken to blend the new structure with the old so that the roof line, the style of the overhang, and other design features complement one another. The new exterior finish products should also be selected to match the style of the house.

First have the existing garage and foundation inspected to determine if the additional load will be supported. This process is required by most building regulations, and your local building department can provide you with information on whom to contact.

The existing garage roof must be removed, and the ceiling joists reinforced to safe standards. If the existing joists are 2 x 8s or larger and are supported by a center beam, the amount of reinforcement will be minimal, whereas 2 x 6s will require substantial reinforcement. After the 5/8" plywood decking has been installed, the walls and roof are framed with standard 2 x 4 and 2 x 6 lumber, and sheathed with 1/2" plywood.

When the exterior is complete, and the structure made weathertight, a doorway or opening can be cut into the new room. (*See the Opening in Load-Bearing Wall project in the* Interior Home Improvement Cost Guide.) The placement of a standard-sized 30" or 33" door requires 2 x 8 headers and jack studs for the rough opening frame, but a 6' or 8' wall opening needs more support and, possibly, temporary ceiling bracing during installation.

Installing the insulation, the rough electrical work, and any partitioning are completed first, and then the ceiling and walls are covered and finished.

Level of Difficulty

Because of its size, this addition is one of the more challenging remodeling projects. Removing the roof and subsequent framing and roofing are difficult operations that must be completed as expediently as possible to protect the interior from the elements. All do-it-yourselfers are encouraged to leave the exterior work to a professional.

Most handy homeowners should be able to handle at least some of the interior work, but should hire professionals for the electrical and carpet installations. Beginners should add 100% to the time estimate for the tasks they undertake. Intermediates and experts should add 50% and 20%, respectively, to the professional time estimate for the interior jobs, and more for the specialty work.

What to Watch Out For

When building above a garage, check local building code requirements regarding fire separation. If not already in place, you probably will have to add

two layers of 5/8" fire–code-rated gypsum board to the garage ceiling. Insulation is also needed especially if the new space will be a bedroom.

See also:
Bow/Bay Window, Room Additions, Skylights, New Window.

Second-Story Addition, 20' x 24'

Description	Quantity/ Unit	Labor-Hours	Material
Demolition, roofing, built-up shingles, asphalt strip	480 S.F.	5.5	
Demolition, roofing, sheathing, per 5 S.F.	96 Ea.	128.0	
Demolition, roofing, rafters, joists, & supports	480 S.F.	7.1	
Frame, floor, 2 x 8 joists, 16" O.C., 12' long	540 L.F.	7.9	537.84
Floor sheathing, 3/4" plywood, 4' x 8' sheets	480 S.F.	6.1	541.44
Frame, walls, 2 x 4 plates, 12' long, 2 x 4 studs, 8' long	264 L.F.	52.8	1,298.88
Headers over windows, 2 x 8 stock	24 L.F.	1.1	23.90
Wall sheathing, 25/32" x 4' x 8' wood fiberboard	704 S.F.	9.4	236.54
Corner bracing, 16 ga. steel straps, 10' long	80 L.F.	1.1	64.32
Housewrap, spun bonded polypropylene	640 S.F.	1.4	122.88
Frame, roof, 2 x 6 rafters, 16" O.C., 14' long	600 L.F.	10.1	597.60
Ridge board, 2 x 8 stock	22 L.F.	0.9	31.15
Sub-fascia, 2 x 6 stock	40 L.F.	0.6	25.92
Ceiling, 2 x 6 joists, 16" O.C.	416 L.F.	5.3	269.57
Roof sheathing, 5/8" x 4' x 8' plywood	960 S.F.	11.8	898.56
Roofing, asphalt shingles, 285 lb. premium	10 Sq.	22.9	654.00
Felt paper, 15 lb., on roof	1,000 S.F.	2.2	36.00
Ridge vent strip, polyethylene	20 L.F.	1.0	61.20
Siding, vinyl panels, 8" - 10" wide	525 S.F.	16.5	434.70
Soffit, 12" wide vented, fascia & drip edge	90 L.F.	12.0	159.84
Gutters, 5" aluminum, 0.027" thick	68 L.F.	4.5	101.18
Downspouts, 2" x 3", embossed aluminum	36 L.F.	1.5	29.38
Wall insulation, fiberglass, 3-1/2" thick, R-11	525 S.F.	2.6	252.00
Ceiling insulation, fiberglass, 9" thick, R-30	480 S.F.	2.4	236.16
Window, double hung, vinyl clad, 3'-0" x 5'-0"	5 Ea.	5.0	1,668.00
Gypsum wallboard, walls, 1/2" thick, taped and finished	1,184 S.F.	19.6	355.20
Carpet, 26-oz. nylon level loop, light to medium traffic	54 S.Y.	8.7	1,067.26
Carpet padding, sponge rubber pad	54 S.Y.	2.9	204.12
Trim, ranch, base & casing	186 L.F.	6.2	341.50
Jamb, 1 x 6 clear pine, at opening to house	19 L.F.	0.8	22.80
Paint, walls & ceiling, primer, oil base, roller	1,200 S.F.	4.7	72.00
Trim, primer, oil base, brushwork	186 L.F.	2.3	4.46
Walls & ceiling, 2 coats, oil base, roller	1,200 S.F.	12.0	144.00
Trim, 2 coats, oil base, brushwork	186 L.F.	3.7	11.16
Wiring, 8 outlets	8 Ea.	12.0	259.20
Wiring, 2 switches	2 Ea.	2.8	69.60
Totals		395.4	$10,832.36

Contractor's Fee, Including Materials: $35,866

Pre-Fabricated Greenhouse

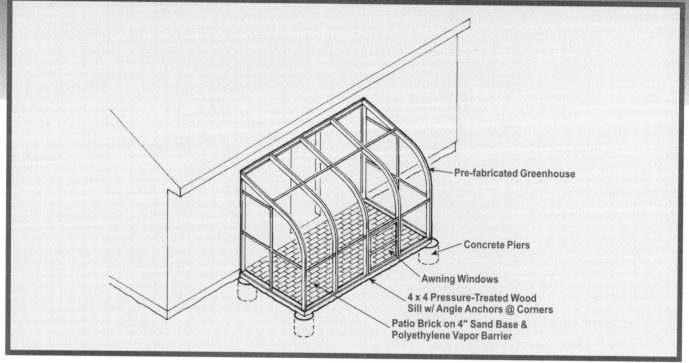

Pre-fabricated Greenhouse

Concrete Piers

Awning Windows

4 x 4 Pressure-Treated Wood
Sill w/ Angle Anchors @ Corners

Patio Brick on 4" Sand Base &
Polyethylene Vapor Barrier

One of the hottest continuing trends in home remodeling is merging the indoors with the outdoors. Greenhouses are one way to accomplish this, and you don't have to be an avid gardener to appreciate them. Greenhouses provide solar heat and natural light—whether for added interior living space or for an indoor garden where you can start seedlings and grow tropical plants.

The new greenhouse should blend with the rest of the house as much as possible, and take advantage of maximum sunlight. The south or southeast side of your home is ideal. Six hours of sunlight are recommended for sun-loving plants, although morning sunlight on the east side is sufficient for most plants.

Materials

The greenhouse kit is the most important and expensive component. There are many styles and sizes. The final choice will depend on how much space you need, your home's layout, and the cost. Prices and quality vary widely depending on the type of materials used in the frame and panels, and the engineering of the structural and tie-in systems. Most greenhouses have aluminum or galvanized steel frames, with glass, fiberglass, or rigid plastic panes.

Like other attached exterior structures, greenhouses require a foundation or some other means of support, a basic framing system, flooring, and inside and outside finish treatments. If you plan to use a factory-prepared greenhouse kit, as in this plan, much of the framing and covering material for the structure has been taken care of by the manufacturer.

The foundation, flooring, and finish materials are standard construction products that can be purchased at most building supply outlets. This project calls for brick flooring, although gravel, stone, concrete, wood, or even sand can be used. It's assumed that there is an existing exterior door that can be used to access the greenhouse. Additional materials and installation time may be needed if the new greenhouse requires an entryway into the house.

What's Involved

Before you can assemble the greenhouse kit, a considerable amount of preliminary work must be completed. Aluminum-framed greenhouses with fiberglass or tempered glass panels require a sturdy and deep foundation or ground support system. If the foundation is not true in its placement, the greenhouse will not sit properly on the slab or against the house wall, and the enclosure may develop leaks.

In climates where deep frost occurs, the recommended slab may not be adequate, and a deeper foundation may be required. Any movement or heaving of the surface can cause the seams along the house and foundation to open, the mullions and weatherstripping in the greenhouse to separate, and, possibly, the glass panels to crack.

Other floor and foundation systems could also be used, such as redwood and pressure-treated pine for the base set in concrete piers, concrete block foundations on footings, or adequate existing supports like decks and patios.

Be sure to fasten the greenhouse correctly to the base frame and tie it into the house with acceptable flashing methods.

Level of Difficulty

Beginners may find the foundation and framing work out of reach, but could install the brick floor and tackle other aspects of the finish work. They should add 100% to the professional time estimate. Intermediate do-it-yourselfers should figure on about 40% more time, and experts, about 10%. All should allow extra time for any work they perform on the assembly and installation of the greenhouse.

What to Watch Out For

Be sure to consider your area's winter weather conditions when selecting a greenhouse. Rain, snow, and ice could potentially accumulate on or slide down the sides of your greenhouse. If your region experiences hailstorms, don't use glass for the panes; choose stronger, more flexible plastic panels with durable glazing.

Options

A greenhouse can collect a considerable amount of passive solar heat in the course of a winter day. The floor, house wall, and plants absorb heat from the sunlight, store it, and then provide a source of radiant warmth after the sun has gone down. One way to collect solar heat is to paint the house wall and other surfaces, like plant racks and floor borders, a dark color. Another is to place portable, dark-colored containers filled with water against the house wall.

To transfer the heat from the greenhouse to the main house, you can install a small exhaust fan in the house wall near the greenhouse ceiling. You can also purchase a thermostatic fan control to turn the fan on and off to ensure a stable temperature.

Other greenhouse add-ons include thermal shades, screens, misters, humidity controls, and vents. You might also want to consider adding an electrical outlet, a small sink, or heating unit—before the finish materials are installed.

See also:

Brick or Flagstone Patio, Sunroom, Interior Doors*, Patio Doors.

*In *Interior Home Improvement Costs*

Pre-Fabricated Greenhouse, 5'-6" x 10'-6"

Description	Quantity/ Unit	Labor- Hours	Material
Excavate, post holes, incl. layout	0.50 C.Y.	0.5	
Forms, round fiber tube, 8" diameter, for posts	12 L.F.	2.5	21.74
Concrete, field mix, 1 C.F. per bag, for posts	4 Bags		27.84
Placing concrete, footings, under 1 C.Y.	0.50 C.Y.	0.4	
Angle anchors for 4 x 4 wood frame	4 Ea.	0.2	2.21
Wood frame for base, 4 x 4 treated lumber	36 L.F.	1.3	58.32
Polyethylene vapor barrier, 0.006" thick	72 S.F.	15.6	237.60
Washed sand for 4" base	1 C.Y.	0.1	31.80
Grade and level sand base	60 S.F.	2.1	
Patio brick, 4" x 8" x 1-1/2" thick, laid flat	60 S.F.	9.6	179.28
Greenhouse, prefab, 5'-6" x 10'-6" x 7' high, w/2 awning windows	58 SF Flr.	16.9	5,916.00
Flashing, aluminum, 0.016" thick	24 S.F.	1.3	18.72
For 8' x 16' greenhouse, add $ 7,100 to the contractor's fee			
Totals		50.5	$6,493.51

Contractor's Fee, Including Materials: $12,027

Sunroom

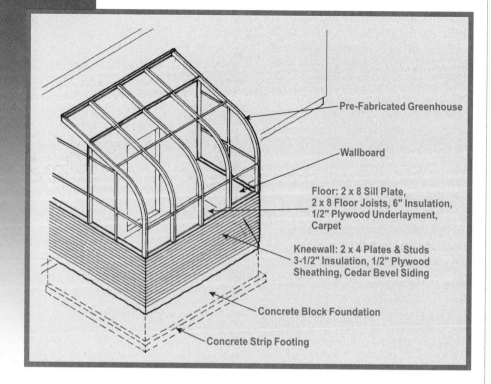

Pre-Fabricated Greenhouse

Wallboard

**Floor: 2 x 8 Sill Plate,
2 x 8 Floor Joists, 6" Insulation,
1/2" Plywood Underlayment,
Carpet**

**Kneewall: 2 x 4 Plates & Studs
3-1/2" Insulation, 1/2" Plywood
Sheathing, Cedar Bevel Siding**

Concrete Block Foundation

Concrete Strip Footing

If the design and layout of your home allow, a new sunroom can provide several benefits: new year-round interior living space (and added value to your home); open, natural light; and passive solar heat. Sunrooms can be custom-designed or built from kits, which are available in many different styles and sizes.

Materials & Tools

This model project is built from a factory-prepared sunroom kit, which consists of an insulated aluminum frame and double-layered tempered glass. The foundation and finish materials are readily available from building supply retailers, but the kit is usually special-order.

Before selecting the type and design of the sunroom kit, consider its thermal and aesthetic impact on your house. Because it will serve as a year-round room, insulating glass and frames are recommended. Shop around, as there are many brand names and variations in quality and cost. Some kits come with screens, shades, and vent windows, while others offer these items at

additional cost. Depending on your budget and climate, you could opt for special features such as reflective shades for hot summers and insulating blankets for winter, both of which can be designed to slide easily in and out of ceiling tracks.

You'll also need materials for the foundation, flooring, and kneewall, and for the entryway into the house. This model project includes a concrete strip footing, concrete block foundation, insulation and flashing, and lumber and wallboard for the kneewall, as well as jambs and casing for the interior doorway.

You'll need basic carpentry tools to assemble the sunroom, including a hammer, drill, caulk and caulk gun, ratchet and socket set, level, safety glasses, and work gloves.

What's Involved

Other than the fabrication and placement of the sunroom itself, the most important operation is constructing the foundation and floor systems. Before the foundation work

begins, a perimeter trench has to be dug to receive the form for the concrete. This may have to be done by hand if there isn't room for a machine to do the excavation.

After the concrete has been allowed to set up for several days, a gravel sub-surface, reinforcing, and expansion joints are installed, and the slab poured. The slab surface is troweled and broom-finished to provide an acceptable base surface for the floor covering. In most cases, a contractor should be hired to complete these operations. Mistakes in concrete work can be difficult and expensive to correct, and a precisely set foundation is a must for the fixed dimensions of the sunroom.

Insulating the foundation and floor is strongly advised. Otherwise, temperature variations and drafts could make your room both uncomfortable and energy-inefficient. When assembling the sunroom kit, work carefully to maintain precise alignment. The base fasteners are usually included, but flashing for the base, roof, and walls is usually extra.

Before the interior finish work can begin, a new doorway from the house must be constructed or the existing opening reconditioned. In most cases, you'll need to remove the exterior siding and possibly enlarge and frame a window opening to receive a 6' sliding glass door or French doors.

If the new sunroom will become part of your indoor, year-round space, you may not need a door. In this case, the opening can be enlarged even more and trimmed with a jamb and facings. You can deduct the expense of the door, but may have to add slightly to the cost of preparing the rough opening.

You have many choices for the finish floor treatment. Wood or laminate flooring or carpet can provide comfort and additional warmth.

the interior wall. A small sink could also be installed. If these electrical and plumbing conveniences are roughed in during the construction process, the procedure will cost less and look better.

See also:
Electrical System Upgrade*, Lighting Fixture Installation/Upgrade*, Interior Doors*, Patio Doors, Pre-Fabricated Greenhouse.

*In *Interior Home Improvement Costs*

outlets and fixtures. Although these items are not listed in this plan, you might consider installing at least one electrical outlet and a light fixture for

Level of Difficulty

Beginners and less experienced intermediate do-it-yourselfers should consider hiring a contractor to install the sunroom, but might tackle the finish work. Beginners should add 100% to the professional time estimates for the work they can do. Intermediates should probably stick to the interior finishing and allow 50% more time. If they attempt to lay a tile or brick floor, they should add 100% to the professional time to allow for slow and careful work.

Experts should leave the foundation and slab installations to a specialist. For the other tasks, they should add 20% to the estimated times, and for the finished floor, about 40%. All do-it-yourselfers should plan to add extra time for any work they do on the assembly and installation of the sunroom kit, as the cost and intricacy of the product demand careful workmanship.

What to Watch Out For

Before finalizing the master plan for your sunroom project, be sure to make provisions for electrical and plumbing

Sunroom, 5'-6" x 10' x 6"

Description	Quantity/Unit	Labor-Hours	Material
Excavation, trench, by hand, pick and shovel, to 6' deep	2 C.Y.	2.0	
Footing, 6" thick x 16" wide x 32' long	2 C.Y.	5.6	242.40
Foundation, 8" concrete block, 2'-8" high, parged 1/2" thick	105 S.F.	9.9	226.80
Perimeter insulation, molded bead board, 1" thick, R-4	100 S.F.	1.2	22.80
Sill plate, 2 x 8 x 12'	22 L.F.	0.8	26.40
Floor joists, 2 x 8, 16" O.C.	132 L.F.	1.9	131.47
Floor insulation, fiberglass blankets, foil-backed, 6" thick, R-19	60 S.F.	0.4	32.40
Underlayment, plywood, 1/2" thick, 4' x 8' sheets	60 S.F.	0.6	44.64
Framing for knee wall, 2 x 4 plates and studs, 3' high, 16" O.C.	108 L.F.	2.7	44.06
Kneewall insulation, fiberglass, foil-backed, 3-1/2" thick, R-11	70 S.F.	0.4	33.60
Sheathing, plywood on knee walls, 1/2" thick	96 S.F.	1.4	71.42
Greenhouse, prefab., 5'-6" x 10'-6" x 6'-6", insul. dbl. glass	58 SF Flr.	9.6	5,533.20
Flashing, aluminum, 0.019" thick	25 S.F.	1.4	19.50
Remove window for opening, to 25 S.F., in existing building	1 Ea.	0.4	
Remove siding for opening in existing building	60 S.F.	1.4	
Frame opening, 2 x 8 stock	24 L.F.	1.1	23.90
Jamb and casing for opening, 9/16" x 3-1/2", interior and exterior	48 L.F.	1.6	88.13
Gypsum wallboard for interior kneewall and sides of opening, 1/2" thick	128 S.F.	2.1	38.40
Paneling, plywood, prefinished, 1/4" thick	128 S.F.	4.1	121.34
Flooring, carpet, 26 oz., nylon level loop, light to medium traffic	60 S.F.	1.1	131.76
Flooring, sponge rubber pad, maximum	60 S.F.	0.4	66.24
Siding for exterior kneewall, cedar, beveled, rough-sawn, stained	80 S.F.	2.7	288.96
For 10'-6" x 18' house, add $21,500 to contractor's fee.			
Totals		52.8	$7,187.42

Contractor's Fee, Including Materials: $13,118

Security Measures

Many thefts are crimes of opportunity, but burglars also observe and select homes they know will be unoccupied. Following are some suggestions to help you reduce your risk.

- Keep doors and windows locked. Remove or prune trees and shrubbery that provide hiding places near doors and windows.

- Keep your garage door closed and locked.

- Use deadbolt locks. With a key on both sides, an intruder cannot break a side window and reach in to open the door. (Keep the key close to the door for household members, but out of sight from the outside.)

- Choose Grade 1 or Grade 2 locks that stand up to prying, twisting, and picking. Look for a 1" throw bolt, a "dead-latch" that will keep the lock from being slipped with a credit card, and a beveled casing that will prevent someone from shearing off the lock's cylinder pins with channel-lock pliers. The strike plate should be heavy-duty with four 3" screws for a wood door frame. This helps prevent the jamb from being kicked open. A minimum of one long screw should also be used in each door hinge.

- Before boring a hole in a door for a deadbolt lock or peephole, check the templates carefully.

- Choose solid core or metal doors.

- Install a wide-angle 160° peephole, mounted at 58" or slightly lower.

- Protect sliding glass doors with a piece of wood or metal placed in the track or by screw-down blocking devices. To prevent loose doors being lifted off the track, keep them properly adjusted and consider an anti-lift device.

- If you have an alarm system, neighborhood watch organization, or watchdog, post a sign on doors or windows.

- If your situation requires them, security shutters, window grilles, or a security gate are usually installed by a specialty distributor. Decorative options blend with different architectural styles.

- Protect windows with wood dowels or pins. Keep windows open for ventilation (up to 6"), as long as the blocking device can't be removed by reaching in from outside. Blocking devices must be easily removable from inside.

- Keep keys for window locks near each window, but hidden from outside view, for quick access in emergencies.

- Don't forget lighting—inside and out. Cover all points of entry, areas where intruders could hide, and walkways. Use timers or photocell lights and motion-detector lights.

- Have your neighbors watch out for you, as you do for them. Pick up one another's mail and newspapers when you're away, and keep an eye on the house.

- Consult your local police department for more tips on crime prevention in your area.

Security Measures

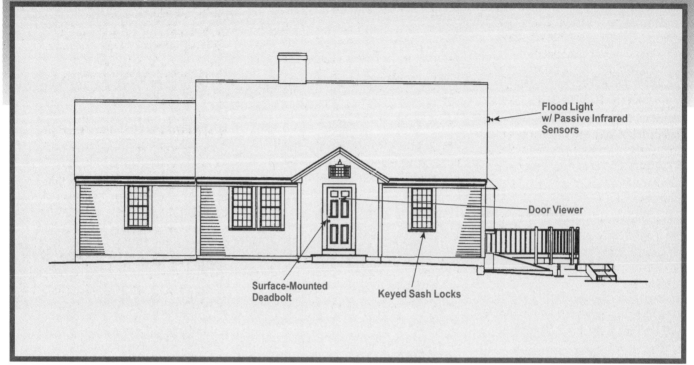

Flood Light
w/ Passive Infrared
Sensors

Door Viewer

Surface-Mounted
Deadbolt

Keyed Sash Locks

Whether you live in the city or a rural area, the risk of crime is a reality, with almost three million homes broken into each year. Protecting your home and family with at least basic security tactics is an inexpensive and prudent course of action.

Materials & Tools

The security measures described in this project are easy to incorporate into any home, but they're also very minimal. A totally integrated security system includes such features as lighting, landscaping, fencing, windows, doors, locks, and alarms.

Setting up a reasonably effective security system for your home does not necessarily mean spending large sums of money on expensive, high tech components. In addition to the suggestions made in this project, your local police department can provide you with valuable advice on home security, and many economical suggestions and tips.

First-floor doors and windows, as well as patio doors, are obvious points of entry that should be equipped with strong locks at the very least. For doors, deadbolt locks are a must. Their prices vary from about $25 to over $300, with a wide variety of features, materials, and quality. The more expensive ones often are made of solid, forged-brass and have a Grade 2 or Grade 1 (commercial duty) security rating. Surface-mounted and cylinder deadbolts are readily available at home supply stores as single cylinders (thumb turn inside and key outside) or double cylinders (key inside and out—to prevent an intruder from breaking glass and reaching in to unlock the door).

Additional security measures for doors include peepholes and brass wrap-around knob reinforcers. Sliding patio doors should be equipped with keyed locks and, if necessary, drop down steel bars to prevent them from opening from the outside.

Windows, especially the double-hung kind, are openings vulnerable to burglars, and you should at the very least have keyed locks or pins on the

sashes. In some locations, homeowners feel the need to install window bars, but it can be difficult to make these look attractive, so they should only be used when absolutely necessary.

In addition to door and window deterrents, your home may warrant added measures, such as an electric security system, which can be costly and complex. A simpler way to obtain some form of a responsive system is to install outdoor passive infrared (PIR) sensors, which detect motion and can switch on lights or activate alarms.

All of these deterrents are fairly easy to install, the most complicated being the electronic sensors. For the bars, locks, and deadbolts, only basic tools are needed—a screwdriver or cordless drill, a wood chisel, and the hardware and screws that will usually come with the unit should be all that is required.

Outdoor wiring can call for more complicated installations and sometimes minor excavation with a shovel to bury the wires. Electrical work should be left to professionals unless you're an expert.

What's Involved

When installing locks on doors and windows, as well as the bars for the patio doors, follow the specific instructions on the units you purchase. You may need to drill holes through doors to get the lock or deadbolt into place. Make sure that you use the correct sized drill bit so the unit fits properly and securely. Screw the locks in place with a screwdriver or drill, and make sure that they close and lock properly before relying on them for protection.

Strengthen a deadbolt or door latch with a reinforcing strike plate attached to the door jamb with 3" screws. Such a plate will reduce the chances of the door's being kicked in or pried open. The length of the screws allows them to go through the jamb and penetrate the studs that form the rough opening.

For double floodlights with sensors mounted between the bulbs, the most common form of exterior PIR security lighting, be sure to adjust the viewing angle to control the size of the area you want to monitor. When the unit detects motion, the sensor will turn the lights on and, after an adjustable delay, shut them off. This type of fixture can be installed in place of a standard floodlight, and can be controlled through an interior switch.

If you opt to mount the PIR sensors separate from the floodlight to detect motion in a wider area range, you'll need to excavate and run wires underground to get to the sensor, which will send radio signals to turn on the lights when triggered. Battery-powered models are also now available, which require no external wiring.

Improvements are being made all the time in PIR technology, and you can now purchase portable units that incorporate a photovoltaic solar panel, a battery, a floodlight, an alarm, and a PIR sensor. During the day, the battery is charged with solar energy, which powers the unit at night.

Level of Difficulty

Hooking up PIR fixtures in place of existing floodlights is not too difficult. If you need to run new wiring, it's best to leave the work to an electrician. You would also be well advised to get an expert's opinion on how best to arrange an outdoor light/alarm system to maximize its effectiveness.

Except for electrical wiring, installation of these basic security devices can be handled successfully by do-it-yourselfers. Security is important enough, however, to justify consideration of hiring professional installers who can design a system that works best for your home and will work within your budget. Beginners should add 100% to the time estimates provided; intermediates, about 50%; and experts, 15%.

What to Watch Out For

When you shop for deadbolt locks, strike plates, and knob reinforcers, be sure you know the material construction and dimensions (thickness) of your door and trim, as well as their existing hardware, to ensure that you come home with the proper items and sizes. Get help if you're unsure of exactly what you need.

If you install double cylinder deadbolts on exterior doors, be sure there is always a key near every lock, and that everyone in the house knows exactly where all the keys are and can locate them, even in the dark. Double-keyed deadbolt locks are effective deterrents against forced entry, but should not prevent your rapid exit in the case of fire or other emergency.

Installing some devices can actually weaken the construction of certain doors or windows. Before purchasing security devices, check on the construction of your doors, jambs, sashes, and frames. Be sure that boring a hole through the door doesn't pose a weakness problem. Also, check the backset (distance from the edge of the door to the center of the hole through the door for the device) to ensure that you have room for all the hardware trim.

Options

Universal keying is a good way to go for door locks if you have several doors and only want to carry one house key. If you're purchasing knob and locksets, dual-torque springs are a nice feature, as they keep knobs from sagging over time.

You can purchase a wireless security system that you can install yourself, including a control panel and sensors for two doors, at your local home center for about $200.

Adding Security Features

Description	Quantity/Unit		Labor-Hours	Material
Locks, surface-mounted, deadbolt, reinf. strike plate, 3" screws	2	Ea.	2.0	331.20
Knob reinforcers, brass, wrap-around	2	Ea.	2.0	52.80
Door viewers	2	Ea.		35.52
Window locks, keyed sash locks, brass	12	Ea.	3.0	146.16
Remove existing floodlights	3	Ea.	1.0	
Floodlights, 2-bulb w/ PIR sensors, to replace exist. floodlights	3	Ea.	3.0	100.80
Totals			11.0	$666.48

Contractor's Fee, Including Materials:	**$1,605**

Part Three
Detailed Costs

This section provides costs for individual construction items. These are helpful if you need to make substitutions in a project featured earlier in the book or want to put together your own project estimate to suit your specific space requirements, budget, style of your house, or personal preferences. This information will also help you determine the total cost of a smaller project, such as adding shutters, a storm door, or an awning.

A variety of materials and their costs are included for comparison, as well as the installation labor-hours involved and corresponding contractors' fees. Everything from foundations to windows and doors is listed in this part of the book, with a range of quality and price in each category.

By knowing the prices of options, as well as their installation times and professional charges, you can make informed choices about the fixtures and materials you want. You can also determine how much of the work you are willing to take on yourself and how much should be done by a contractor.

This section's illustrations help to identify the various components, which makes planning and purchasing easier and helps you to visualize the renovation in progress.

Applying the Detailed Costs can be a bit more challenging than the projects in Part Two. For example, different standards of measure (such as cubic yard, linear foot, each, and set) are used according to the type of material. The quantity of building components must be calculated with the measurements you make yourself. A "How To Use" page follows this introduction and explains the format in more detail. Some organization is required when you use the Detailed Costs to draw up plans and make estimates, and a bit of imagination is helpful, too, when it comes to putting the separate elements together.

The benefits of this independent approach are the ability to create your own remodeling project estimate, expand or reduce the size of an estimate in Part Two, and change certain aspects to fit your budget or design.

How to Use the Detailed Costs

In these pages, basic building elements are grouped together to create a complete "system." These systems can be used to make up a proposed project. For example, the "Three Fixture Bathroom System" lists all the basic elements required to plumb a bathroom. The result is the total cost for the fixtures and installation.

These systems can also be combined to create an estimate for a more complex project. For example, if you need to install drywall and flooring as part of

your project, you can consult the "Drywall and Thincoat Systems" and the "Flooring Systems" pages to calculate the cost per square foot for redoing the interior walls and floor, and then add in the cost of the plumbing fixtures.

There are two types of pages in the Detailed Costs section. One has an illustration of a building system and lists the building elements that go into it. Many have different material options at varying costs. For example, the "Shingle

Siding Systems" page lists the cost per square foot for various types of building shingles.

The second type of page lists the cost for items that are not part of a system. For example, the "Insulation Systems" page lists the cost per square foot for various types of building insulation.

The terms that appear in this section are explained below. The illustration on the facing page shows how this part of the book is organized.

Key to Abbreviations
C.L.F. – 100 linear feet
C.Y. – cubic yard
Ea. – each
Gal. – gallon
Lb. – pound
L.F. – linear foot
Pr. – pair
Sq. – square
S.F. – square foot
S.Y. – square yard
V.L.F. – vertical linear foot

System Descriptions: A detailed definition of what is in the system being priced.

Quantity: The number of each component per system.

Unit: The standard unit of measure for each item. *(A Key to Abbreviations appears to the left on this page.)*

Material Cost: The cost for each item in the system. In systems that have several components, the total material cost for one unit of the whole system appears in a box at the bottom of the column.

Total Cost: The total cost one "unit" (such as one square foot) of each system. This includes the price of materials (without sales tax), labor costs, and the contractor's overhead and profit.

System Description	QUAN.	UNIT	LABOR-HOURS	MAT. COST	COST PER S.F.
WHITE CEDAR SHINGLES, 5″ EXPOSURE					
White cedar shingles, 16″ long, grade "A", 5″ exposure	1.000	S.F.	.033	1.35	
Building wrap, Spunbonded polypropylene	1.100	S.F.	.002	.12	
Trim, cedar	.125	S.F.	.005	.21	
Paint, primer & 1 coat	1.000	S.F.	.017	.09	**Contractor's Fee**
TOTAL			.057	1.77	**5.97**

Labor-Hours: The time it takes to install one unit of the system.

Table of Contents

The Detailed Costs are organized into six divisions, listed here. The schematic diagram pictured below shows how the components of a typical house fit into these divisions.

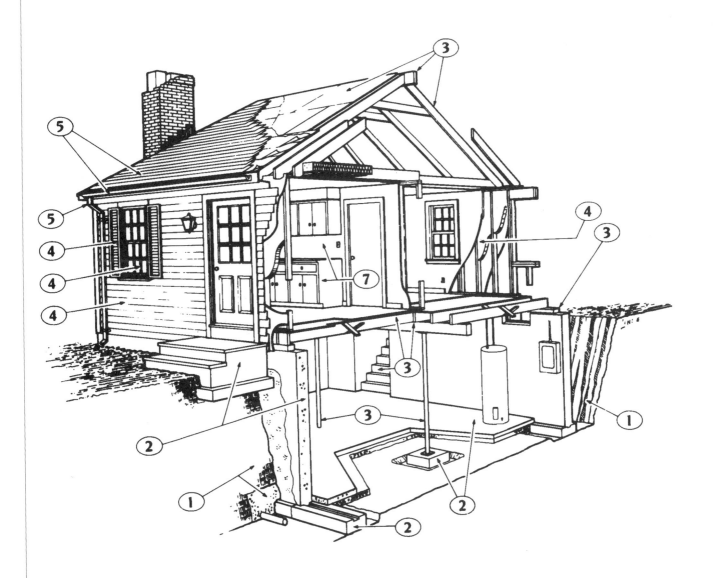

Division 1
SITE WORK

Footing Excavation Systems

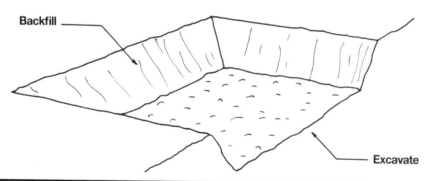

Backfill

Excavate

System Description	QUAN.	UNIT	LABOR-HOURS	MAT. COST	COST EACH
BUILDING, 24' X 38', 4' DEEP					
Cut & chip light trees to 6" diam.	.190	Acre	9.120		
Excavate, backhoe	174.000	C.Y.	4.641		
Backfill, dozer, 4" lifts, no compaction	87.000	C.Y.	.580		
Rough grade, dozer, 30' from building	87.000	C.Y.	.580		**Contractor's Fee**
TOTAL			14.921		**1,472.79**

Foundation Excavation Systems

Backfill

Excavate

System Description	QUAN.	UNIT	LABOR-HOURS	MAT. COST	COST EACH
BUILDING, 24' X 38', 8' DEEP					
Clear & grub, dozer, medium brush, 30' from building	.190	Acre	2.027		
Excavate, track loader, 1-1/2 C.Y. bucket	550.000	C.Y.	7.860		
Backfill, dozer, 8" lifts, no compaction	180.000	C.Y.	1.201		
Rough grade, dozer, 30' from building	280.000	C.Y.	1.868		**Contractor's Fee**
TOTAL			12.956		**1,915.16**

Be sure to read pages 178-179 for proper use of this section.

Utility Trenching Systems

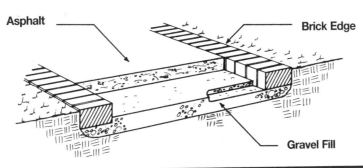

Backfill · Bedding · Sewer Pipe · Excavation

System Description	QUAN.	UNIT	LABOR-HOURS	MAT. COST	COST PER L.F.
2′ DEEP					
Excavation, backhoe	.296	C.Y.	.032		
Bedding, sand	.111	C.Y.	.044	1.44	
Utility, sewer, 6″ cast iron	1.000	L.F.	.283	12.99	Contractor's Fee
Backfill, incl. compaction	.185	C.Y.	.044		**43.98**
TOTAL			.403	14.43	

Sidewalk Systems

Asphalt · Brick Edge · Gravel Fill

System Description	QUAN.	UNIT	LABOR-HOURS	MAT. COST	COST PER S.F.
ASPHALT SIDEWALK SYSTEM, 3′ WIDE WALK					
Gravel fill, 4″ deep	1.000	S.F.	.001	.28	
Compact fill	.012	C.Y.	.001		
Handgrade	1.000	S.F.	.004		
Walking surface, bituminous paving, 2″ thick	1.000	S.F.	.007	.45	Contractor's Fee
Edging, brick, laid on edge	.670	L.F.	.079	1.37	**7.93**
TOTAL			.092	2.10	

Be sure to read pages 178-179 for proper use of this section.

Driveway Systems

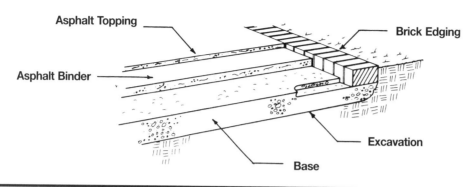

System Description	QUAN.	UNIT	LABOR-HOURS	MAT. COST	COST PER S.F.
ASPHALT DRIVEWAY TO 10' WIDE					
Excavation, driveway to 10' wide, 6" deep	.019	C.Y.	.001		
Base, 6" crushed stone	1.000	S.F.	.001	.65	
Handgrade base	1.000	S.F.	.004		
2" thick base	1.000	S.F.	.002	.45	
1" topping	1.000	S.F.	.001	.26	
Edging, brick pavers	.200	L.F.	.024	.41	**Contractor's Fee**
TOTAL			.033	1.77	**4.66**

Septic Systems

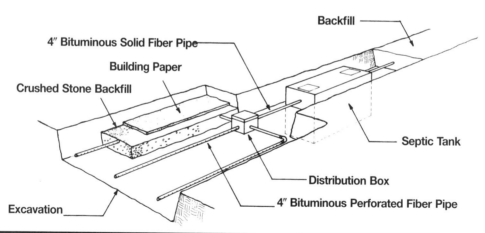

System Description	QUAN.	UNIT	LABOR-HOURS	MAT. COST	COST EACH
SEPTIC SYSTEM WITH 1000 S.F. LEACHING FIELD, 1000 GALLON TANK					
Tank, 1000 gallon, concrete	1.000	Ea.	3.500	615	
Distribution box, concrete	1.000	Ea.	1.000	114	
4" PVC pipe	25.000	L.F.	1.600	52.75	
Tank and field excavation	119.000	C.Y.	13.130		
Crushed stone backfill	76.000	C.Y.	12.160	1,976	
Backfill with excavated material	36.000	C.Y.	.240		
Building paper	125.000	S.Y.	2.430	45	
4" PVC perforated pipe	145.000	L.F.	9.280	305.95	
4" pipe fittings	2.000	Ea.	1.939	19.90	**Contractor's Fee**
TOTAL			45.279	3,128.60	**7,862.20**

Be sure to read pages 178-179 for proper use of this section.

Chain Link Fence Price Sheet

SYSTEM DESCRIPTION	QUAN.	UNIT	LABOR-HOURS	COST PER UNIT	
				MAT.	TOTAL
Chain link fence					
Galv.9ga. wire, 1-5/8"post 10'O.C., 1-3/8"top rail, 2"corner post, 3'hi	1.000	L.F.	.130	4.87	13.06
4' high	1.000	L.F.	.141	7.25	17.23
6' high	1.000	L.F.	.209	8.20	21.60
Add for gate 3' wide 1-3/8" frame 3' high	1.000	Ea.	2.000	43.50	152.05
4' high	1.000	Ea.	2.400	54	185.39
6' high	1.000	Ea.	2.400	97.50	252.85
Add for gate 4' wide 1-3/8" frame 3' high	1.000	Ea.	2.667	51	192.14
4' high	1.000	Ea.	2.667	67	216.95
6' high	1.000	Ea.	3.000	124	319.45
Alum.9ga. wire, 1-5/8"post, 10'O.C., 1-3/8"top rail, 2"corner post,3'hi	1.000	L.F.	.130	5.85	14.58
4' high	1.000	L.F.	.141	6.70	16.38
6' high	1.000	L.F.	.209	8.60	22.22
Add for gate 3' wide 1-3/8" frame 3' high	1.000	Ea.	2.000	57	173.02
4' high	1.000	Ea.	2.400	78	222.64
6' high	1.000	Ea.	2.400	117	283.06
Add for gate 4' wide 1-3/8" frame 3' high	1.000	Ea.	2.400	78	222.64
4' high	1.000	Ea.	2.667	104	274.31
6' high	1.000	Ea.	3.000	163	380.02
Vinyl 9ga. wire, 1-5/8"post 10'O.C., 1-3/8"top rail, 2"corner post,3'hi	1.000	L.F.	.130	5.20	13.57
4' high	1.000	L.F.	.141	8.55	19.24
6' high	1.000	L.F.	.209	9.75	24.01
Add for gate 3' wide 1-3/8" frame 3' high	1.000	Ea.	2.000	65	185.39
4' high	1.000	Ea.	2.400	84.50	232.73
6' high	1.000	Ea.	2.400	130	303.25
Add for gate 4' wide 1-3/8" frame 3' high	1.000	Ea.	2.400	88.50	238.91
4' high	1.000	Ea.	2.667	117	294.43
6' high	1.000	Ea.	3.000	169	389.40
Tennis court, chain link fence, 10' high					
Galv.11ga.wire, 2"post 10'O.C., 1-3/8"top rail, 2-1/2"corner post	1.000	L.F.	.253	13	30.89
Add for gate 3' wide 1-3/8" frame	1.000	Ea.	2.400	163	354.43
Alum.11ga.wire, 2"post 10'O.C., 1-3/8"top rail, 2-1/2"corner post	1.000	L.F.	.253	18.20	38.96
Add for gate 3' wide 1-3/8" frame	1.000	Ea.	2.400	208	424.23
Vinyl 11ga.wire,2"post 10' O.C.,1-3/8"top rail,2-1/2"corner post	1.000	L.F.	.253	15.60	34.92
Add for gate 3' wide 1-3/8" frame	1.000	Ea.	2.400	234	464.61

Be sure to read pages 178-179 for proper use of this section.

Wood Fence Price Sheet

SYSTEM DESCRIPTION	QUAN.	UNIT	LABOR-HOURS	COST PER UNIT MAT.	COST PER UNIT TOTAL
Basketweave, 3/8"x4" boards, 2"x4" stringers on spreaders, 4"x4" posts					
No. 1 cedar, 6' high	1.000	L.F.	.150	8.60	19.69
Treated pine, 6' high	1.000	L.F.	.160	10.45	22.99
Board fence, 1"x4" boards, 2"x4" rails, 4"x4" posts					
Preservative treated, 2 rail, 3' high	1.000	L.F.	.166	6.40	16.95
4' high	1.000	L.F.	.178	7	18.39
3 rail, 5' high	1.000	L.F.	.185	7.90	20.07
6' high	1.000	L.F.	.192	9.05	22.21
Western cedar, No. 1, 2 rail, 3' high	1.000	L.F.	.166	6.95	17.81
3 rail, 4' high	1.000	L.F.	.178	8.25	20.33
5' high	1.000	L.F.	.185	9.50	22.56
6' high	1.000	L.F.	.192	10.40	24.30
No. 1 cedar, 2 rail, 3' high	1.000	L.F.	.166	10.45	23.23
4' high	1.000	L.F.	.178	11.85	25.92
3 rail, 5' high	1.000	L.F.	.185	13.75	29.14
6' high	1.000	L.F.	.192	15.30	31.90
Shadow box, 1"x6" boards, 2"x4" rails, 4"x4" posts					
Fir, pine or spruce, treated, 3 rail, 6' high	1.000	L.F.	.160	11.70	24.93
No. 1 cedar, 3 rail, 4' high	1.000	L.F.	.185	14.40	30.15
6' high	1.000	L.F.	.192	17.75	35.71
Open rail, split rails, No. 1 cedar, 2 rail, 3' high	1.000	L.F.	.150	5.80	15.35
3 rail, 4' high	1.000	L.F.	.160	7.80	18.87
No. 2 cedar, 2 rail, 3' high	1.000	L.F.	.150	4.50	13.33
3 rail, 4' high	1.000	L.F.	.160	5.10	14.69
Open rail, rustic rails, No. 1 cedar, 2 rail, 3' high	1.000	L.F.	.150	3.61	11.95
3 rail, 4' high	1.000	L.F.	.160	4.84	14.29
No. 2 cedar, 2 rail, 3' high	1.000	L.F.	.150	3.45	11.71
3 rail, 4' high	1.000	L.F.	.160	3.65	12.44
Rustic picket, molded pine pickets, 2 rail, 3' high	1.000	L.F.	.171	5.10	15.16
3 rail, 4' high	1.000	L.F.	.197	5.85	17.39
No. 1 cedar, 2 rail, 3' high	1.000	L.F.	.171	6.95	18.03
3 rail, 4' high	1.000	L.F.	.197	8	20.72
Picket fence, fir, pine or spruce, preserved, treated					
2 rail, 3' high	1.000	L.F.	.171	4.44	14.14
3 rail, 4' high	1.000	L.F.	.185	5.25	15.96
Western cedar, 2 rail, 3' high	1.000	L.F.	.171	5.55	15.86
3 rail, 4' high	1.000	L.F.	.185	5.70	16.66
No. 1 cedar, 2 rail, 3' high	1.000	L.F.	.171	11.10	24.47
3 rail, 4' high	1.000	L.F.	.185	12.95	27.90
Stockade, No. 1 cedar, 3-1/4" rails, 6' high	1.000	L.F.	.150	10.45	22.56
8' high	1.000	L.F.	.155	13.55	27.58
No. 2 cedar, treated rails, 6' high	1.000	L.F.	.150	10.45	22.56
Treated pine, treated rails, 6' high	1.000	L.F.	.150	10.25	22.25
Gates, No. 2 cedar, picket, 3'-6" wide 4' high	1.000	Ea.	2.667	55.50	199.11
No. 2 cedar, rustic round, 3' wide, 3' high	1.000	Ea.	2.667	70.50	222.42
No. 2 cedar, stockade screen, 3'-6" wide, 6' high	1.000	Ea.	3.000	61.50	222.64
General, wood, 3'-6" wide, 4' high	1.000	Ea.	2.400	53.50	184.68
6' high	1.000	Ea.	3.000	67	231.17

Be sure to read pages 178-179 for proper use of this section.

Division 2
FOUNDATIONS

Footing Systems

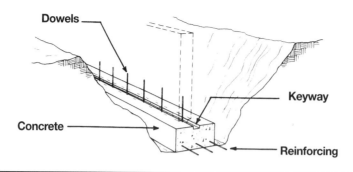

System Description	QUAN.	UNIT	LABOR-HOURS	MAT. COST	COST PER L.F.
8″ THICK BY 18″ WIDE FOOTING					
Concrete, 3000 psi	.040	C.Y.		3.56	
Place concrete, direct chute	.040	C.Y.	.016		
Forms, footing, 4 uses	1.330	SFCA	.103	.78	
Reinforcing, 1/2″ diameter bars, 2 each	1.380	Lb.	.011	.63	
Keyway, 2″ x 4″, beveled, 4 uses	1.000	L.F.	.015	.20	
Dowels, 1/2″ diameter bars, 2′ long, 6′ O.C.	.166	Ea.	.006	.11	**Contractor's Fee**
TOTAL			.151	5.28	**15.99**

Block Wall Systems

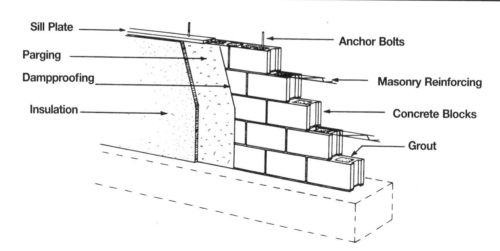

System Description	QUAN.	UNIT	LABOR-HOURS	MAT. COST	COST PER S.F.
8″ WALL, GROUTED, FULL HEIGHT					
Concrete block, 8″ x 16″ x 8″	1.000	S.F.	.094	1.97	
Masonry reinforcing, every second course	.750	L.F.	.002	.14	
Parging, plastering with portland cement plaster, 1 coat	1.000	S.F.	.014	.23	
Dampproofing, bituminous coating, 1 coat	1.000	S.F.	.012	.08	
Insulation, 1″ rigid polystyrene	1.000	S.F.	.010	.42	
Grout, solid, pumped	1.000	S.F.	.059	.94	
Anchor bolts, 1/2″ diameter, 8″ long, 4′ O.C.	.060	Ea.	.002	.04	
Sill plate, 2″ x 4″, treated	.250	L.F.	.007	.14	**Contractor's Fee**
TOTAL			.200	3.96	**16.84**

Be sure to read pages 178-179 for proper use of this section.

Concrete Wall Systems

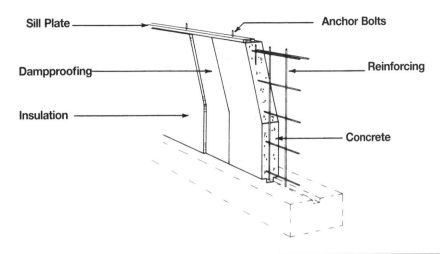

System Description	QUAN.	UNIT	LABOR-HOURS	MAT. COST	COST PER S.F.
8″ THICK, POURED CONCRETE WALL					
Concrete, 8″ thick , 3000 psi	.025	C.Y.		2.23	
Forms, prefabricated plywood, 4 uses per month	2.000	SFCA	.076	1.22	
Reinforcing, light	.670	Lb.	.004	.31	
Placing concrete, direct chute	.025	C.Y.	.013		
Dampproofing, brushed on, 2 coats	1.000	S.F.	.016	.11	
Rigid insulation, 1″ polystyrene	1.000	S.F.	.010	.42	
Anchor bolts, 1/2″ diameter, 12″ long, 4′ O.C.	.060	Ea.	.003	.08	
Sill plates, 2″ x 4″, treated	.250	L.F.	.007	.14	Contractor's Fee
TOTAL			.129	4.51	**13.70**

Wood Wall Foundation Systems

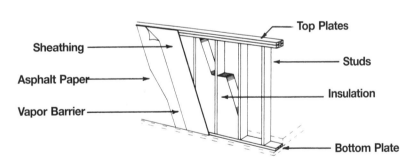

System Description	QUAN.	UNIT	LABOR-HOURS	MAT. COST	COST PER S.F.
2″ X 4″ STUDS, 16″ O.C., WALL					
Studs, 2″ x 4″, 16″ O.C., treated	1.000	L.F.	.015	.54	
Plates, double top plate, single bottom plate, treated, 2″ x 4″	.750	L.F.	.011	.41	
Sheathing, 1/2″, exterior grade, CDX, treated	1.000	S.F.	.014	.88	
Asphalt paper, 15# roll	1.100	S.F.	.002	.04	
Vapor barrier, 4 mil polyethylene	1.000	S.F.	.002	.03	
Insulation, batts, fiberglass, 3-1/2″ thick, R 11	1.000	S.F.	.005	.30	Contractor's Fee
TOTAL			.049	2.20	**6.25**

Be sure to read pages 178-179 for proper use of this section.

Floor Slab Systems

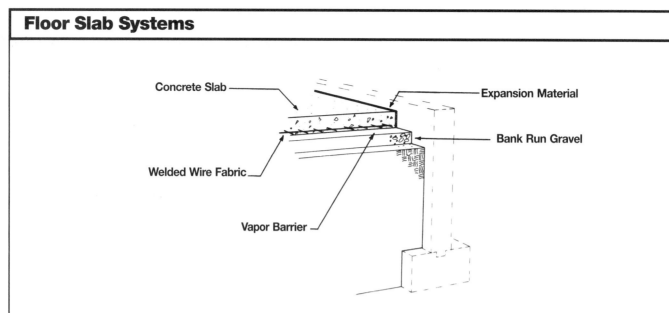

Concrete Slab —
Expansion Material —
Bank Run Gravel —
Welded Wire Fabric —
Vapor Barrier —

System Description	QUAN.	UNIT	LABOR-HOURS	MAT. COST	COST PER S.F.
4″ THICK SLAB					
Concrete, 4″ thick, 3000 psi concrete	.012	C.Y.		1.07	
Place concrete, direct chute	.012	C.Y.	.005		
Bank run gravel, 4″ deep	1.000	S.F.	.001	.32	
Polyethylene vapor barrier, .006″ thick	1.000	S.F.	.002	.03	
Edge forms, expansion material	.100	L.F.	.005	.03	
Welded wire fabric, 6 x 6, 10/10 (W1.4/W1.4)	1.100	S.F.	.005	.23	
Steel trowel finish	1.000	S.F.	.015		Contractor's Fee
TOTAL			.033	1.68	**4.39**

Be sure to read pages 178-179 for proper use of this section.

Division 3
FRAMING

Floor Framing Systems

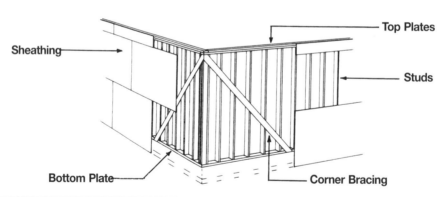

Box Sill · Sheathing · Bridging · Furring · Girder · Wood Joists · Box Sill

System Description	QUAN.	UNIT	LABOR-HOURS	MAT. COST	COST PER S.F.
2" X 8", 16" O.C.					
Wood joists, 2" x 8", 16" O.C.	1.000	L.F.	.015	.91	
Bridging, 1" x 3", 6' O.C.	.080	Pr.	.005	.04	
Box sills, 2" x 8"	.150	L.F.	.002	.14	
Concrete filled steel column, 4" diameter	.125	L.F.	.002	.11	
Girder, built up from three 2" x 8"	.125	L.F.	.013	.34	
Sheathing, plywood, subfloor, 5/8" CDX	1.000	S.F.	.012	.86	Contractor's Fee
Furring, 1" x 3", 16" O.C.	1.000	L.F.	.023	.31	
TOTAL			.072	2.71	**8.37**

Exterior Wall Framing Systems

Sheathing · Top Plates · Studs · Bottom Plate · Corner Bracing

System Description	QUAN.	UNIT	LABOR-HOURS	MAT. COST	COST PER S.F.
2" X 4", 16" O.C.					
2" x 4" studs, 16" O.C.	1.000	L.F.	.015	.37	
Plates, 2" x 4", double top, single bottom	.375	L.F.	.005	.14	
Corner bracing, let-in, 1" x 6"	.063	L.F.	.003	.04	
Sheathing, 1/2" plywood, CDX	1.000	S.F.	.011	.68	Contractor's Fee
TOTAL			.034	1.23	**3.92**

Be sure to read pages 178-179 for proper use of this section.

Gable End Roof Framing Systems

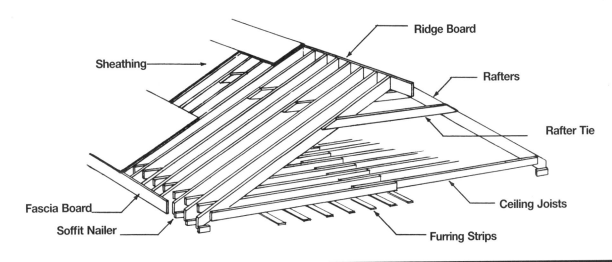

System Description	QUAN.	UNIT	LABOR-HOURS	MAT. COST	COST PER S.F.
2″ X 6″ RAFTERS, 16″ O.C., 4/12 PITCH					
Rafters, 2″ x 6″, 16″ O.C., 4/12 pitch	1.170	L.F.	.019	.69	
Ceiling joists, 2″ x 4″, 16″ O.C.	1.000	L.F.	.013	.37	
Ridge board, 2″ x 6″	.050	L.F.	.002	.03	
Fascia board, 2″ x 6″	.100	L.F.	.005	.07	
Rafter tie, 1″ x 4″, 4′ O.C.	.060	L.F.	.001	.03	
Soffit nailer (outrigger), 2″ x 4″, 24″ O.C.	.170	L.F.	.004	.06	
Sheathing, exterior, plywood, CDX, 1/2″ thick	1.170	S.F.	.013	.80	Contractor's Fee
Furring strips, 1″ x 3″, 16″ O.C.	1.000	L.F.	.023	.31	**8.32**
TOTAL			.080	2.36	

Truss Roof Framing Systems

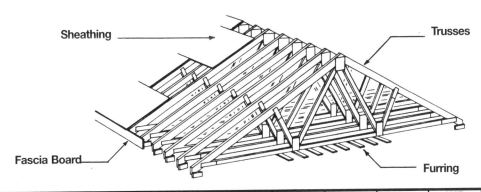

System Description	QUAN.	UNIT	LABOR-HOURS	MAT. COST	COST PER S.F.
TRUSS, 16″ O.C., 4/12 PITCH, 1′ OVERHANG, 26′ SPAN					
Truss, 40# loading, 16″ O.C., 4/12 pitch, 26′ span	.030	Ea.	.021	2.34	
Fascia board, 2″ x 6″	.100	L.F.	.005	.07	
Sheathing, exterior, plywood, CDX, 1/2″ thick	1.170	S.F.	.013	.80	Contractor's Fee
Furring, 1″ x 3″, 16″ O.C.	1.000	L.F.	.023	.31	**9.46**
TOTAL			.062	3.52	

Be sure to read pages 178-179 for proper use of this section.

Hip Roof Framing Systems

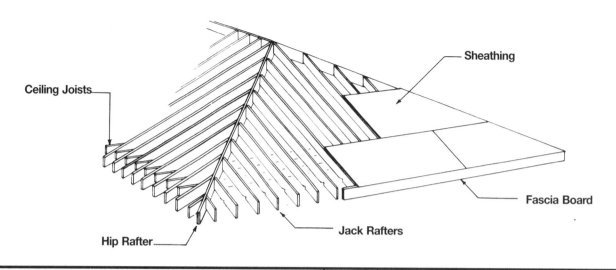

System Description	QUAN.	UNIT	LABOR-HOURS	MAT. COST	COST PER S.F.
2" X 6", 16" O.C., 4/12 PITCH					
Hip rafters, 2" x 8", 4/12 pitch	.160	L.F.	.004	.15	
Jack rafters, 2" x 6", 16" O.C., 4/12 pitch	1.430	L.F.	.038	.84	
Ceiling joists, 2" x 6", 16" O.C.	1.000	L.F.	.013	.59	
Fascia board, 2" x 8"	.220	L.F.	.016	.20	
Soffit nailer (outrigger), 2" x 4", 24" O.C.	.220	L.F.	.006	.08	
Sheathing, 1/2" exterior plywood, CDX	1.570	S.F.	.018	1.07	**Contractor's Fee**
Furring strips, 1" x 3", 16" O.C.	1.000	L.F.	.023	.31	
TOTAL			.118	3.24	**11.79**

Gambrel Roof Framing Systems

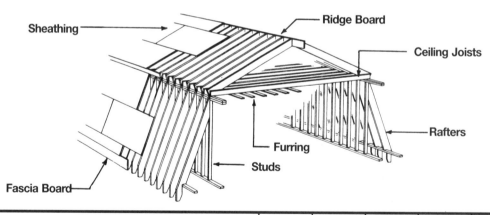

System Description	QUAN.	UNIT	LABOR-HOURS	MAT. COST	COST PER S.F.
2" X 6" RAFTERS, 16" O.C.					
Roof rafters, 2" x 6", 16" O.C.	1.430	L.F.	.029	.84	
Ceiling joists, 2" x 6", 16" O.C.	.710	L.F.	.009	.42	
Stud wall, 2" x 4", 16" O.C., including plates	.790	L.F.	.012	.29	
Furring strips, 1" x 3", 16" O.C.	.710	L.F.	.016	.22	
Ridge board, 2" x 8"	.050	L.F.	.002	.05	
Fascia board, 2" x 6"	.100	L.F.	.006	.07	**Contractor's Fee**
Sheathing, exterior grade plywood, 1/2" thick	1.450	S.F.	.017	.99	
TOTAL			.091	2.88	**9.73**

Be sure to read pages 178-179 for proper use of this section.

Mansard Roof Framing Systems

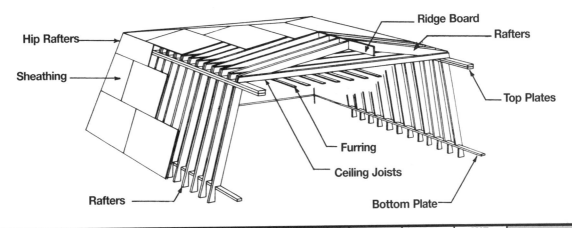

System Description	QUAN.	UNIT	LABOR-HOURS	MAT. COST	COST PER S.F.
2″ X 6″ RAFTERS, 16″ O.C.					
Roof rafters, 2″ x 6″, 16″ O.C.	1.210	L.F.	.033	.71	
Rafter plates, 2″ x 6″, double top, single bottom	.364	L.F.	.010	.21	
Ceiling joists, 2″ x 4″, 16″ O.C.	.920	L.F.	.012	.34	
Hip rafter, 2″ x 6″	.070	L.F.	.002	.04	
Jack rafter, 2″ x 6″, 16″ O.C.	1.000	L.F.	.039	.59	
Ridge board, 2″ x 6″	.018	L.F.	.001	.01	
Sheathing, exterior grade plywood, 1/2″ thick	2.210	S.F.	.025	1.50	
Furring strips, 1″ x 3″, 16″ O.C.	.920	L.F.	.021	.29	Contractor's Fee
TOTAL			.143	3.69	**14.01**

Shed/Flat Roof Framing Systems

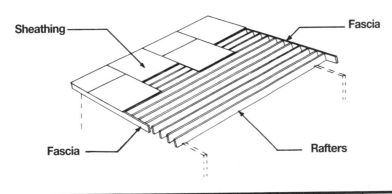

System Description	QUAN.	UNIT	LABOR-HOURS	MAT. COST	COST PER S.F.
2″ X 6″, 16″ O.C., 4/12 PITCH					
Rafters, 2″ x 6″, 16″ O.C., 4/12 pitch	1.170	L.F.	.019	.69	
Fascia, 2″ x 6″	.100	L.F.	.006	.07	
Bridging, 1″ x 3″, 6′ O.C.	.080	Pr.	.005	.04	
Sheathing, exterior grade plywood, 1/2″ thick	1.230	S.F.	.014	.84	Contractor's Fee
TOTAL			.044	1.64	**5.07**

Be sure to read pages 178-179 for proper use of this section.

Gable Dormer Framing Systems

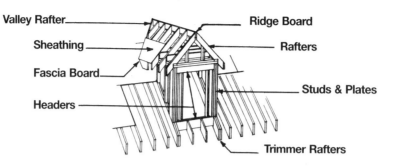

System Description	QUAN.	UNIT	LABOR-HOURS	MAT. COST	COST PER S.F.
2″ X 6″, 16″ O.C.					
Dormer rafter, 2″ x 6″, 16″ O.C.	1.330	L.F.	.036	.78	
Ridge board, 2″ x 6″	.280	L.F.	.009	.17	
Trimmer rafters, 2″ x 6″	.880	L.F.	.014	.52	
Wall studs & plates, 2″ x 4″, 16″ O.C.	3.160	L.F.	.056	1.17	
Fascia, 2″ x 6″	.220	L.F.	.012	.15	
Valley rafter, 2″ x 6″, 16″ O.C.	.280	L.F.	.009	.17	
Cripple rafter, 2″ x 6″, 16″ O.C.	.560	L.F.	.022	.33	
Headers, 2″ x 6″, doubled	.670	L.F.	.030	.40	
Ceiling joist, 2″ x 4″, 16″ O.C.	1.000	L.F.	.013	.37	
Sheathing, exterior grade plywood, 1/2″ thick	3.610	S.F.	.041	2.45	Contractor's Fee
TOTAL			.242	6.51	**24.12**

Shed Dormer Framing Systems

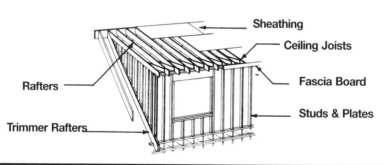

System Description	QUAN.	UNIT	LABOR-HOURS	MAT. COST	COST PER S.F.
2″ X 6″ RAFTERS, 16″ O.C.					
Dormer rafter, 2″ x 6″, 16″ O.C.	1.080	L.F.	.029	.64	
Trimmer rafter, 2″ x 6″	.400	L.F.	.006	.24	
Studs & plates, 2″ x 4″, 16″ O.C.	2.750	L.F.	.049	1.02	
Fascia, 2″ x 6″	.250	L.F.	.014	.17	
Ceiling joist, 2″ x 4″, 16″ O.C.	1.000	L.F.	.013	.37	
Sheathing, exterior grade plywood, CDX, 1/2″ thick	2.940	S.F.	.034	2	Contractor's Fee
TOTAL			.145	4.44	**15.26**

Be sure to read pages 178-179 for proper use of this section.

Partition Framing Systems

System Description	QUAN.	UNIT	LABOR-HOURS	MAT. COST	COST PER S.F.
2″ X 4″, 16″ O.C.					
2″ x 4″ studs, #2 or better, 16″ O.C.	1.000	L.F.	.015	.37	
Plates, double top, single bottom	.375	L.F.	.005	.14	**Contractor's Fee**
Cross bracing, let-in, 1″ x 6″	.080	L.F.	.004	.05	
TOTAL			.024	.56	**2.26**
2″ X 4″, 24″ O.C.					
2″ x 4″ studs, #2 or better, 24″ O.C.	.800	L.F.	.012	.30	
Plates, double top, single bottom	.375	L.F.	.005	.14	**Contractor's Fee**
Cross bracing, let-in, 1″ x 6″	.080	L.F.	.003	.05	
TOTAL			.020	.49	**1.89**
2″ X 6″, 16″ O.C.					
2″ x 6″ studs, #2 or better, 16″ O.C.	1.000	L.F.	.016	.59	
Plates, double top, single bottom	.375	L.F.	.006	.22	**Contractor's Fee**
Cross bracing, let-in, 1″ x 6″	.080	L.F.	.004	.05	
TOTAL			.026	.86	**2.84**
2″ X 6″, 24″ O.C.					
2″ x 6″ studs, #2 or better, 24″ O.C.	.800	L.F.	.013	.47	
Plates, double top, single bottom	.375	L.F.	.006	.22	**Contractor's Fee**
Cross bracing, let-in, 1″ x 6″	.080	L.F.	.003	.05	
TOTAL			.022	.74	**2.38**

Be sure to read pages 178-179 for proper use of this section.

Division 4
EXTERIOR WALLS

Block Masonry Systems

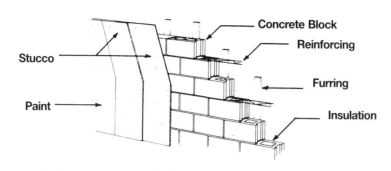

Stucco

Paint

Concrete Block

Reinforcing

Furring

Insulation

System Description	QUAN.	UNIT	LABOR-HOURS	MAT. COST	COST PER S.F.
8″ THICK CONCRETE BLOCK WALL					
8″ thick concrete block, 8″ x 8″ x 16″	1.000	S.F.	.107	1.67	
Masonry reinforcing, truss strips every other course	.625	L.F.	.002	.12	
Furring, 1″ x 3″, 16″ O.C.	1.000	L.F.	.016	.31	
Masonry insulation, poured vermiculite	1.000	S.F.	.018	.80	
Stucco, 2 coats	1.000	S.F.	.069	.20	
Masonry paint, 2 coats	1.000	S.F.	.016	.19	**Contractor's Fee**
TOTAL			.228	3.29	**17.26**

Brick/Stone Veneer Systems

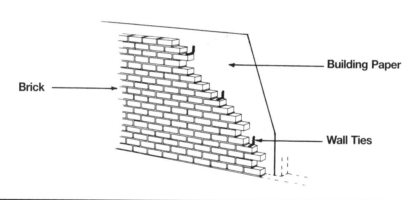

Brick

Building Paper

Wall Ties

System Description	QUAN.	UNIT	LABOR-HOURS	MAT. COST	COST PER S.F.
SELECT COMMON BRICK					
Brick, select common, running bond	1.000	S.F.	.174	3.17	
Wall ties, 7/8″ x 7″, 22 gauge	1.000	Ea.	.008	.06	
Building paper, spunbonded polypropylene	1.100	S.F.	.002	.12	
Trim, pine, painted	.125	L.F.	.004	.09	**Contractor's Fee**
TOTAL			.188	3.44	**15.32**

Be sure to read pages 178-179 for proper use of this section.

Wood Siding Systems

Trim → Building Paper → Beveled Cedar Siding

System Description	QUAN.	UNIT	LABOR-HOURS	MAT. COST	COST PER S.F.
1/2″ X 6″ BEVELED CEDAR SIDING, "A" GRADE					
1/2″ x 6″ beveled cedar siding	1.000	S.F.	.032	3.61	
Building wrap, Spundbonded polypropylene	1.100	S.F.	.002	.12	
Trim, cedar	.125	L.F.	.005	.21	
Paint, primer & 2 coats	1.000	S.F.	.017	.18	Contractor's Fee
TOTAL			.056	4.12	**9.52**
1/2″ X 8″ BEVELED CEDAR SIDING, "A" GRADE					
1/2″ x 8″ beveled cedar siding	1.000	S.F.	.029	3.21	
Building wrap, Spundbonded polypropylene	1.100	S.F.	.002	.12	
Trim, cedar	.125	L.F.	.005	.21	
Paint, primer & 2 coats	1.000	S.F.	.017	.18	Contractor's Fee
TOTAL			.053	3.72	**8.74**
1″ X 4″ TONGUE & GROOVE, REDWOOD, VERTICAL GRAIN					
Redwood, clear, vertical grain, 1″ x 10″	1.000	S.F.	.018	3.62	
Building Wrap, Spundbonded polypropylene	1.100	S.F.	.002	.12	
Trim, redwood	.125	L.F.	.005	.21	
Sealer, 1 coat, stain, 1 coat	1.000	S.F.	.013	.12	Contractor's Fee
TOTAL			.038	4.07	**8.47**
1″ X 6″ TONGUE & GROOVE, REDWOOD, VERTICAL GRAIN					
Redwood, clear, vertical grain, 1″ x 10″	1.000	S.F.	.019	3.73	
Building wrap, Spundbonded polypropylene	1.100	S.F.	.002	.12	
Trim, redwood	.125	L.F.	.005	.21	
Sealer, 1 coat, stain, 1 coat	1.000	S.F.	.013	.12	Contractor's Fee
TOTAL			.039	4.18	**8.67**

Be sure to read pages 178-179 for proper use of this section.

Shingle Siding Systems

Trim · Building Paper · White Cedar Shingles

System Description	QUAN.	UNIT	LABOR-HOURS	MAT. COST	COST PER S.F.
WHITE CEDAR SHINGLES, 5″ EXPOSURE					
White cedar shingles, 16″ long, grade "A", 5″ exposure	1.000	S.F.	.033	1.35	
Building wrap, Spunbonded polypropylene	1.100	S.F.	.002	.12	
Trim, cedar	.125	S.F.	.005	.21	
Paint, primer & 1 coat	1.000	S.F.	.017	.09	Contractor's Fee
TOTAL			.057	1.77	**5.97**
NO. 1 PERFECTIONS, 5-1/2″ EXPOSURE					
No. 1 perfections, red cedar, 5-1/2″ exposure	1.000	S.F.	.029	1.76	
Building wrap, Spunbonded polypropylene	1.100	S.F.	.002	.12	
Trim, cedar	.125	S.F.	.005	.21	
Stain, sealer & 1 coat	1.000	S.F.	.017	.09	Contractor's Fee
TOTAL			.053	2.18	**6.36**
RESQUARED & REBUTTED PERFECTIONS, 5-1/2″ EXPOSURE					
Resquared & rebutted perfections, 5-1/2″ exposure	1.000	S.F.	.027	2.19	
Building wrap, Spunbonded polypropylene	1.100	S.F.	.002	.12	
Trim, cedar	.125	S.F.	.005	.21	
Stain, sealer & 1 coat	1.000	S.F.	.017	.09	Contractor's Fee
TOTAL			.051	2.61	**6.89**
HAND-SPLIT SHAKES, 8-1/2″ EXPOSURE					
Hand-split red cedar shakes, 18″ long, 8-1/2″ exposure	1.000	S.F.	.040	1.07	
Building wrap, Spunbonded polypropylene	1.100	S.F.	.002	.12	
Trim, cedar	.125	S.F.	.005	.21	
Stain, sealer & 1 coat	1.000	S.F.	.017	.09	Contractor's Fee
TOTAL			.064	1.49	**5.92**

Be sure to read pages 178-179 for proper use of this section.

Metal & Plastic Siding Systems

Aluminum Trim →

Building Paper

Alum. Horizontal Siding

Backer Insulation Board

System Description	QUAN.	UNIT	LABOR-HOURS	MAT. COST	COST PER S.F.
ALUMINUM CLAPBOARD SIDING, 8″ WIDE, WHITE					
Aluminum horizontal siding, 8″ clapboard	1.000	S.F.	.031	1.38	
Backer, insulation board	1.000	S.F.	.008	.45	
Trim, aluminum	.600	L.F.	.016	.62	Contractor's Fee
Building wrap, Spunbonded polypropylene	1.100	S.F.	.002	.12	
TOTAL			.057	2.57	**7.30**
ALUMINUM VERTICAL BOARD & BATTEN, WHITE					
Aluminum vertical board & batten	1.000	S.F.	.027	1.54	
Backer insulation board	1.000	S.F.	.008	.45	
Trim, aluminum	.600	L.F.	.016	.62	Contractor's Fee
Building wrap, Spunbonded polypropylene	1.100	S.F.	.002	.12	
TOTAL			.053	2.73	**7.32**
VINYL CLAPBOARD SIDING, 8″ WIDE, WHITE					
PVC vinyl horizontal siding, 8″ clapboard	1.000	S.F.	.032	.68	
Backer, insulation board	1.000	S.F.	.008	.45	
Trim, vinyl	.600	L.F.	.014	.45	Contractor's Fee
Building wrap, Spunbonded polypropylene	1.100	S.F.	.002	.12	
TOTAL			.056	1.70	**5.92**
VINYL VERTICAL BOARD & BATTEN, WHITE					
PVC vinyl vertical board & batten	1.000	S.F.	.029	1.51	
Backer, insulation board	1.000	S.F.	.008	.45	
Trim, vinyl	.600	L.F.	.014	.45	Contractor's Fee
Building wrap, Spunbonded polypropylene	1.100	S.F.	.002	.12	
TOTAL			.053	2.53	**7.02**

Be sure to read pages 178-179 for proper use of this section.

Double Hung Window Systems

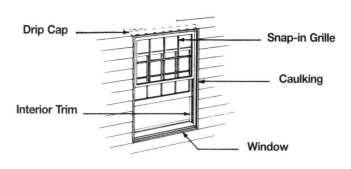

Drip Cap — Snap-in Grille — Caulking — Interior Trim — Window

System Description	QUAN.	UNIT	LABOR-HOURS	MAT. COST	COST EACH
BUILDER'S QUALITY WOOD WINDOW 2' X 3', DOUBLE HUNG					
Window, primed, builder's quality, 2' x 3', insulating glass	1.000	Ea.	.800	198	
Trim, interior casing	11.000	L.F.	.367	10.34	
Paint, interior & exterior, primer & 2 coats	2.000	Face	1.778	2.10	
Caulking	10.000	L.F.	.323	1.70	
Snap-in grille	1.000	Set	.333	46	
Drip cap, metal	2.000	L.F.	.040	.56	Contractor's Fee
TOTAL			3.641	258.70	**600.20**
PLASTIC CLAD WOOD WINDOW 3' X 4', DOUBLE HUNG					
Window, plastic clad, premium, 3' x 4', insulating glass	1.000	Ea.	.889	355	
Trim, interior casing	15.000	L.F.	.500	14.10	
Paint, interior, primer & 2 coats	1.000	Face	.889	1.05	
Caulking	14.000	L.F.	.452	2.38	
Snap-in grille	1.000	Set	.333	46	Contractor's Fee
TOTAL			3.063	418.53	**820.84**
METAL CLAD WOOD WINDOW, 3' X 5', DOUBLE HUNG					
Window, metal clad, deluxe, 3' x 5', insulating glass	1.000	Ea.	1.000	297	
Trim, interior casing	17.000	L.F.	.567	15.98	
Paint, interior, primer & 2 coats	1.000	Face	.889	1.05	
Caulking	16.000	L.F.	.516	2.72	
Snap-in grille	1.000	Set	.235	128	
Drip cap, metal	3.000	L.F.	.060	.84	Contractor's Fee
TOTAL			3.267	445.59	**875.06**

Be sure to read pages 178-179 for proper use of this section.

Casement Window Systems

System Description	QUAN.	UNIT	LABOR-HOURS	MAT. COST	COST EACH
BUILDER'S QUALITY WINDOW, WOOD, 2' BY 3', CASEMENT					
Window, primed, builder's quality, 2' x 3', insulating glass	1.000	Ea.	.800	310	
Trim, interior casing	11.000	L.F.	.367	10.34	
Paint, interior & exterior, primer & 2 coats	2.000	Face	1.778	2.10	
Caulking	10.000	L.F.	.323	1.70	
Snap-in grille	1.000	Ea.	.267	26	
Drip cap, metal	2.000	L.F.	.040	.56	Contractor's Fee
TOTAL			3.575	350.70	**739.03**
PLASTIC CLAD WOOD WINDOW, 2' X 4', CASEMENT					
Window, plastic clad, premium, 2' x 4', insulating glass	1.000	Ea.	.889	305	
Trim, interior casing	13.000	L.F.	.433	12.22	
Paint, interior, primer & 2 coats	1.000	Ea.	.889	1.05	
Caulking	12.000	L.F.	.387	2.04	
Snap-in grille	1.000	Ea.	.267	26	Contractor's Fee
TOTAL			2.865	346.31	**696.96**
METAL CLAD WOOD WINDOW, 2' X 5', CASEMENT					
Window, metal clad, deluxe, 2' x 5', insulating glass	1.000	Ea.	1.000	263	
Trim, interior casing	15.000	L.F.	.500	14.10	
Paint, interior, primer & 2 coats	1.000	Ea.	.889	1.05	
Caulking	14.000	L.F.	.452	2.38	
Snap-in grille	1.000	Ea.	.250	37	
Drip cap, metal	12.000	L.F.	.040	.56	Contractor's Fee
TOTAL			3.131	318.09	**669.58**

Be sure to read pages 178-179 for proper use of this section.

Awning Window Systems

Drip Cap — Interior Trim

Snap-in Grille

Caulking

Window

System Description	QUAN.	UNIT	LABOR-HOURS	MAT. COST	COST EACH
BUILDER'S QUALITY WINDOW, WOOD, 34" X 22", AWNING					
Window, builder quality, 34" x 22", insulating glass	1.000	Ea.	.800	243	
Trim, interior casing	10.500	L.F.	.350	9.87	
Paint, interior & exterior, primer & 2 coats	2.000	Face	1.778	2.10	
Caulking	9.500	L.F.	.306	1.62	
Snap-in grille	1.000	Ea.	.267	21	
Drip cap, metal	3.000	L.F.	.060	.84	**Contractor's Fee**
TOTAL			3.561	278.43	**626.21**
PLASTIC CLAD WOOD WINDOW, 40" X 28", AWNING					
Window, plastic clad, premium, 40" x 28", insulating glass	1.000	Ea.	.889	315	
Trim interior casing	13.500	L.F.	.450	12.69	
Paint, interior, primer & 2 coats	1.000	Face	.889	1.05	
Caulking	12.500	L.F.	.403	2.13	
Snap-in grille	1.000	Ea.	.267	21	**Contractor's Fee**
TOTAL			2.898	351.87	**707.80**
METAL CLAD WOOD WINDOW, 48" X 36", AWNING					
Window, metal clad, deluxe, 48" x 36", insulating glass	1.000	Ea.	1.000	340	
Trim, interior casing	15.000	L.F.	.500	14.10	
Paint, interior, primer & 2 coats	1.000	Face	.889	1.05	
Caulking	14.000	L.F.	.452	2.38	
Snap-in grille	1.000	Ea.	.250	30.50	
Drip cap, metal	4.000	L.F.	.080	1.12	**Contractor's Fee**
TOTAL			3.171	389.15	**782.11**

Be sure to read pages 178-179 for proper use of this section.

Sliding Window Systems

Drip Cap — Snap-in Grille — Caulking — Interior Trim — Window

System Description	QUAN.	UNIT	LABOR-HOURS	MAT. COST	COST EACH
BUILDER'S QUALITY WOOD WINDOW, 3' X 2', SLIDING					
Window, primed, builder's quality, 3' x 2', insul. glass	1.000	Ea.	.800	200	
Trim, interior casing	11.000	L.F.	.367	10.34	
Paint, interior & exterior, primer & 2 coats	2.000	Face	1.778	2.10	
Caulking	10.000	L.F.	.323	1.70	
Snap-in grille	1.000	Set	.333	25.50	
Drip cap, metal	3.000	L.F.	.060	.84	Contractor's Fee
TOTAL			3.661	240.48	**573.16**
PLASTIC CLAD WOOD WINDOW, 4' X 3'-6", SLIDING					
Window, plastic clad, premium, 4' x 3'-6", insulating glass	1.000	Ea.	.889	705	
Trim, interior casing	16.000	L.F.	.533	15.04	
Paint, interior, primer & 2 coats	1.000	Face	.889	1.05	
Caulking	17.000	L.F.	.548	2.89	Contractor's Fee
Snap-in grille	1.000	Set	.333	25.50	
TOTAL			3.192	749.48	**1,341.31**
METAL CLAD WOOD WINDOW, 6' X 5', SLIDING					
Window, metal clad, deluxe, 6' x 5', insulating glass	1.000	Ea.	1.000	710	
Trim, interior casing	23.000	L.F.	.767	21.62	
Paint, interior, primer & 2 coats	1.000	Face	.889	1.05	
Caulking	22.000	L.F.	.710	3.74	
Snap-in grille	1.000	Set	.364	38	Contractor's Fee
Drip cap, metal	6.000	L.F.	.120	1.68	
TOTAL			3.850	776.09	**1,422.23**

Be sure to read pages 178-179 for proper use of this section.

Bow/Bay Window Systems

Drip Cap

Caulking

Snap-in Grille

Window

System Description	QUAN.	UNIT	LABOR-HOURS	MAT. COST	COST EACH
AWNING TYPE BOW WINDOW, BUILDER'S QUALITY, 8' X 5'					
Window, primed, builder's quality, 8' x 5', insulating glass	1.000	Ea.	1.600	1,325	
Trim, interior casing	27.000	L.F.	.900	25.38	
Paint, interior & exterior, primer & 1 coat	2.000	Face	3.200	11.20	
Drip cap, vinyl	1.000	Ea.	.533	82.50	
Caulking	26.000	L.F.	.839	4.42	
Snap-in grilles	1.000	Set	1.067	104	Contractor's Fee
TOTAL			8.139	1,552.50	**2,855.19**
CASEMENT TYPE BOW WINDOW, PLASTIC CLAD, 10' X 6'					
Window, plastic clad, premium, 10' x 6', insulating glass	1.000	Ea.	2.286	1,875	
Trim, interior casing	33.000	L.F.	1.100	31.02	
Paint, interior, primer & 1 coat	1.000	Face	1.778	2.10	
Drip cap, vinyl	1.000	Ea.	.615	90	
Caulking	32.000	L.F.	1.032	5.44	
Snap-in grilles	1.000	Set	1.333	130	Contractor's Fee
TOTAL			8.144	2,133.56	**3,768.35**
DOUBLE HUNG TYPE, METAL CLAD, 9' X 5'					
Window, metal clad, deluxe, 9' x 5', insulating glass	1.000	Ea.	2.667	1,200	
Trim, interior casing	29.000	L.F.	.967	27.26	
Paint, interior, primer & 1 coat	1.000	Face	1.778	2.10	
Drip cap, vinyl	1.000	Set	.615	90	
Caulking	28.000	L.F.	.903	4.76	
Snap-in grilles	1.000	Set	1.067	104	Contractor's Fee
TOTAL			7.997	1,428.12	**2,664.70**

Be sure to read pages 178-179 for proper use of this section.

Fixed Window Systems

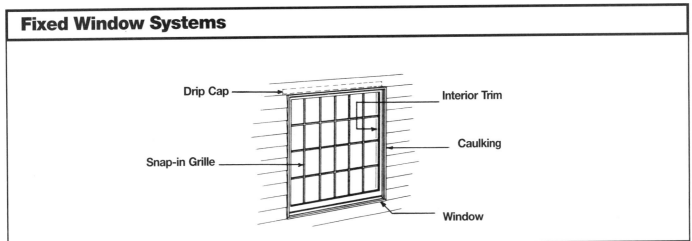

System Description	QUAN.	UNIT	LABOR-HOURS	MAT. COST	COST EACH
BUILDER'S QUALITY PICTURE WINDOW, 4' X 4'					
Window, primed, builder's quality, 4' x 4', insulating glass	1.000	Ea.	1.333	320	
Trim, interior casing	17.000	L.F.	.567	15.98	
Paint, interior & exterior, primer & 2 coats	2.000	Face	1.778	2.10	
Caulking	16.000	L.F.	.516	2.72	
Snap-in grille	1.000	Ea.	.267	146	Contractor's Fee
Drip cap, metal	4.000	L.F.	.080	1.12	**1,008.58**
TOTAL			4.541	487.92	
PLASTIC CLAD WOOD WINDOW, 4'-6" X 6'-6"					
Window, plastic clad, prem., 4'-6" x 6'-6", insul. glass	1.000	Ea.	1.455	700	
Trim, interior casing	23.000	L.F.	.767	21.62	
Paint, interior, primer & 2 coats	1.000	Face	.889	1.05	
Caulking	22.000	L.F.	.710	3.74	Contractor's Fee
Snap-in grille	1.000	Ea.	.267	146	**1,584.44**
TOTAL			4.088	872.41	
METAL CLAD WOOD WINDOW, 6'-6" X 6'-6"					
Window, metal clad, deluxe, 6'-6" x 6'-6", insulating glass	1.000	Ea.	1.600	565	
Trim interior casing	27.000	L.F.	.900	25.38	
Paint, interior, primer & 2 coats	1.000	Face	1.600	5.60	
Caulking	26.000	L.F.	.839	4.42	
Snap-in grille	1.000	Ea.	.267	146	Contractor's Fee
Drip cap, metal	6.500	L.F.	.130	1.82	**1,458.45**
TOTAL			5.336	748.22	

Be sure to read pages 178-179 for proper use of this section.

Entrance Door Systems

System Description	QUAN.	UNIT	LABOR-HOURS	MAT. COST	COST EACH
COLONIAL, 6 PANEL, 3' X 6'-8", WOOD					
Door, 3' x 6'-8" x 1-3/4" thick, pine, 6 panel colonial	1.000	Ea.	1.067	390	
Frame, 5-13/16" deep, incl. exterior casing & drip cap	17.000	L.F.	.725	141.95	
Interior casing, 2-1/2" wide	18.000	L.F.	.600	16.92	
Sill, 8/4 x 8" deep	3.000	L.F.	.480	43.95	
Butt hinges, brass, 4-1/2" x 4-1/2"	1.500	Pr.		20.40	
Lockset	1.000	Ea.	.571	34.50	
Weatherstripping, metal, spring type, bronze	1.000	Set	1.053	17.70	
Paint, interior & exterior, primer & 2 coats	2.000	Face	1.778	11.10	**Contractor's Fee**
TOTAL			6.274	676.52	**1,401.29**
SOLID CORE BIRCH, FLUSH, 3' X 6'-8"					
Door, 3' x 6'-8", 1-3/4" thick, birch, flush solid core	1.000	Ea.	1.067	104	
Frame, 5-13/16" deep, incl. exterior casing & drip cap	17.000	L.F.	.725	141.95	
Interior casing, 2-1/2" wide	18.000	L.F.	.600	16.92	
Sill, 8/4 x 8" deep	3.000	L.F.	.480	43.95	
Butt hinges, brass, 4-1/2" x 4-1/2"	1.500	Pr.		20.40	
Lockset	1.000	Ea.	.571	34.50	
Weatherstripping, metal, spring type, bronze	1.000	Set	1.053	17.70	
Paint, Interior & exterior, primer & 2 coats	2.000	Face	1.778	10.30	**Contractor's Fee**
TOTAL			6.274	389.72	**956.35**

Be sure to read pages 178-179 for proper use of this section.

Sliding Door Systems

Drip Cap

Interior Casing

Frame & Exterior Casing

Door

Sill

System Description	QUAN.	UNIT	LABOR-HOURS	MAT. COST	COST EACH
WOOD SLIDING DOOR, 8' WIDE, PREMIUM					
Wood, 5/8" thick tempered insul. glass, 8' wide, premium	1.000	Ea.	5.333	1,250	
Interior casing	22.000	L.F.	.733	20.68	
Exterior casing	22.000	L.F.	.733	20.68	
Sill, oak, 8/4 x 8" deep	8.000	L.F.	1.280	117.20	
Drip cap	8.000	L.F.	.160	2.24	
Paint, interior & exterior, primer & 2 coats	2.000	Face	2.816	15.84	**Contractor's Fee**
TOTAL			11.055	1,426.64	**2,833.93**
ALUMINUM SLIDING DOOR, 8' WIDE, PREMIUM					
Aluminum, 5/8" tempered insul. glass, 8' wide, premium	1.000	Ea.	5.333	1,425	
Interior casing	22.000	L.F.	.733	20.68	
Exterior casing	22.000	L.F.	.733	20.68	
Sill, oak, 8/4 x 8" deep	8.000	L.F.	1.280	117.20	
Drip cap	8.000	L.F.	.160	2.24	
Paint, interior & exterior, primer & 2 coats	2.000	Face	2.816	15.84	**Contractor's Fee**
TOTAL			11.055	1,601.64	**3,104.05**

Be sure to read pages 178-179 for proper use of this section.

Residential Garage Door Systems

System Description	QUAN.	UNIT	LABOR-HOURS	MAT. COST	COST EACH
OVERHEAD, SECTIONAL GARAGE DOOR, 9' X 7'					
Wood, overhead sectional door, std., incl. hardware, 9' x 7'	1.000	Ea.	2.000	485	
Jamb & header blocking, 2" x 6"	25.000	L.F.	.901	14.75	
Exterior trim	25.000	L.F.	.833	23.50	
Paint, interior & exterior, primer & 2 coats	2.000	Face	3.556	22.20	
Weatherstripping, molding type	1.000	Set	.767	21.62	Contractor's Fee
Drip cap	9.000	L.F.	.180	2.52	
TOTAL			8.237	569.59	**1,336.64**
OVERHEAD, SECTIONAL GARAGE DOOR, 16' X 7'					
Wood, overhead sectional, std., incl. hardware, 16' x 7'	1.000	Ea.	2.667	980	
Jamb & header blocking, 2" x 6"	30.000	L.F.	1.081	17.70	
Exterior trim	30.000	L.F.	1.000	28.20	
Paint, interior & exterior, primer & 2 coats	2.000	Face	5.333	33.30	
Weatherstripping, molding type	1.000	Set	1.000	28.20	Contractor's Fee
Drip cap	16.000	L.F.	.320	4.48	
TOTAL			11.401	1,091.88	**2,318.55**
OVERHEAD, SWING-UP TYPE, GARAGE DOOR, 16' X 7'					
Wood, overhead, swing-up, std., incl. hardware, 16' x 7'	1.000	Ea.	2.667	670	
Jamb & header blocking, 2" x 6"	30.000	L.F.	1.081	17.70	
Exterior trim	30.000	L.F.	1.000	28.20	
Paint, interior & exterior, primer & 2 coats	2.000	Face	5.333	33.30	
Weatherstripping, molding type	1.000	Set	1.000	28.20	Contractor's Fee
Drip cap	16.000	L.F.	.320	4.48	
TOTAL			11.401	781.88	**1,838.02**

Be sure to read pages 178-179 for proper use of this section.

Aluminum Window Systems

System Description	QUAN.	UNIT	LABOR-HOURS	MAT. COST	COST EACH
SINGLE HUNG, 2' X 3' OPENING					
Window, 2' x 3' opening, enameled, insulating glass	1.000	Ea.	1.600	191	
Blocking, 1" x 3" furring strip nailers	10.000	L.F.	.146	3.10	
Drywall, 1/2" thick, standard	5.000	S.F.	.040	1.20	
Corner bead, 1" x 1", galvanized steel	8.000	L.F.	.160	1.04	
Finish drywall, tape and finish corners inside and outside	16.000	L.F.	.269	1.12	Contractor's Fee
Sill, slate	2.000	L.F.	.400	17.30	
TOTAL			2.615	214.76	**501.55**
SLIDING, 3' X 2' OPENING					
Window, 3' x 2' opening, enameled, insulating glass	1.000	Ea.	1.600	197	
Blocking, 1" x 3" furring strip nailers	10.000	L.F.	.146	3.10	
Drywall, 1/2" thick, standard	5.000	S.F.	.040	1.20	
Corner bead, 1" x 1", galvanized steel	7.000	L.F.	.140	.91	
Finish drywall, tape and finish corners inside and outside	14.000	L.F.	.236	.98	Contractor's Fee
Sill, slate	3.000	L.F.	.600	25.95	
TOTAL			2.762	229.14	**530.83**
AWNING, 3'-1" X 3'-2"					
Window, 3'-1" x 3'-2" opening, enameled, insul. glass	1.000	Ea.	1.600	212	
Blocking, 1" x 3" furring strip, nailers	12.500	L.F.	.182	3.88	
Drywall, 1/2" thick, standard	4.500	S.F.	.036	1.08	
Corner bead, 1" x 1", galvanized steel	9.250	L.F.	.185	1.20	
Finish drywall, tape and finish corners, inside and outside	18.500	L.F.	.312	1.30	Contractor's Fee
Sill, slate	3.250	L.F.	.650	28.11	
TOTAL			2.965	247.57	**570.94**

Be sure to read pages 178-179 for proper use of this section.

Storm Door & Window Systems

Aluminum Window →

← Aluminum Door

SYSTEM DESCRIPTION	QUAN.	UNIT	LABOR-HOURS	COST EACH	
				MAT.	TOTAL
Storm door, aluminum, combination, storm & screen, anodized, 2'-6" x 6'-8"	1.000	Ea.	1.067	167	318.74
2'-8" x 6'-8"	1.000	Ea.	1.143	191	360.26
3'-0" x 6'-8"	1.000	Ea.	1.143	191	360.26
Mill finish, 2'-6" x 6'-8"	1.000	Ea.	1.067	222	404.04
2'-8" x 6'-8"	1.000	Ea.	1.143	222	408.31
3'-0" x 6'-8"	1.000	Ea.	1.143	240	436.17
Painted, 2'-6" x 6'-8"	1.000	Ea.	1.067	222	404.04
2'-8" x 6'-8"	1.000	Ea.	1.143	225	412.86
3'-0" x 6'-8"	1.000	Ea.	1.143	235	428.50
Wood, combination, storm & screen, crossbuck, 2'-6" x 6'-9"	1.000	Ea.	1.455	288	527.73
2'-8" x 6'-9"	1.000	Ea.	1.600	281	524.74
3'-0" x 6'-9"	1.000	Ea.	1.778	287	543.94
Full lite, 2'-6" x 6'-9"	1.000	Ea.	1.455	292	533.99
2'-8" x 6'-9"	1.000	Ea.	1.600	292	541.81
3'-0" x 6'-9"	1.000	Ea.	1.778	300	564.13
Windows, aluminum, combination storm & screen, basement, 1'-10" x 1'-0"	1.000	Ea.	.533	31	77.91
2'-9" x 1'-6"	1.000	Ea.	.533	33.50	81.82
3'-4" x 2'-0"	1.000	Ea.	.533	40.50	92.69
Double hung, anodized, 2'-0" x 3'-5"	1.000	Ea.	.533	80	153.97
2'-6" x 5'-0"	1.000	Ea.	.571	107	197.97
4'-0" x 6'-0"	1.000	Ea.	.640	227	387.55
Painted, 2'-0" x 3'-5"	1.000	Ea.	.533	95.50	177.92
2'-6" x 5'-0"	1.000	Ea.	.571	153	269.27
4'-0" x 6'-0"	1.000	Ea.	.640	275	462.05
Fixed window, anodized, 4'-6" x 4'-6"	1.000	Ea.	.640	122	224.77
5'-8" x 4'-6"	1.000	Ea.	.800	139	260.31
Painted, 4'-6" x 4'-6"	1.000	Ea.	.640	122	224.77
5'-8" x 4'-6"	1.000	Ea.	.800	139	260.31

Be sure to read pages 178-179 for proper use of this section.

Shutters/Blinds Systems

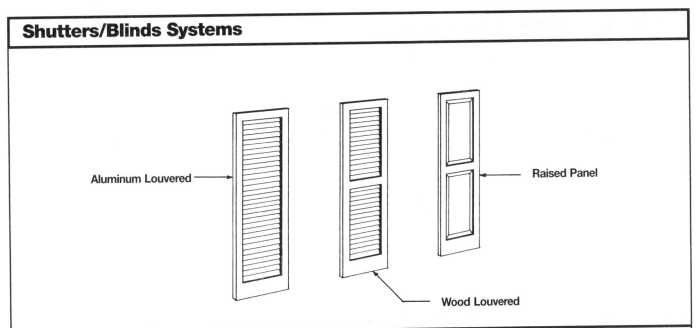

Aluminum Louvered

Wood Louvered

Raised Panel

SYSTEM DESCRIPTION	QUAN.	UNIT	LABOR-HOURS	COST PER PAIR	
				MAT.	TOTAL
Shutters, exterior blinds, aluminum, louvered, 1'-4" wide, 3"-0" long	1.000	Set	.800	46	116.15
4'-0" long	1.000	Set	.800	55	130.08
5'-4" long	1.000	Set	.800	72.50	157.24
6'-8" long	1.000	Set	.889	92.50	193.21
Wood, louvered, 1'-2" wide, 3'-3" long	1.000	Set	.800	87.50	180.48
4'-7" long	1.000	Set	.800	118	227.75
5'-3" long	1.000	Set	.800	134	252.63
1'-6" wide, 3'-3" long	1.000	Set	.800	93	189.01
4'-7" long	1.000	Set	.800	130	246.52
Polystyrene, solid raised panel, 3'-3" wide, 3'-0" long	1.000	Set	.800	182	327.13
3'-11" long	1.000	Set	.800	194	345.75
5'-3" long	1.000	Set	.800	255	440.15
6'-8" long	1.000	Set	.889	287	494.89
Polystyrene, louvered, 1'-2" wide, 3'-3" long	1.000	Set	.800	44	113.02
4'-7" long	1.000	Set	.800	54.50	129.30
5'-3" long	1.000	Set	.800	58	134.78
6'-8" long	1.000	Set	.889	95.50	197.90
Vinyl, louvered, 1'-2" wide, 4'-7" long	1.000	Set	.720	51.50	120.42
1'-4" x 6'-8" long	1.000	Set	.889	87	184.68

Be sure to read pages 178-179 for proper use of this section.

Insulation Systems

System Description	QUAN.	UNIT	LABOR-HOURS	COST PER S.F.	
				MAT.	TOTAL
Poured insulation, cellulose fiber, R3.8 per inch (1" thick)	1.000	S.F.	.003	.04	.25
Fiberglass , R4.0 per inch (1" thick)	1.000	S.F.	.003	.03	.23
Mineral wool, R3.0 per inch (1" thick)	1.000	S.F.	.003	.03	.23
Polystyrene, R4.0 per inch (1" thick)	1.000	S.F.	.003	.20	.50
Vermiculite, R2.7 per inch (1" thick)	1.000	S.F.	.003	.15	.42
Perlite, R2.7 per inch (1" thick)	1.000	S.F.	.003	.15	.42
Reflective insulation, aluminum foil reinforced with scrim	1.000	S.F.	.004	.15	.47
Reinforced with woven polyolefin	1.000	S.F.	.004	.19	.54
With single bubble air space, R8.8	1.000	S.F.	.005	.30	.76
With double bubble air space, R9.8	1.000	S.F.	.005	.32	.79
Rigid insulation, fiberglass, unfaced,					
1-1/2" thick, R6.2	1.000	S.F.	.008	.45	1.14
2" thick, R8.3	1.000	S.F.	.008	.51	1.23
2-1/2" thick, R10.3	1.000	S.F.	.010	.63	1.53
3" thick, R12.4	1.000	S.F.	.010	.63	1.53
Foil faced, 1" thick, R4.3	1.000	S.F.	.008	.88	1.81
1-1/2" thick, R6.2	1.000	S.F.	.008	1.19	2.29
2" thick, R8.7	1.000	S.F.	.009	1.49	2.81
2-1/2" thick, R10.9	1.000	S.F.	.010	1.76	3.28
3" thick, R13.0	1.000	S.F.	.010	1.91	3.52
Foam glass, 1-1/2" thick R2.64	1.000	S.F.	.010	1.43	2.77
2" thick R5.26	1.000	S.F.	.011	3.22	5.61
Perlite, 1" thick R2.77	1.000	S.F.	.010	.29	1
2" thick R5.55	1.000	S.F.	.011	.55	1.46
Polystyrene, extruded, blue, 2.2#/C.F., 3/4" thick R4	1.000	S.F.	.010	.40	1.17
1-1/2" thick R8.1	1.000	S.F.	.011	.78	1.82
2" thick R10.8	1.000	S.F.	.011	1.08	2.29
Molded bead board, white, 1" thick R3.85	1.000	S.F.	.010	.18	.83
1-1/2" thick, R5.6	1.000	S.F.	.011	.44	1.29
2" thick, R7.7	1.000	S.F.	.011	.57	1.50
Non-rigid insulation, batts					
Fiberglass, kraft faced, 3-1/2" thick, R11, 11" wide	1.000	S.F.	.005	.25	.67
15" wide	1.000	S.F.	.005	.25	.67
23" wide	1.000	S.F.	.005	.25	.67
6" thick, R19, 11" wide	1.000	S.F.	.006	.34	.85
15" wide	1.000	S.F.	.006	.34	.85
23" wide	1.000	S.F.	.006	.34	.85
9" thick, R30, 15" wide	1.000	S.F.	.006	.66	1.35
23" wide	1.000	S.F.	.006	.66	1.35
12" thick, R38, 15" wide	1.000	S.F.	.006	.84	1.63
23" wide	1.000	S.F.	.006	.84	1.63
Fiberglass, foil faced, 3-1/2" thick, R11, 15" wide	1.000	S.F.	.005	.37	.86
23" wide	1.000	S.F.	.005	.37	.86
6" thick, R19, 15" thick	1.000	S.F.	.005	.45	.98
23" wide	1.000	S.F.	.005	.45	.98
9" thick, R30, 15" wide	1.000	S.F.	.006	.78	1.54
23" wide	1.000	S.F.	.006	.78	1.54

Be sure to read pages 178-179 for proper use of this section.

Division 5
ROOFING

Gable End Roofing Systems

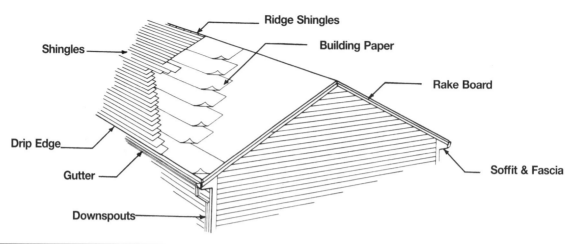

System Description	QUAN.	UNIT	LABOR-HOURS	MAT. COST	COST PER S.F.
ASPHALT, ROOF SHINGLES, CLASS A					
Shingles, inorganic class A, 210-235 lb./sq., 4/12 pitch	1.160	S.F.	.017	.41	
Drip edge, metal, 5″ wide	.150	L.F.	.003	.04	
Building paper, #15 felt	1.300	S.F.	.002	.05	
Ridge shingles, asphalt	.042	L.F.	.001	.04	
Soffit & fascia, white painted aluminum, 1′ overhang	.083	L.F.	.012	.20	
Rake trim, 1″ x 6″	.040	L.F.	.002	.04	
Rake trim, prime and paint	.040	L.F.	.002	.01	
Gutter, seamless, aluminum painted	.083	L.F.	.006	.10	
Downspouts, aluminum painted	.035	L.F.	.002	.04	**Contractor's Fee**
TOTAL			.047	.93	**4.03**
WOOD, CEDAR SHINGLES NO. 1 PERFECTIONS, 18″ LONG					
Shingles, wood, cedar, No. 1 perfections, 4/12 pitch	1.160	S.F.	.035	2.11	
Drip edge, metal, 5″ wide	.150	L.F.	.003	.04	
Building paper, #15 felt	1.300	S.F.	.002	.05	
Ridge shingles, cedar	.042	L.F.	.001	.11	
Soffit & fascia, white painted aluminum, 1′ overhang	.083	L.F.	.012	.20	
Rake trim, 1″ x 6″	.040	L.F.	.002	.04	
Rake trim, prime and paint	.040	L.F.	.002	.01	
Gutter, seamless, aluminum, painted	.083	L.F.	.006	.10	
Downspouts, aluminum, painted	.035	L.F.	.002	.04	**Contractor's Fee**
TOTAL			.065	2.70	**7.90**

Be sure to read pages 178-179 for proper use of this section.

Hip Roof Roofing Systems

System Description	QUAN.	UNIT	LABOR-HOURS	MAT. COST	COST PER S.F.
ASPHALT, ROOF SHINGLES, CLASS A					
Shingles, inorganic, class A, 210-235 lb./sq. 4/12 pitch	1.570	S.F.	.023	.54	
Drip edge, metal, 5" wide	.122	L.F.	.002	.03	
Building paper, #15 asphalt felt	1.800	S.F.	.002	.07	
Ridge shingles, asphalt	.075	L.F.	.002	.07	
Soffit & fascia, white painted aluminum, 1' overhang	.120	L.F.	.017	.29	
Gutter, seamless, aluminum, painted	.120	L.F.	.008	.14	**Contractor's Fee**
Downspouts, aluminum, painted	.035	L.F.	.002	.04	
TOTAL			.056	1.18	**5.07**
WOOD, CEDAR SHINGLES, NO. 1 PERFECTIONS, 18" LONG					
Shingles, red cedar, No. 1 perfections, 5" exp., 4/12 pitch	1.570	S.F.	.047	2.82	
Drip edge, metal, 5" wide	.122	L.F.	.002	.03	
Building paper, #15 asphalt felt	1.800	S.F.	.002	.07	
Ridge shingles, wood, cedar	.075	L.F.	.002	.20	
Soffit & fascia, white painted aluminum, 1' overhang	.120	L.F.	.017	.29	
Gutter, seamless, aluminum, painted	.120	L.F.	.008	.14	**Contractor's Fee**
Downspouts, aluminum, painted	.035	L.F.	.002	.04	
TOTAL			.080	3.59	**10.29**

Be sure to read pages 178-179 for proper use of this section.

Gambrel Roofing Systems

System Description	QUAN.	UNIT	LABOR-HOURS	MAT. COST	COST PER S.F.
ASPHALT, ROOF SHINGLES, CLASS A					
Shingles, asphalt, inorganic, class A, 210-235 lb./sq.	1.450	S.F.	.022	.51	
Drip edge, metal, 5" wide	.146	L.F.	.003	.04	
Building paper, #15 asphalt felt	1.500	S.F.	.002	.06	
Ridge shingles, asphalt	.042	L.F.	.001	.04	
Soffit & fascia, painted aluminum, 1' overhang	.083	L.F.	.012	.20	
Rake trim, 1" x 6"	.063	L.F.	.003	.07	
Rake trim, prime and paint	.063	L.F.	.003	.02	
Gutter, seamless, aluminum, painted	.083	L.F.	.006	.10	
Downspouts, aluminum, painted	.042	L.F.	.002	.05	Contractor's Fee
TOTAL			.054	1.09	**4.65**
WOOD, CEDAR SHINGLES, NO. 1 PERFECTIONS, 18" LONG					
Shingles, wood, red cedar, No. 1 perfections, 5" exposure	1.450	S.F.	.044	2.64	
Drip edge, metal, 5" wide	.146	L.F.	.003	.04	
Building paper, #15 asphalt felt	1.500	S.F.	.002	.06	
Ridge shingles, wood	.042	L.F.	.001	.11	
Soffit & fascia, white painted aluminum, 1' overhang	.083	L.F.	.012	.20	
Rake trim, 1" x 6"	.063	L.F.	.003	.07	
Rake trim, prime and paint	.063	L.F.	.001	.01	
Gutter, seamless, aluminum, painted	.083	L.F.	.006	.10	
Downspouts, aluminum, painted	.042	L.F.	.002	.05	Contractor's Fee
TOTAL			.074	3.28	**9.37**

Be sure to read pages 178-179 for proper use of this section.

Mansard Roofing Systems

System Description	QUAN.	UNIT	LABOR-HOURS	MAT. COST	COST PER S.F.
ASPHALT, ROOF SHINGLES, CLASS A					
Shingles, standard inorganic class A 210-235 lb./sq.	2.210	S.F.	.032	.75	
Drip edge, metal, 5″ wide	.122	L.F.	.002	.03	
Building paper, #15 asphalt felt	2.300	S.F.	.003	.09	
Ridge shingles, asphalt	.090	L.F.	.002	.08	
Soffit & fascia, white painted aluminum, 1′ overhang	.122	L.F.	.018	.29	
Gutter, seamless, aluminum, painted	.122	L.F.	.008	.15	Contractor's Fee
Downspouts, aluminum, painted	.042	L.F.	.002	.05	
TOTAL			.067	1.44	**6.03**
WOOD, CEDAR SHINGLES, NO. 1 PERFECTIONS, 18″ LONG					
Shingles, wood, red cedar, No. 1 perfections, 5″ exposure	2.210	S.F.	.064	3.87	
Drip edge, metal, 5″ wide	.122	L.F.	.002	.03	
Building paper, #15 asphalt felt	2.300	S.F.	.003	.09	
Ridge shingles, wood	.090	L.F.	.003	.24	
Soffit & fascia, white painted aluminum, 1′ overhang	.122	L.F.	.018	.29	
Gutter, seamless, aluminum, painted	.122	L.F.	.008	.15	Contractor's Fee
Downspouts, aluminum, painted	.042	L.F.	.002	.05	
TOTAL			.100	4.72	**13.16**

Be sure to read pages 178-179 for proper use of this section.

Shed Roofing Systems

Building Paper — Drip Edge — Shingles — Soffit & Fascia — Gutter — Rake Boards — Downspouts

System Description	QUAN.	UNIT	LABOR-HOURS	MAT. COST	COST PER S.F.
ASPHALT, ROOF SHINGLES, CLASS A					
Shingles, inorganic class A 210-235 lb./sq. 4/12 pitch	1.230	S.F.	.019	.44	
Drip edge, metal, 5" wide	.100	L.F.	.002	.02	
Building paper, #15 asphalt felt	1.300	S.F.	.002	.05	
Soffit & fascia, white painted aluminum, 1' overhang	.080	L.F.	.012	.19	
Rake trim, 1" x 6"	.043	L.F.	.002	.05	
Rake trim, prime and paint	.043	L.F.	.002	.01	
Gutter, seamless, aluminum, painted	.040	L.F.	.003	.05	Contractor's Fee
Downspouts, painted aluminum	.020	L.F.	.001	.02	**3.64**
TOTAL			.043	.83	
WOOD, CEDAR SHINGLES, NO. 1 PERFECTIONS, 18" LONG					
Shingles, red cedar, No. 1 perfections, 5" exp., 4/12 pitch	1.230	S.F.	.035	2.11	
Drip edge, metal, 5" wide	.100	L.F.	.002	.02	
Building paper, #15 asphalt felt	1.300	S.F.	.002	.05	
Soffit & fascia, white painted aluminum, 1' overhang	.080	L.F.	.012	.19	
Rake trim, 1" x 6"	.043	L.F.	.002	.05	
Rake trim, prime and paint	.043	L.F.	.001	.01	
Gutter, seamless, aluminum, painted	.040	L.F.	.003	.05	Contractor's Fee
Downspouts, painted aluminum	.020	L.F.	.001	.02	**7.20**
TOTAL			.058	2.50	

Be sure to read pages 178-179 for proper use of this section.

Gable Dormer Roofing Systems

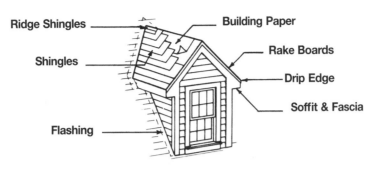

Ridge Shingles — Building Paper — Shingles — Rake Boards — Drip Edge — Soffit & Fascia — Flashing

System Description	QUAN.	UNIT	LABOR-HOURS	MAT. COST	COST PER S.F.
ASPHALT, ROOF SHINGLES, CLASS A					
Shingles, standard inorganic class A 210-235 lb./sq	1.400	S.F.	.020	.48	
Drip edge, metal, 5" wide	.220	L.F.	.004	.05	
Building paper, #15 asphalt felt	1.500	S.F.	.002	.06	
Ridge shingles, asphalt	.280	L.F.	.007	.24	
Soffit & fascia, aluminum, vented	.220	L.F.	.032	.52	Contractor's Fee
Flashing, aluminum, mill finish, .013" thick	1.500	S.F.	.083	.59	
TOTAL			.148	1.94	**11.10**
WOOD, CEDAR, NO. 1 PERFECTIONS					
Shingles, red cedar, No.1 perfections, 18" long, 5" exp.	1.400	S.F.	.041	2.46	
Drip edge, metal, 5" wide	.220	L.F.	.004	.05	
Building paper, #15 asphalt felt	1.500	S.F.	.002	.06	
Ridge shingles, wood	.280	L.F.	.008	.74	
Soffit & fascia, aluminum, vented	.220	L.F.	.032	.52	Contractor's Fee
Flashing, aluminum, mill finish, .013" thick	1.500	S.F.	.083	.59	
TOTAL			.170	4.42	**16.34**
SLATE, BUCKINGHAM, BLACK					
Shingles, Buckingham, Virginia, black	1.400	S.F.	.064	8.47	
Drip edge, metal, 5" wide	.220	L.F.	.004	.05	
Building paper, #15 asphalt felt	1.500	S.F.	.002	.06	
Ridge shingles, slate	.280	L.F.	.011	2.62	
Soffit & fascia, aluminum, vented	.220	L.F.	.032	.52	Contractor's Fee
Flashing, copper, 16 oz.	1.500	S.F.	.104	4.32	
TOTAL			.217	16.04	**36.72**

Be sure to read pages 178-179 for proper use of this section.

Shed Dormer Roofing Systems

Building Paper — Shingles

Drip Edge

Rake Boards

Soffit & Fascia

Flashing

System Description	QUAN.	UNIT	LABOR-HOURS	MAT. COST	COST PER S.F.
ASPHALT, ROOF SHINGLES, CLASS A					
Shingles, standard inorganic class A 210-235 lb./sq.	1.100	S.F.	.016	.37	
Drip edge, aluminum, 5" wide	.250	L.F.	.005	.07	
Building paper, #15 asphalt felt	1.200	S.F.	.002	.05	
Soffit & fascia, aluminum, vented, 1' overhang	.250	L.F.	.036	.60	Contractor's Fee
Flashing, aluminum, mill finish, 0.013" thick	.800	L.F.	.044	.31	
TOTAL			.103	1.40	**7.89**
WOOD, CEDAR, NO. 1 PERFECTIONS, 18" LONG					
Shingles, wood, red cedar, #1 perfections, 5" exposure	1.100	S.F.	.032	1.94	
Drip edge, aluminum, 5" wide	.250	L.F.	.005	.07	
Building paper, #15 asphalt felt	1.200	S.F.	.002	.05	
Soffit & fascia, aluminum, vented, 1' overhang	.250	L.F.	.036	.60	Contractor's Fee
Flashing, aluminum, mill finish, 0.013" thick	.800	L.F.	.044	.31	
TOTAL			.119	2.97	**11.33**
SLATE, BUCKINGHAM, BLACK					
Shingles, slate, Buckingham, black	1.100	S.F.	.050	6.66	
Drip edge, aluminum, 5" wide	.250	L.F.	.005	.07	
Building paper, #15 asphalt felt	1.200	S.F.	.002	.05	
Soffit & fascia, aluminum, vented, 1' overhang	.250	L.F.	.036	.60	Contractor's Fee
Flashing, copper, 16 oz.	.800	L.F.	.056	2.30	
TOTAL			.149	9.68	**23.19**

Be sure to read pages 178-179 for proper use of this section.

Built-up Roofing Systems

System Description	QUAN.	UNIT	LABOR-HOURS	MAT. COST	COST PER S.F.
ASPHALT, ORGANIC, 4-PLY, INSULATED DECK					
Membrane, asphalt, 4-plies #15 felt, gravel surfacing	1.000	S.F.	.025	.71	
Insulation board, 2-layers of 1-1/16" glass fiber	2.000	S.F.	.016	1.68	
Wood blocking, 2" x 6"	.040	L.F.	.004	.07	
Treated 4" x 4" cant strip	.040	L.F.	.001	.06	Contractor's Fee
Flashing, aluminum, 0.040" thick	.050	S.F.	.003	.08	
TOTAL			.049	2.60	**6.83**
ASPHALT, INORGANIC, 3-PLY, INSULATED DECK					
Membrane, asphalt, 3-plies type IV glass felt, gravel surfacing	1.000	S.F.	.028	.66	
Insulation board, 2-layers of 1-1/16" glass fiber	2.000	S.F.	.016	1.68	
Wood blocking, 2" x 6"	.040	L.F.	.004	.07	
Treated 4" x 4" cant strip	.040	L.F.	.001	.06	Contractor's Fee
Flashing, aluminum, 0.040" thick	.050	S.F.	.003	.08	
TOTAL			.052	2.55	**6.90**
COAL TAR, ORGANIC, 4-PLY, INSULATED DECK					
Membrane, coal tar, 4-plies #15 felt, gravel surfacing	1.000	S.F.	.027	1.17	
Insulation board, 2-layers of 1-1/16" glass fiber	2.000	S.F.	.016	1.68	
Wood blocking, 2" x 6"	.040	L.F.	.004	.07	
Treated 4" x 4" cant strip	.040	L.F.	.001	.06	Contractor's Fee
Flashing, aluminum, 0.040" thick	.050	S.F.	.003	.08	
TOTAL			.051	3.06	**7.62**
COAL TAR, INORGANIC, 3-PLY, INSULATED DECK					
Membrane, coal tar, 3-plies type IV glass felt, gravel surfacing	1.000	S.F.	.029	.96	
Insulation board, 2-layers of 1-1/16" glass fiber	2.000	S.F.	.016	1.68	
Wood blocking, 2" x 6"	.040	L.F.	.004	.07	
Treated 4" x 4" cant strip	.040	L.F.	.001	.06	Contractor's Fee
Flashing, aluminum, 0.040" thick	.050	S.F.	.003	.08	
TOTAL			.053	2.85	**7.45**

Be sure to read pages 178-179 for proper use of this section.

Division 7
SPECIALTIES

Greenhouse Systems

SYSTEM DESCRIPTION	QUAN.	UNIT	LABOR-HOURS	COST EACH	
				MAT.	TOTAL
Economy, lean to, shell only, not including 2' stub wall, fndtn, flrs, heat					
4' x 16'	1.000	Ea.	26.212	2,000	4,563.62
4' x 24'	1.000	Ea.	30.259	2,325	5,278.01
6' x 10'	1.000	Ea.	16.552	1,675	3,522.23
6' x 16'	1.000	Ea.	23.034	2,350	4,922.59
6' x 24'	1.000	Ea.	29.793	3,025	6,362.05
8' x 10'	1.000	Ea.	22.069	2,250	4,720
8' x 16'	1.000	Ea.	38.400	3,900	8,181.81
8' x 24'	1.000	Ea.	49.655	5,050	10,605.79
Free standing, 8' x 8'	1.000	Ea.	17.356	2,625	5,036.33
8' x 16'	1.000	Ea.	30.211	4,575	8,761.15
8' x 24'	1.000	Ea.	39.051	5,900	11,313.08
10' x 10'	1.000	Ea.	18.824	3,150	5,935.55
10' x 16'	1.000	Ea.	24.095	4,025	7,584.70
10' x 24'	1.000	Ea.	31.624	5,300	9,994.46
14' x 10'	1.000	Ea.	20.741	3,925	7,239.94
14' x 16'	1.000	Ea.	24.889	4,700	8,675.85
14' x 24'	1.000	Ea.	33.349	6,300	11,622.30
Standard, lean to, shell only, not incl. 2' stub wall, fndtn, flrs, heat 4'x10'	1.000	Ea.	28.235	2,150	4,901.27
4' x 16'	1.000	Ea.	39.341	3,000	6,859.64
4' x 24'	1.000	Ea.	45.412	3,475	7,915.25
6' x 10'	1.000	Ea.	24.827	2,525	5,295.79
6' x 16'	1.000	Ea.	34.538	3,500	7,350.12
6' x 24'	1.000	Ea.	44.689	4,525	9,507.54
8' x 10'	1.000	Ea.	33.103	3,350	7,044.46
8' x 16'	1.000	Ea.	57.600	5,850	12,272.72
8' x 24'	1.000	Ea.	74.482	7,550	15,869.59
Free standing, 8' x 8'	1.000	Ea.	26.034	3,925	7,542.05
8' x 16'	1.000	Ea.	45.316	6,850	13,139.95
8' x 24'	1.000	Ea.	58.577	8,850	16,989.17
10' x 10'	1.000	Ea.	28.236	4,725	8,889.10
10' x 16'	1.000	Ea.	36.142	6,050	11,405.49
10' x 24'	1.000	Ea.	47.436	7,950	14,988.14
14' x 10'	1.000	Ea.	31.112	5,875	10,854.58
14' x 16'	1.000	Ea.	37.334	7,050	12,997.78
14' x 24'	1.000	Ea.	50.030	9,450	17,429.89
Deluxe, lean to, shell only, not incl. 2' stub wall, fndtn, flrs or heat, 4'x10'	1.000	Ea.	20.645	3,750	6,952.05
4' x 16'	1.000	Ea.	33.032	5,975	11,082.05
4' x 24'	1.000	Ea.	49.548	8,975	16,655.07
6' x 10'	1.000	Ea.	30.968	5,600	10,392.54
6' x 16'	1.000	Ea.	49.548	8,975	16,655.07
6' x 24'	1.000	Ea.	74.323	13,500	25,014.59
8' x 10'	1.000	Ea.	41.290	7,475	13,868.56
8' x 16'	1.000	Ea.	66.065	12,000	22,235.19
8' x 24'	1.000	Ea.	99.097	18,000	33,381.22
Freestanding, 8' x 8'	1.000	Ea.	18.618	5,150	9,024.16
8' x 16'	1.000	Ea.	37.236	10,300	18,034.11
8' x 24'	1.000	Ea.	55.855	15,500	27,154.23
10' x 10'	1.000	Ea.	29.091	8,050	14,117.36
10' x 16'	1.000	Ea.	46.546	12,900	22,597.72
10' x 24'	1.000	Ea.	69.818	19,300	33,800.62
14' x 10'	1.000	Ea.	40.727	11,300	19,789.89
14' x 16'	1.000	Ea.	65.164	18,000	31,547.25
14' x 24'	1.000	Ea.	97.746	27,000	47,306.65

Be sure to read pages 178-179 for proper use of this section.

Wood Deck Systems

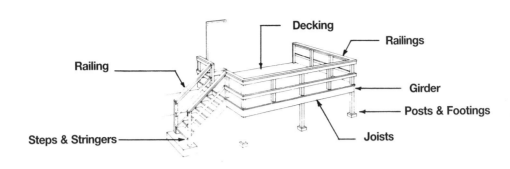

System Description	QUAN.	UNIT	LABOR-HOURS	MAT. COST	COST PER S.F.
8' X 12' DECK, PRESSURE TREATED LUMBER, JOISTS 16" O.C.					
Decking, 2" x 6" lumber	2.080	L.F.	.027	1.23	
Lumber preservative	2.080	L.F.		.27	
Joists, 2" x 8", 16" O.C.	1.000	L.F.	.015	.91	
Lumber preservative	1.000	L.F.		.17	
Girder, 2" x 10"	.125	L.F.	.002	.16	
Lumber preservative	.125	L.F.		.03	
Hand excavation for footings	.250	L.F.	.006		
Concrete footings	.250	L.F.	.006	.22	
4" x 4" Posts	.250	L.F.	.010	.35	
Lumber preservative	.250	L.F.		.04	
Framing, pressure treated wood stairs, 3' wide, 8 closed risers	1.000	Set	.080	1.65	
Railings, 2" x 4"	1.000	L.F.	.026	.37	Contractor's Fee
Lumber preservative	1.000	L.F.		.09	
TOTAL			.172	5.49	**18.28**
12' X 16' DECK, PRESSURE TREATED LUMBER, JOISTS 24" O.C.					
Decking, 2" x 6"	2.080	L.F.	.027	1.23	
Lumber preservative	2.080	L.F.		.27	
Joists, 2" x 10", 24" O.C.	.800	L.F.	.014	1.03	
Lumber preservative	.800	L.F.		.17	
Girder, 2" x 10"	.083	L.F.	.001	.11	
Lumber preservative	.083	L.F.		.02	
Hand excavation for footings	.122	L.F.	.006		
Concrete footings	.122	L.F.	.006	.22	
4" x 4" Posts	.122	L.F.	.005	.17	
Lumber preservative	.122	L.F.		.02	
Framing, pressure treated wood stairs, 3' wide, 8 closed risers	1.000	Set	.040	.83	
Railings, 2" x 4"	.670	L.F.	.017	.25	Contractor's Fee
Lumber preservative	.670	L.F.		.06	
TOTAL			.116	4.38	**13.37**
12' X 24' DECK, REDWOOD OR CEDAR, JOISTS 16" O.C.					
Decking, 2" x 6" redwood	2.080	L.F.	.027	7.55	
Joists, 2" x 10", 16" O.C.	1.000	L.F.	.018	6.95	
Girder, 2" x 10"	.083	L.F.	.001	.58	
Hand excavation for footings	.111	L.F.	.006		
Concrete footings	.111	L.F.	.006	.22	
Lumber preservative	.111	L.F.		.02	
Post, 4" x 4", including concrete footing	.111	L.F.	.009	.73	
Framing, redwood or cedar stairs, 3' wide, 8 closed risers	1.000	Set	.028	2	Contractor's Fee
Railings, 2" x 4"	.540	L.F.	.005	1.31	
TOTAL			.100	19.36	**35.62**

Be sure to read pages 178-179 for proper use of this section.

Adjusting Project Costs to Your Location

Adjusting Project Costs to Your Location

Costs shown in *Exterior Home Improvement Costs* are based on National Averages for materials and installation. To adjust these costs to a specific location, simply multiply the base cost by the factor for that city. The data is arranged alphabetically by state and postal zip code numbers. For a city not listed, use the factor for a nearby city with similar economic characteristics.

STATE	CITY	Residential
ALABAMA		
350-352	Birmingham	.87
354	Tuscaloosa	.73
355	Jasper	.71
356	Decatur	.77
357-358	Huntsville	.84
359	Gadsden	.73
360-361	Montgomery	.76
362	Anniston	.68
363	Dothan	.75
364	Evergreen	.71
365-366	Mobile	.80
367	Selma	.72
368	Phenix City	.73
369	Butler	.71
ALASKA		
995-996	Anchorage	1.26
997	Fairbanks	1.28
998	Juneau	1.26
999	Ketchikan	1.27
ARIZONA		
850,853	Phoenix	.87
852	Mesa/Tempe	.84
855	Globe	.80
856-857	Tucson	.84
859	Show Low	.83
860	Flagstaff	.85
863	Prescott	.82
864	Kingman	.81
865	Chambers	.81
ARKANSAS		
716	Pine Bluff	.75
717	Camden	.65
718	Texarkana	.70
719	Hot Springs	.64
720-722	Little Rock	.81
723	West Memphis	.74
724	Jonesboro	.73
725	Batesville	.71
726	Harrison	.72
727	Fayetteville	.68
728	Russellville	.70
729	Fort Smith	.76
CALIFORNIA		
900-902	Los Angeles	1.06
903-905	Inglewood	1.04
906-908	Long Beach	1.03
910-912	Pasadena	1.04
913-916	Van Nuys	1.07
917-918	Alhambra	1.08
919-921	San Diego	1.04
922	Palm Springs	1.02
923-924	San Bernardino	1.04
925	Riverside	1.06
926-927	Santa Ana	1.04
928	Anaheim	1.07
930	Oxnard	1.07
931	Santa Barbara	1.07
932-933	Bakersfield	1.04

STATE	CITY	Residential
934	San Luis Obispo	1.07
935	Mojave	1.05
936-938	Fresno	1.10
939	Salinas	1.12
940-941	San Francisco	1.21
942,956-958	Sacramento	1.11
943	Palo Alto	1.16
944	San Mateo	1.19
945	Vallejo	1.13
946	Oakland	1.18
947	Berkeley	1.21
948	Richmond	1.22
949	San Rafael	1.20
950	Santa Cruz	1.14
951	San Jose	1.18
952	Stockton	1.09
953	Modesto	1.08
954	Santa Rosa	1.14
955	Eureka	1.09
959	Marysville	1.10
960	Redding	1.10
961	Susanville	1.11
COLORADO		
800-802	Denver	.96
803	Boulder	.94
804	Golden	.92
805	Fort Collins	.91
806	Greeley	.80
807	Fort Morgan	.94
808-809	Colorado Springs	.92
810	Pueblo	.93
811	Alamosa	.90
812	Salida	.91
813	Durango	.93
814	Montrose	.88
815	Grand Junction	.93
816	Glenwood Springs	.92
CONNECTICUT		
060	New Britain	1.06
061	Hartford	1.05
062	Willimantic	1.06
063	New London	1.05
064	Meriden	1.05
065	New Haven	1.06
066	Bridgeport	1.06
067	Waterbury	1.06
068	Norwalk	1.06
069	Stamford	1.07
D.C.		
200-205	Washington	.93
DELAWARE		
197	Newark	1.00
198	Wilmington	1.01
199	Dover	1.00
FLORIDA		
320,322	Jacksonville	.79
321	Daytona Beach	.86
323	Tallahassee	.72

STATE	CITY	Residential
324	Panama City	.67
325	Pensacola	.75
326,344	Gainesville	.77
327-328,347	Orlando	.85
329	Melbourne	.87
330-332,340	Miami	.83
333	Fort Lauderdale	.84
334,349	West Palm Beach	.84
335-336,346	Tampa	.87
337	St. Petersburg	.77
338	Lakeland	.83
339,341	Fort Myers	.81
342	Sarasota	.85
GEORGIA		
300-303,399	Atlanta	.89
304	Statesboro	.67
305	Gainesville	.74
306	Athens	.74
307	Dalton	.70
308-309	Augusta	.76
310-312	Macon	.78
313-314	Savannah	.79
315	Waycross	.71
316	Valdosta	.71
317	Albany	.75
318-319	Columbus	.80
HAWAII		
967	Hilo	1.22
968	Honolulu	1.23
STATES & POSS.		
969	Guam	1.62
IDAHO		
832	Pocatello	.88
833	Twin Falls	.72
834	Idaho Falls	.72
835	Lewiston	.97
836-837	Boise	.89
838	Coeur d'Alene	.84
ILLINOIS		
600-603	North Suburban	1.09
604	Joliet	1.11
605	South Suburban	1.09
606-608	Chicago	1.15
609	Kankakee	1.00
610-611	Rockford	1.03
612	Rock Island	.96
613	La Salle	1.01
614	Galesburg	.98
615-616	Peoria	1.01
617	Bloomington	.96
618-619	Champaign	.98
620-622	East St. Louis	.97
623	Quincy	.97
624	Effingham	.98
625	Decatur	.97
626-627	Springfield	.98
628	Centralia	.96
629	Carbondale	.96
INDIANA		
460	Anderson	.91
461-462	Indianapolis	.96

STATE	CITY	Residential
463-464	Gary	1.00
465-466	South Bend	.91
467-468	Fort Wayne	.91
469	Kokomo	.92
470	Lawrenceburg	.86
471	New Albany	.85
472	Columbus	.92
473	Muncie	.91
474	Bloomington	.94
475	Washington	.90
476-477	Evansville	.90
478	Terre Haute	.90
479	Lafayette	.91
IOWA		
500-503,509	Des Moines	.92
504	Mason City	.77
505	Fort Dodge	.76
506-507	Waterloo	.80
508	Creston	.81
510-511	Sioux City	.86
512	Sibley	.73
513	Spencer	.75
514	Carroll	.75
515	Council Bluffs	.81
516	Shenandoah	.74
520	Dubuque	.84
521	Decorah	.77
522-524	Cedar Rapids	.93
525	Ottumwa	.83
526	Burlington	.87
527-528	Davenport	.98
KANSAS		
660-662	Kansas City	.95
664-666	Topeka	.79
667	Fort Scott	.85
668	Emporia	.72
669	Belleville	.74
670-672	Wichita	.81
673	Independence	.75
674	Salina	.73
675	Hutchinson	.68
676	Hays	.74
677	Colby	.76
678	Dodge City	.73
679	Liberal	.68
KENTUCKY		
400-402	Louisville	.93
403-405	Lexington	.84
406	Frankfort	.82
407-409	Corbin	.67
410	Covington	.93
411-412	Ashland	.94
413-414	Campton	.68
415-416	Pikeville	.77
417-418	Hazard	.67
420	Paducah	.89
421-422	Bowling Green	.89
423	Owensboro	.82
424	Henderson	.91
425-426	Somerset	.67
427	Elizabethtown	.86
LOUISIANA		
700-701	New Orleans	.85

STATE	CITY	Residential
703	Thibodaux	.80
704	Hammond	.78
705	Lafayette	.77
706	Lake Charles	.79
707-708	Baton Rouge	.78
710-711	Shreveport	.78
712	Monroe	.73
713-714	Alexandria	.73
MAINE		
039	Kittery	.80
040-041	Portland	.88
042	Lewiston	.88
043	Augusta	.83
044	Bangor	.86
045	Bath	.81
046	Machias	.82
047	Houlton	.86
048	Rockland	.80
049	Waterville	.75
MARYLAND		
206	Waldorf	.84
207-208	College Park	.84
209	Silver Spring	.85
210-212	Baltimore	.89
214	Annapolis	.85
215	Cumberland	.86
216	Easton	.68
217	Hagerstown	.85
218	Salisbury	.75
219	Elkton	.81
MASSACHUSETTS		
010-011	Springfield	1.05
012	Pittsfield	1.02
013	Greenfield	1.01
014	Fitchburg	1.08
015-016	Worcester	1.11
017	Framingham	1.12
018	Lowell	1.13
019	Lawrence	1.12
020-022, 024	Boston	1.19
023	Brockton	1.11
025	Buzzards Bay	1.10
026	Hyannis	1.09
027	New Bedford	1.11
MICHIGAN		
480,483	Royal Oak	1.03
481	Ann Arbor	1.03
482	Detroit	1.10
484-485	Flint	.97
486	Saginaw	.94
487	Bay City	.95
488-489	Lansing	.96
490	Battle Creek	.93
491	Kalamazoo	.92
492	Jackson	.94
493,495	Grand Rapids	.83
494	Muskegon	.89
496	Traverse City	.81
497	Gaylord	.84
498-499	Iron Mountain	.90
MINNESOTA		
550-551	Saint Paul	1.13

STATE	CITY	Residential
553-555	Minneapolis	1.18
556-558	Duluth	1.10
559	Rochester	1.04
560	Mankato	1.00
561	Windom	.84
562	Willmar	.85
563	St. Cloud	1.07
564	Brainerd	.96
565	Detroit Lakes	.97
566	Bemidji	.94
567	Thief River Falls	.92
MISSISSIPPI		
386	Clarksdale	.61
387	Greenville	.68
388	Tupelo	.64
389	Greenwood	.65
390-392	Jackson	.73
393	Meridian	.66
394	Laurel	.62
395	Biloxi	.75
396	McComb	.74
397	Columbus	.64
MISSOURI		
630-631	St. Louis	1.01
633	Bowling Green	.90
634	Hannibal	.87
635	Kirksville	.80
636	Flat River	.93
637	Cape Girardeau	.86
638	Sikeston	.83
639	Poplar Bluff	.83
640-641	Kansas City	1.02
644-645	St. Joseph	.93
646	Chillicothe	.85
647	Harrisonville	.94
648	Joplin	.82
650-651	Jefferson City	.88
652	Columbia	.87
653	Sedalia	.85
654-655	Rolla	.88
656-658	Springfield	.84
MONTANA		
590-591	Billings	.88
592	Wolf Point	.84
593	Miles City	.86
594	Great Falls	.89
595	Havre	.81
596	Helena	.88
597	Butte	.83
598	Missoula	.83
599	Kalispell	.82
NEBRASKA		
680-681	Omaha	.89
683-685	Lincoln	.78
686	Columbus	.69
687	Norfolk	.77
688	Grand Island	.77
689	Hastings	.76
690	Mccook	.69
691	North Platte	.75
692	Valentine	.66
693	Alliance	.65

Adjusting Project Costs to Your Location

STATE	CITY	Residential		STATE	CITY	Residential
NEVADA				125-126	Poughkeepsie	1.06
889-891	Las Vegas	1.00		127	Monticello	1.03
893	Ely	.90		128	Glens Falls	.88
894-895	Reno	.97		129	Plattsburgh	.92
897	Carson City	.97		130-132	Syracuse	.96
898	Elko	.97		133-135	Utica	.93
				136	Watertown	.91
NEW HAMPSHIRE				137-139	Binghamton	.92
030	Nashua	.90		140-142	Buffalo	1.06
031	Manchester	.90		143	Niagara Falls	1.01
032-033	Concord	.86		144-146	Rochester	.98
034	Keene	.73		147	Jamestown	.88
035	Littleton	.81		148-149	Elmira	.87
036	Charleston	.70				
037	Claremont	.72		**NORTH CAROLINA**		
038	Portsmouth	.84		270,272-274	Greensboro	.73
				271	Winston-Salem	.73
NEW JERSEY				275-276	Raleigh	.74
070-071	Newark	1.12		277	Durham	.73
072	Elizabeth	1.15		278	Rocky Mount	.64
073	Jersey City	1.12		279	Elizabeth City	.61
074-075	Paterson	1.12		280	Gastonia	.74
076	Hackensack	1.11		281-282	Charlotte	.74
077	Long Branch	1.12		283	Fayetteville	.71
078	Dover	1.12		284	Wilmington	.72
079	Summit	1.12		285	Kinston	.62
080,083	Vineland	1.09		286	Hickory	.62
081	Camden	1.10		287-288	Asheville	.72
082,084	Atlantic City	1.13		289	Murphy	.66
085-086	Trenton	1.11				
087	Point Pleasant	1.10		**NORTH DAKOTA**		
088-089	New Brunswick	1.12		580-581	Fargo	.80
				582	Grand Forks	.77
NEW MEXICO				583	Devils Lake	.80
870-872	Albuquerque	.85		584	Jamestown	.75
873	Gallup	.85		585	Bismarck	.80
874	Farmington	.85		586	Dickinson	.77
875	Santa Fe	.85		587	Minot	.80
877	Las Vegas	.85		588	Williston	.77
878	Socorro	.85				
879	Truth/Consequences	.84		**OHIO**		
880	Las Cruces	.83		430-432	Columbus	.96
881	Clovis	.85		433	Marion	.93
882	Roswell	.85		434-436	Toledo	1.01
883	Carrizozo	.85		437-438	Zanesville	.91
884	Tucumcari	.86		439	Steubenville	.96
				440	Lorain	1.01
NEW YORK				441	Cleveland	1.02
100-102	New York	1.36		442-443	Akron	.99
103	Staten Island	1.27		444-445	Youngstown	.96
104	Bronx	1.29		446-447	Canton	.94
105	Mount Vernon	1.15		448-449	Mansfield	.95
106	White Plains	1.19		450	Hamilton	.96
107	Yonkers	1.21		451-452	Cincinnati	.96
108	New Rochelle	1.21		453-454	Dayton	.92
109	Suffern	1.13		455	Springfield	.94
110	Queens	1.27		456	Chillicothe	.96
111	Long Island City	1.30		457	Athens	.90
112	Brooklyn	1.32		458	Lima	.92
113	Flushing	1.29				
114	Jamaica	1.29		**OKLAHOMA**		
115,117,118	Hicksville	1.19		730-731	Oklahoma City	.80
116	Far Rockaway	1.28		734	Ardmore	.78
119	Riverhead	1.20		735	Lawton	.81
120-122	Albany	.95		736	Clinton	.77
123	Schenectady	.96		737	Enid	.77
124	Kingston	1.02		738	Woodward	.76

Adjusting Project Costs to Your Location

STATE	CITY	Residential		STATE	CITY	Residential
739	Guymon	.67		293	Spartanburg	.71
740-741	Tulsa	.79		294	Charleston	.72
743	Miami	.82		295	Florence	.66
744	Muskogee	.72		296	Greenville	.70
745	Mcalester	.74		297	Rock Hill	.65
746	Ponca City	.77		298	Aiken	.83
747	Durant	.76		299	Beaufort	.67
748	Shawnee	.76				
749	Poteau	.77		**SOUTH DAKOTA**		
				570-571	Sioux Falls	.77
OREGON				572	Watertown	.73
970-972	Portland	1.02		573	Mitchell	.75
973	Salem	1.01		574	Aberdeen	.76
974	Eugene	1.01		575	Pierre	.76
975	Medford	.99		576	Mobridge	.73
976	Klamath Falls	1.00		577	Rapid City	.76
977	Bend	1.02				
978	Pendleton	.99		**TENNESSEE**		
979	Vale	.99		370-372	Nashville	.85
				373-374	Chattanooga	.77
PENNSYLVANIA				375,380-381	Memphis	.85
150-152	Pittsburgh	.99		376	Johnson City	.72
153	Washington	.94		377-379	Knoxville	.75
154	Uniontown	.91		382	Mckenzie	.70
155	Bedford	.89		383	Jackson	.71
156	Greensburg	.95		384	Columbia	.72
157	Indiana	.91		385	Cookeville	.68
158	Dubois	.90				
159	Johnstown	.90		**TEXAS**		
160	Butler	.93		750	Mckinney	.75
161	New Castle	.93		751	Waxahackie	.76
162	Kittanning	.94		752-753	Dallas	.83
163	Oil City	.89		754	Greenville	.69
164-165	Erie	.97		755	Texarkana	.74
166	Altoona	.89		756	Longview	.68
167	Bradford	.89		757	Tyler	.74
168	State College	.92		758	Palestine	.67
169	Wellsboro	.88		759	Lufkin	.72
170-171	Harrisburg	.94		760-761	Fort Worth	.83
172	Chambersburg	.89		762	Denton	.77
173-174	York	.88		763	Wichita Falls	.79
175-176	Lancaster	.91		764	Eastland	.72
177	Williamsport	.85		765	Temple	.75
178	Sunbury	.89		766-767	Waco	.77
179	Pottsville	.89		768	Brownwood	.68
180	Lehigh Valley	.98		769	San Angelo	.72
181	Allentown	1.01		770-772	Houston	.85
182	Hazleton	.89		773	Huntsville	.69
183	Stroudsburg	.92		774	Wharton	.70
184-185	Scranton	.95		775	Galveston	.83
186-187	Wilkes-Barre	.91		776-777	Beaumont	.82
188	Montrose	.89		778	Bryan	.74
189	Doylestown	1.04		779	Victoria	.74
190-191	Philadelphia	1.14		780	Laredo	.73
193	Westchester	1.07		781-782	San Antonio	.80
194	Norristown	1.03		783-784	Corpus Christi	.77
195-196	Reading	.96		785	Mc Allen	.75
				786-787	Austin	.79
PUERTO RICO				788	Del Rio	.66
009	San Juan	.84		789	Giddings	.70
				790-791	Amarillo	.78
RHODE ISLAND				792	Childress	.76
028	Newport	1.08		793-794	Lubbock	.76
029	Providence	1.08		795-796	Abilene	.75
				797	Midland	.76
SOUTH CAROLINA				798-799,885	El Paso	.75
290-292	Columbia	.73				

Adjusting Project Costs to Your Location

STATE	CITY	Residential
UTAH		
840-841	Salt Lake City	.83
842,844	Ogden	.81
843	Logan	.82
845	Price	.72
846-847	Provo	.83
VERMONT		
050	White River Jct.	.73
051	Bellows Falls	.75
052	Bennington	.74
053	Brattleboro	.75
054	Burlington	.80
056	Montpelier	.82
057	Rutland	.81
058	St. Johnsbury	.75
059	Guildhall	.74
VIRGINIA		
220-221	Fairfax	.85
222	Arlington	.87
223	Alexandria	.90
224-225	Fredericksburg	.75
226	Winchester	.71
227	Culpeper	.77
228	Harrisonburg	.67
229	Charlottesville	.72
230-232	Richmond	.81
233-235	Norfolk	.81
236	Newport News	.79
237	Portsmouth	.77
238	Petersburg	.78
239	Farmville	.68
240-241	Roanoke	.72
242	Bristol	.67
243	Pulaski	.66
244	Staunton	.68
245	Lynchburg	.69
246	Grundy	.67
WASHINGTON		
980-981,987	Seattle	1.01
982	Everett	1.03
983-984	Tacoma	1.00
985	Olympia	1.00
986	Vancouver	.97
988	Wenatchee	.90
989	Yakima	.95
990-992	Spokane	1.00
993	Richland	.97
994	Clarkston	.96
WEST VIRGINIA		
247-248	Bluefield	.88
249	Lewisburg	.89
250-253	Charleston	.97
254	Martinsburg	.84
255-257	Huntington	.96
258-259	Beckley	.90
260	Wheeling	.92
261	Parkersburg	.91
262	Buckhannon	.91
263-264	Clarksburg	.90
265	Morgantown	.91
266	Gassaway	.92
267	Romney	.86
268	Petersburg	.88

STATE	CITY	Residential
WISCONSIN		
530,532	Milwaukee	1.05
531	Kenosha	1.03
534	Racine	1.01
535	Beloit	.99
537	Madison	.99
538	Lancaster	.97
539	Portage	.96
540	New Richmond	.98
541-543	Green Bay	1.01
544	Wausau	.95
545	Rhinelander	.95
546	La Crosse	.94
547	Eau Claire	.98
548	Superior	.97
549	Oshkosh	.95
WYOMING		
820	Cheyenne	.75
821	Yellowstone Nat. Pk.	.70
822	Wheatland	.71
823	Rawlins	.69
824	Worland	.69
825	Riverton	.70
826	Casper	.76
827	Newcastle	.68
828	Sheridan	.73
829-831	Rock Springs	.73
ALBERTA		
	Calgary	1.06
	Edmonton	1.05
	Fort McMurray	1.03
	Lethbridge	1.04
	Lloydminster	1.03
	Medicine Hat	1.04
	Red Deer	1.04
BRITISH COLUMBIA		
	Kamloops	1.01
	Prince George	1.01
	Vancouver	1.08
	Victoria	1.01
MANITOBA		
	Brandon	1.00
	Portage la Prairie	1.00
	Winnipeg	1.01
NEW BRUNSWICK		
	Bathurst	.91
	Dalhousie	.91
	Fredericton	.99
	Moncton	.91
	Newcastle	.91
	Saint John	.99
NEWFOUNDLAND		
	Corner Brook	.92
	St. John's	.93
NORTHWEST TERRITORIES		
	Yellowknife	.98

STATE	CITY	Residential
NOVA SCOTIA		
	Dartmouth	.93
	Halifax	.93
	New Glasgow	.93
	Sydney	.91
	Yarmouth	.93
ONTARIO		
	Barrie	1.10
	Brantford	1.12
	Cornwall	1.11
	Hamilton	1.12
	Kingston	1.11
	Kitchener	1.06
	London	1.10
	North Bay	1.08
	Oshawa	1.10
	Ottawa	1.12
	Owen Sound	1.09
	Peterborough	1.09
	Sarnia	1.12
	Sudbury	1.03
	Thunder Bay	1.08
	Toronto	1.15
	Windsor	1.09
PRINCE EDWARD ISLAND		
	Charlottetown	.88
	Summerside	.88
QUEBEC		
	Cap-de-la-Madeleine	1.11
	Charlesbourg	1.11
	Chicoutimi	1.11
	Gatineau	1.10
	Laval	1.11
	Montreal	1.11
	Quebec	1.13
	Sherbrooke	1.10
	Trois Rivieres	1.11
SASKATCHEWAN		
	Moose Jaw	.91
	Prince Albert	.90
	Regina	.92
	Saskatoon	.91
YUKON		
	Whitehorse	.90

Part Four
Additional Resources

Safety Tips

Safety is an essential component of a successful home improvement project. Knowing the risks and doing your best to eliminate them will lead not only to a job well done, but also to a safe new space in your home—built correctly. In addition to the guidelines listed here, each project in this book features a "What To Watch Out For" section, which often includes specific safety tips. Be sure to also comply with the requirements of your local building codes.

Tackle only the projects you can handle. If you're unsure of how to do something, consult or hire a professional. If you're undertaking a project yourself, follow the safety guidelines listed below.

- **Protect yourself:** Wear the proper clothing and gear to protect yourself from possible hazards. Avoid loose or torn clothes, especially when working with power tools or insulation. Wear heavy shoes or boots, safety glasses when working with power tools, a hardhat if materials or tools could fall on your head, and work gloves when practical. Use hearing protection when operating loud machinery or when hammering in an enclosed space. Wear a dust mask to keep from inhaling sawdust, insulation fibers, or other airborne particles.

- **Organize the work area:** Keep your work space neat and organized. Eliminate tripping hazards and clutter, especially in areas used for access. Take the time to clean up and reorganize as you go. This means not only keeping materials neat, but also remembering not to leave nails sticking out; pull them or bend them over.

- **Don't strain yourself:** When lifting equipment or materials, always try to let your leg muscles do the work, not your back. Seek assistance when moving heavy or awkward objects, and, remember, if an object is on wheels, it is easier to push than to pull it.

- **Check equipment for safety:** When working from a ladder, scaffold, or temporary platform, make sure it is stable and well braced. When walking on joists, trusses, or rafters, always watch each step to see that what you are stepping on is secure. Do not use or store gasoline-fueled equipment inside a building or enclosed area.

- **Follow product manufacturers' recommendations:** When working with adhesives, protective coatings, or other volatile products, be sure to follow their installation and ventilation guidelines. Pay particular attention to drying times and fire hazards associated with the product. If possible, obtain from the supplier a Material Safety Data Sheet, which will clearly describe any associated hazards.

- **Shut off affected utilities:** When working with electricity or gas, be sure you know how to shut off the supply when needed. It may be wise to invest in a simple current-testing device to determine when electric current is present. If you don't already have one, purchase a fire extinguisher, learn how to use it, and keep it handy.

- **Use tools correctly:** Keep in mind that you'll need special tools for some jobs. Study how to use them, and practice with them beforehand. When using power tools, never pin back safety guards. Choose the correct cutting blade for the material you are using.

Keep children or bystanders away from the work area, and never interrupt someone using a power tool or actively performing an operation. Keep drill bits, blades, and cutters sharp; dull tools require extra force and can be dangerous. Always unplug tools when leaving them unattended or when servicing or changing blades.

A few tips on hand tools:

- Do not use any tool for a purpose other than the one for which it was designed. In other words, do not use a screwdriver as a pry bar, pliers as a hammer, etc.

- Do not use any striking tool (such as a hammer or sledgehammer) that has dents or cracks, shows excessive wear, or has a damaged or loose handle. Also, do not strike a hammer with another hammer in an attempt to remove a stubborn nail, get at an awkward spot, etc. Do not strike hard objects (such as concrete or steel), which could chip the tool, causing personal injury.

- If you rent or borrow tools and equipment, take time to read the instructions or have an experienced person demonstrate proper usage.

Home remodeling can be a satisfying and rewarding experience. Proper planning, common sense, and good safety practices go a long way to ensure a successful project. Take your time, know your limitations, get some good advice, and have fun!

CAUTION: The projects outlined in this book are intended as a general overview only. Consult a skilled professional to learn proper, complete, and detailed procedures for all home improvement work.

Material Shopping List

Use this form to develop a shopping list of materials needed to complete your home improvement projects. Materials have been grouped under headings that will make it easier for you to find items at your lumberyard or home center. Use the blank lines under each category for other items you may select.

The shaded section of the form is provided to help you track multiple prices, if you do comparison shopping. Simply enter the name of each store at the top of the column, and put the prices obtained from each one in the spaces provided. Circle the best price for each item.

This form can become a Project Estimate by performing the following steps:

1. Enter the quantity and the cost of each item in the appropriate spaces in the **Project Estimate** section. Multiply **Quantity** times **Cost** to determine the **Total Cost** for each item.

2. Add all the numbers in the Total Cost column *on each page* to determine **Page Totals**. Enter each Page Total in the space provided at the bottom of the page.

3. On the last page, add all Page Totals to determine the **Project Subtotal**.

4. Multiply the Project Subtotal by your local sales tax (if necessary) to determine the **Project Total**. (For example, if the sales tax is 5%, multiply the Project Subtotal by 1.05.)

ITEM DESCRIPTION	Unit	PRICE COMPARISON			PROJECT ESTIMATE		
		Store 1	Store 2	Store 3	Quantity ×	Cost Used =	Total Cost
Building Materials							
Concrete Field Mix							
Round Fiber Tube							
Roof Shingles							
Asphalt							
Wood							
Roll Roofing							
Roofing Cement							
Cement Board							
Roof Accessories							
Flashing							
Drip Edge							
Soffit Vent							
Ridge Vent							
Gutter							
Downspout							
Elbows							
Siding							
Vinyl							
Cedar Bevel							

PAGE TOTAL _____

ITEM DESCRIPTION	Unit	PRICE COMPARISON			PROJECT ESTIMATE		
		Store 1	Store 2	Store 3	Quantity ×	Cost Used =	Total Cost
1/2″ × 6″							
1/2″ × 8″							
Cedar Shingles							
Rough-sawn Cedar Boards, Random Width							
Exterior Trim							
Molding							
1 × 3							
Door Casings							
Window Casings							
Exterior Deck/Stair Parts							
Treads							
Railing							
Balusters							
Insulation							
Fiberglass Batt, Kraft Faced, 3-1/2″							
Fiberglass Batt, Kraft Faced, 6″							
Fiberglass Batt, Foil Faced, 6″							
Fiberglass Batt, Unfaced, 6″							
Fiberglass Batt, Unfaced, 8″							
Rigid Molded Beadboard, 1/2″							
Polyethylene Vapor Barrier							
Asphalt Dampproofing							
Asphalt Felt Building Paper							
Air Infiltration Barrier (House-wrap)							
Metal Bulkhead							
Caulking Compound and Sealants							
Masonry Materials							
Cement							
Lime							
Mortar							
Gypsum Board, 1/2″							
Lumber							
Framing Lumber							
1 × 4							
2 × 3							
2 × 4							
2 × 6							

PAGE TOTAL _____

ITEM DESCRIPTION	Unit	PRICE COMPARISON			PROJECT ESTIMATE		
		Store 1	Store 2	Store 3	Quantity ×	Cost Used =	Total Cost
2 × 8							
2 × 10							
2 × 12							
4 × 4							
Pressure-Treated							
1 × 4							
2 × 4							
2 × 6							
2 × 8							
2 × 10							
2 × 12							
4 × 4							
Redwood or specialty wood							
2 × 4							
2 × 6							
2 × 8							
Plywood Sheathing							
1/2″ CDX							
5/8″ CDX							
3/4″ CDX							
Plywood Underlayment, 5/8″							
Finish Plywood							
1/2″ AA							
3/4″ AA							
3/4″ MDO							
Exterior Trim, Pine							
1 × 4							
1 × 6							
1 × 8							
1 × 10							
3″ Finials							
Composite Material							
Doors & Windows							
Doors							
Sliding Door							
Vinyl Clad							
Wood							
French Door							
Pre-Hung Entrance							
Combination Screen/Storm							
Garage—Overhead							

PAGE TOTAL _____

ITEM DESCRIPTION	Unit	PRICE COMPARISON			PROJECT ESTIMATE		
		Store 1	Store 2	Store 3	Quantity ×	Cost Used =	Total Cost
Door Grilles							
Windows							
Bow/Bay							
Double-Hung							
Wood							
Vinyl Clad							
Sliding							
Vinyl Replacement							
Storm—Aluminum							
Skylights							
Fixed							
Operable							
Hardware and Fasteners							
Nails							
Screws							
Hinges							
Joist Hangers							
Timber Connectors							
Door Hardware							
Entrance Lock, Cylinder, Grip Handle							
Lockset							
Post Base, 4 × 4							
Post Cap, 4 × 4							
Window Locks							
Door Viewer							
Knob Reinforcer							
Plumbing							
Rough Plumbing							
Supply Pipe							
Drain Pipe							
Vent Pipe							
Traps							
Exterior Hose Bibb							
Tubing							
Irrigation Plumbing							
Drip Irrigation Supply Manifold							
Fittings							
Drip Tubing							
Sprinkler Head							
Painting and Staining							
Paint							
Stain							
Paint Thinner							
Wood Preservative							
Drop Cloth							

PAGE TOTAL _____

ITEM DESCRIPTION	Unit	PRICE COMPARISON			PROJECT ESTIMATE		
		Store 1	Store 2	Store 3	Quantity ×	Cost Used =	Total Cost
Brushes							
Sandpaper							
Masking Tape							
Electrical							
Switches							
Light Switch							
Receptacles							
Duplex							
GFI							
Conduit							
Wire							
Boxes							
Fittings							
Lighting							
Light Fixtures							
Exterior Entrance							
Post							
Floodlight							
Lamps							
Landscaping							
Landscape Timbers							
Wood Edging							
Retaining Wall Stones							
Plastic Drainage Pipe							
Brick Pavers							
Brick							
Flagstone or Paving Stones							
Concrete, Ready-Mix							
Sand							
Topsoil							
Mulch							
Fertilizer							
Lime							
Grass Seed							
Shrubs							
Trees							
Sod							
Ground Cover							
Peat Moss							

PAGE TOTAL _____

ITEM DESCRIPTION	Unit	PRICE COMPARISON			PROJECT ESTIMATE		
		Store 1	Store 2	Store 3	Quantity ×	Cost Used =	Total Cost
Bituminous Asphalt							
Pea Gravel							
Outdoor Living Accessories							
Refrigerator							
Fireplace/Grill							
Sink							
Countertop							
Cabinets							
Awning							
Fountain							
Tools/Equipment							
Eye and Ear Protection							
Dust Mask							
Fire Extinguisher							
Hard Hat							
Work Boots							
Gloves							
Rubbish Barrels, Trash Bags							
Rental Equipment							
Saw Blades							
Flat Bar, Cat's Paw							
Drill Bits							
Extension Cords and Lights							
Saw Horses							
Ladders							
Hammer							
Wrench							
Drill							
Screwdriver							
Level							
Plumb							
Tape Measure							
Shovel							
Trowel							
Screws							
Post Hole Digger							
Saws							
Wire Cutters							
Sidewall Stripper							

PAGE TOTAL _____

TOTAL FROM PREVIOUS PAGES _____

PROJECT SUBTOTAL _____

TAX @ _____ %

PROJECT TOTAL _____

Glossary

A

Access door or panel

A means of access for the inspection, repair, or service of concealed systems, such as air conditioning equipment.

Anchor bolt

A threaded bolt, usually embedded in a foundation, for securing a sill, framework, or machinery. Also called a foundation bolt or hold-down bolt.

Apron

(1) A piece of finished trim placed under a window stool. (2) A slab of concrete extending beyond the entrance to a building, particularly at an entrance for vehicular traffic.

Architectural millwork

Millwork manufactured to meet specifications of a particular job, as distinguished from stock millwork. Also called custom millwork.

Asphalt shingles

Roofing felt, saturated with asphalt and coated on the weather side with a harder asphalt and aggregate particles, which has been cut into shingles for application to a sloped roof. Also called composition shingles or strip slates.

B

Backer

Three studs nailed together in a U-shape, to which a partition is attached.

Backfill

Earth, soil, or other material used to replace previously excavated material, as around a newly constructed foundation wall.

Baluster, banister

One of a series of short, vertical supporting elements for a handrail or a coping.

Base flashing

In roofing, the flashing supplied by the upturned edges of a watertight membrane.

Batt insulation

Thermal- or sound-insulating material, such as fiberglass or expanded shale, which has been fashioned into a flexible, blanket-like form. It often has a vapor barrier on one side. It is installed between the studs or joists of a frame construction.

Bead

Any molding, stop, or caulking used around a glass or panel to hold it in position.

Beam

A large horizontal structure of wood or steel.

Bearing

(1) The section of a structural member, such as a beam or truss, that rests on the supports. (2) Descriptive of any wall that provides support to the floor and/or roof of a building.

Bearing wall

Any wall that supports a vertical load, as well as its own weight.

Bed

(1) The mortar into which masonry units are set. (2) Sand or other aggregate on which pipe or conduit is laid in a trench.

Blocking

(1) Small pieces of wood used to secure, join, or reinforce members, or to fill spaces between members. (2) Small wood blocks used for shimming.

Board and batten

A method of siding in which the joints between vertically placed boards or plywood are covered by narrow strips of wood.

Board foot

The basic unit of measure for lumber. One board foot is equal to a 1" thick board 1' in width and 1' in length.

Bonding agent

A substance applied to a suitable substrate to create a bond between it and a succeeding layer.

Box nail

Nail similar to a common nail, but with a smaller diameter shank.

Box sill

A common method of frame construction using a header nailed across the ends of floor joists where they rest on the sill.

Brace

(1) A diagonal tie that interconnects scaffold members. (2) A temporary support for aligning vertical concrete formwork. (3) A horizontal or inclined member used to hold sheeting in place. (4) A hand tool with a handle, crank, and chuck used for turning a bit or auger.

Brick

A solid masonry unit of clay or shale, formed into a rectangular prism while plastic, and then burned or fired in a kiln.

Building paper

A heavy, asphalt-impregnated paper used as a lining and/or vapor barrier between sheathing and an outside wall covering, or as a lining between rough and finish flooring.

Butt hinge

A common form of hinge consisting of two plates, each with one meshing knuckle edge and connected by means of a removable or fixed pin through the knuckles.

Butt joint

(1) A square joint between two members at right angles to each other. The contact surface of the outstanding member is cut square and is flush to the surface of the other member. (2) A joint in which the ends of two members butt each other so only tensile or compressive loads are transferred.

C

Casement window

A window assembly having at least one casement or vertically hinged sash.

Casing

A piece of wood or metal trim that finishes off the frame of a door or window.

Caulk

(1) To fill a joint, crack, or opening with a sealer material. (2) The material used for that application.

Cement

Any chemical binder that makes bodies adhere to it or to each other, such as glue, paste, or Portland cement.

Common nail

Nail used in framing and rough carpentry having a flat head about twice the diameter of its shank.

Concrete

A composite material that consists essentially of a binding medium within which are embedded particles or fragments of aggregate.

Conduit

(1) A pipe, tube, or channel used to direct the flow of a fluid. (2) A pipe or tube used to enclose electric wires to protect them from damage.

Cornice

The horizontal projection of a roof overhang at the eaves, consisting of lookout, soffit, and fascia.

Crown

The high point of a piece of lumber with a curve in it.

Curb appeal

Exterior improvements to a home, such as landscaping, stone walls, and paint and trim details in order to make the home more visually appealing from the street.

D

d

Abbreviation for penny. Refers only to nail size.

Dampproofing

An application of a water-resisting treatment or material to the surface of a concrete or masonry wall to prevent passage or absorption of water or moisture.

Deck

(1) An uncovered wood platform usually attached to or on the roof of a structure. (2) The flooring of a building. (3) The structural assembly to which a roof covering is applied.

Dressed lumber

Lumber that has been processed through a planing machine for the purpose of attaining a smooth surface and uniformity of size on at least one side or edge.

Drip edge

The edge of a roof that drips into a gutter or into the open.

E

Eave

The part of a roof that projects beyond its supporting walls.

Edging

An edge molding or strip used to decoratively define the boundary of a patio, garden, or walkway.

ENERGY STAR®

A voluntary U.S. Environmental Protection Agency and Department of Energy whole building standard for rating the energy efficiency of appliances, homes, and commercial buildings.

Estimate

The anticipated cost of materials, labor, services, or any combination of these for a proposed construction project.

F

Fascia

(1) A board used on the outside vertical face of a cornice. (2) The board connecting the top of the siding with the bottom of a soffit. (3) A board nailed across the ends of the rafters at the eaves. (4) The edge beam of a bridge. (5) A flat member or band at the surface of a building.

Finish flooring

The material used to make the wearing surface of a floor, such as hardwood, tile, or terrazzo.

Flagstone

A flat, irregular-shaped stone, used as pedestrian paving or flooring.

Flashing

A thin, impervious sheet of material placed in construction to prevent water penetration or direct the flow of water. Flashing is used especially at roof hips and valleys, roof penetrations, joints between a roof and a vertical wall, and in masonry walls to direct the flow of water and moisture.

Foundation

The material or materials through which the load of a structure is transmitted to the earth.

Frame

The wood skeleton of a building. Also called framing.

Framing

(1) Structural timbers assembled into a given construction system. (2) Any construction work involving and incorporating a frame, as around a window or door opening. (3) The unfinished structure, or underlying rough timbers of a building, including walls, roofs, and floors.

Furring

(1) Strips of wood or metal fastened to a wall or other surface to even it, to form an air space, to give appearance of greater thickness, or for the application of an interior finish such as plaster. (2) Lumber 1" in thickness and less than 4" in width, frequently the product of resawing a wider piece.

G

Gable

The portion of the end of a building that extends from the eaves upward to the peak or ridge of the roof.

Gable roof

A roof shape characterized by two sections of roof of constant slope that meet at a ridge; peaked roof.

Girder

A large principal beam of steel, reinforced concrete, wood, or combination of these, used to support other structural members at isolated points along its length.

Glue laminated, glu-lam

The result of a process in which individual pieces of lumber or veneer are bonded together with adhesives to make a single piece in which the grain of all the constituent pieces is parallel.

Grade

(1) A designation of quality, especially of lumber and plywood. (2) Ground level. (3) The slope of the ground on a building site.

Grain

The direction of fibers in wood.

Ground

(1) A strip of wood that is fixed in a wall of concrete or masonry to provide a place for attaching wood trim or furring strips. (2) A screed, strip of wood, or bead of metal fastened around an opening in a wall and acting as a thickness guide for plastering or as a fastener for trim.

Ground Fault Interrupter (GFI)

A code required safety feature installed on a circuit that senses a leak of current or a false ground and immediately cuts the power to that circuit. GFI outlets are required on any circuit near water, such as those found outdoors or in bathrooms.

Grout

(1) A hydrous mortar whose consistency allows it to be placed or pumped into small joints or cavities, as between pieces of ceramic clay, slate, and floor tile. (2) Various mortar mixes used in foundation work to fill voids in soils, usually through successive injections through drilled holes.

Gutter

(1) A shallow channel of wood, metal, or PVC positioned just below and following along the eaves of a building for the purpose of collecting and diverting water from a roof. (2) In electrical wiring, the rectangular space allowed around the interior of an electrical panel for the installation of feeder and branch wiring conductors.

Gypsum

A naturally occurring, soft, whitish mineral (hydrous calcium sulfate) that, after processing, is used as a retarding agent in Portland cement and as the primary ingredient in plaster, gypsum board, and related products.

H

Header

(1) A rectangular masonry unit laid across the thickness of a wall, so as to expose its end(s). (2) A lintel. (3) A member extending horizontally between two joists to support tailpieces. (4) In piping, a chamber, pipe, or conduit having several openings through which it collects or distributes material from other pipes or conduits. (5) The wood surrounding an area of asphaltic concrete paving.

Hip

(1) The exterior inclining angle created by the junction of the sides of adjacent sloping roofs, excluding the ridge angle. (2) The rafter at this angle. (3) In a truss, the joint at which the upper chord meets an inclined end post.

Hip roof

A roof shape characterized by four or more sections of constant slope, all of which run from a uniform eave height to the ridge.

J

Jamb

An exposed upright member on each side of a window frame, door frame, or door lining. In a window, these jambs outside the frame are called reveals.

Joint

(1) The point, area, position, or condition at which two or more things are jointed. (2) The space, however small, where two surfaces meet. (3) The mortar-filled space between adjacent masonry units. (4) The place where separate but adjacent timbers are connected, as by nails or screws, or by mortises and tenons, glue, etc.

Joist

A piece of lumber 2" or 4" thick and 6" or more wide, used horizontally as a support for a ceiling or floor. Also, such a support made from steel, aluminum, or other material.

K

Keyway

A recess or groove in one lift or placement of concrete that is filled with concrete of the next lift, giving shear strength to the joint. Also called a key.

Kneewall

A short wall under a slope, usually in attic space.

L

Labor-hour

A unit describing the work performed by one person in one hour.

Lally column

A trade name for a pipe column 3" to 6" in diameter, sometimes filled with concrete.

Landing

An intermediate platform between flights of stairs, or the platform at the top or bottom of a staircase.

Layout

A design scheme or plan showing the proposed arrangement of objects and spaces within and outside a structure.

Level

(1) A term used to describe any horizontal surface that has all points at the same elevation and thus does not tilt or slope. (2) In surveying, an instrument that measures heights from an established reference. (3) A spirit level, consisting of small tubes of liquid with bubbles in each. The small tubes are positioned in a length of wood or metal that is hand held and, by observing the position of the bubbles, used to find and check level surfaces.

Lintel

A horizontal supporting member, installed above an opening such as a window or a door, that serves to carry the weight of the wall above it.

Low-emissivity (Low-e) glass

Insulated glass with a special coating to reflect internal long wave radiation, like that from a heat source, back into a home while keeping the short wave radiation of the sun out, resulting in the reduction of the home's net heating and cooling costs.

M

Main beam

A structural beam that transmits its load directly to columns, rather than to another beam.

Millwork

All the building products made of wood that are produced in a planing mill such as moldings, door and window frames, doors, windows, blinds, and stairs. Millwork does not include flooring, ceilings, and siding.

Mortar

A plastic mixture used in masonry construction that can be troweled and hardens in place.

N

Nonbearing wall

A dividing wall that supports none of the structure above it.

Nosing

The rounded front edge of a stair tread that extends over the riser.

O

On center (O.C.)

Layout spacing designation that refers to distance from the center of one framing member to another.

Outlet

The point in an electrical wiring circuit at which the current is supplied to an appliance or device.

P

Paint

(1) A mixture of a solid pigment in a liquid vehicle that dries to a protective and decorative coating. (2) The resultant dry coating.

Panel door

A door constructed with panels, usually shaped to a pattern, installed between the stiles and rails that form the outside frame of the door.

Paneling

The material used to cover an interior wall. Paneling may be made from a 4 x 4 sheet milled to a pattern and may be either hardwood or softwood plywood, often prefinished or overlaid with a decorative finish, or hardboard, also usually prefinished.

Particle board

A generic term used to describe panel products made from discrete particles of wood or other ligno-cellulosic material rather than from fibers. The wood particles are mixed with resins and formed into a solid board under heat and pressure.

Partition

An interior wall that divides a building into rooms or areas, usually nonload-bearing.

Pavers

Blocks or tiles used as a wearing surface, such as on patios and walkways.

Pitch

The angle or inclination of a roof, which varies according to the climate and roofing materials used.

Plaster

A cementitious material (or combination of cementitious material and aggregate) that, when mixed with a suitable amount of water, forms a plastic mass or paste.

Plumb

Straight up and down, perfectly vertical.

Plumbing system

The water supply and distribution pipes; plumbing fixtures and traps; soil, waste, and vent pipes; building drains and sewers; and respective devices and appurtenances within a building.

Plywood

A flat panel made up of a number of thin sheets, or veneers, of wood, in which the grain direction of each ply, or layer, is at right angles to the one adjacent to it. The veneer sheets are united under pressure by a bonding agent.

Polystyrene foam

A low-cost, foamed plastic weighing about 1 lb. per cu. ft., with good insulating properties; it is resistant to grease.

Polyurethane

Reaction product of an isocyanate with any of a wide variety of other compounds containing an active hydrogen group. Polyurethane is used to formulate tough, abrasion-resistant coatings.

Polyvinyl chloride (PVC)

A synthetic resin prepared by the polymerization of vinyl chloride, used in the manufacture of nonmetallic waterstops for concrete, floor coverings, pipe and fittings.

Post-and-beam framing

Framing in which the horizontal members are supported by a distinct column, as opposed to a wall.

Pounds per square inch (PSI)

A unit to measure water pressure.

Pre-hung door

A packaged unit consisting of a finished door on a frame with all necessary hardware and trim.

Purlin

In roofs, a horizontal member supporting the common rafters.

Q

Quotation

A price quoted by a contractor, subcontractor, material supplier, or vendor to furnish materials, labor, or both.

R

Rail

A horizontal member supported by vertical posts.

Resilient flooring

A durable floor covering that has the ability to resume its original shape, such as linoleum.

Retaining wall

(1) A structure used to sustain the pressure of the earth behind it.
(2) Any wall subjected to lateral pressure other than wind pressure.

Rib

One of a number of parallel structural members backing sheathing.

Rise and run

The angle of inclination or slope of a member or structure, expressed as the ratio of the vertical rise to the horizontal run.

Riser

A vertical member between two stair treads.

Rough flooring

Any materials used to construct an unfinished floor.

Rough opening (R.O.)

Any opening formed by the framing members to accommodate doors or windows.

R-value

A measure of a material's resistance to heat flow given a thickness of material. The term is the reciprocal of the U-value. The higher the R-value, the more effective the particular insulation.

S

Sheathing

(1) The material forming the contact face of forms. Also called lagging or sheeting. (2) Plywood, waferboard, oriented strand board, or lumber used to close up side walls, floors, or roofs preparatory to the installation of finish materials on the surface.

Shed dormer

A dormer window having vertical framing projecting from a sloping roof, and an eave line parallel to the eave line of the principal roof.

Shim

A thin piece of material, often tapered (such as a wood shingle) inserted between building materials for the purpose of straightening or making their surfaces flush at a joint.

Shingle

A roof-covering unit made of asphalt, wood, slate, cement, or other material cut into stock sizes and applied on sloping roofs in an overlapping pattern.

Shoe

Any piece of timber, metal, or stone receiving the lower end of virtually any member.

Sill

(1) The horizontal member forming the bottom of a window or exterior door frame. (2) As applied to general construction, the lowest member of the frame of the structure, resting on the foundation and supporting the frame.

Sized lumber

Lumber uniformly manufactured to net surfaced sizes. Sized lumber may be rough, surfaced, or partly surfaced on one or more faces.

Slab

A flat, horizontal (or nearly so) molded layer of plain or reinforced concrete, usually of uniform but sometimes of variable thickness, either on the ground or supported by beams, columns, walls, or other framework.

Sleeper

Lumber laid on a concrete floor as a nailing base for wood flooring.

Slope

The pitch of a roof, expressed as inches of rise per 12" of run.

Soffit

The underside of a projection, such as a cornice.

Span

(1) The distance between supports of a member. (2) The measure of distance between two supporting members.

Spread footing

A generally rectangular prism of concrete, larger in lateral dimensions than the column or wall it supports, that distributes the load of a column or wall to the subgrade.

Stair

(1) A single step. (2) A series of steps or flights of steps connected by landings, used for passage from one level to another.

Stepped footing

A wall footing with horizontal steps to accommodate a sloping grade or bearing stratum.

Stock lumber

Lumber cut to standard sizes and readily available from suppliers.

Strapping

(1) Flexible metal bands used to bind units for ease of handling and storage. (2) Another name for furring.

Stringer

(1) A secondary flexural member parallel to the longitudinal axis of a bridge or other structure. (2) A horizontal timber used to support joists or other cross members.

Structural lumber

Any lumber with nominal dimensions of 2" or more in thickness and 4" or more in width that is intended for use where working stresses are required.

Stud

(1) A vertical member of appropriate size and spacing to support sheathing or concrete forms. (2) A framing member, usually cut to a precise length at the mill, designed to be used in framing building walls with little or no trimming before it is set in place.

Sub-floor

The rough floor, usually plywood, laid across floor joists and under finish flooring.

T

Thermal barrier

An element of low conductivity placed between two conductive materials to limit heat flow, for use in metal windows or curtain walls that are to be used in cold climates.

Thermopane

Trade name of a double-glazed panel for doors or windows, used for thermal insulation.

Threshold

A shaped strip on the floor between the jambs of a door, used to separate different types of flooring or to provide weather protection at an exterior door.

Toe

(1) Any projection from the base of a construction or object to give it increased bearing and stability.
(2) That part of the base of a retaining wall that projects beyond the face, away from the retained material.

Toenailing

To drive a nail at an angle to join two pieces of wood.

Tongue-and-groove

Lumber machined to have a groove on one side and a protruding tongue on the other so that pieces will fit snugly together, with the tongue of one fitting into the groove of the other.

Top plate

A member on top of a stud wall on which joists rest to support an additional floor or form a ceiling.

Tread

The horizontal part of a stair.

Truss

A structural component composed of a combination of members, usually in a triangular arrangement, to form a rigid framework; often used to support a roof.

V

Vapor barrier

Material used to prevent the passage of vapor or moisture into a structure or another material, thus preventing condensation within them.

Veneer

(1) A masonry facing that is attached to the backup but not so bonded as to act with it under load. (2) Wood peeled, sawn, or sliced into sheets of a given constant thickness and combined with glue to produce plywood.

Ventilation

A natural or mechanical process by which air is introduced to or removed from a space, with or without heating, cooling, or purification treatment.

W

Wallboard

The term commonly applied to preformed sheets used for interior finish construction.

Waterstop

A thin sheet of metal, rubber, plastic, or other material inserted across a joint to obstruct the seeping of water through the joint.

Wood preservative

Any chemical preservative for wood, applied by washing on or pressure-impregnating. Products used include creosote, sodium fluoride, copper sulfate, and tar or pitch.

X

Xeriscaping

Using native, drought-tolerant, or well-adapted plants in dry climates to avoid the need for irrigation.

Z

Z-bar

A Z-shaped member that is used as a main runner in some types of acoustical ceiling.

Zoning permit

A permit issued by appropriate government officials authorizing land to be used for a specific purpose.

Index

Notes

Notes

Notes

Notes